T0270890

2D Electrostatic Fields

2D Electrostatic Fields
A Complex Variable Approach

Robert L. Coffie

CRC Press
Taylor & Francis Group
Boca Raton London New York

CRC Press is an imprint of the
Taylor & Francis Group, an **informa** business

First edition published 2022
by CRC Press
6000 Broken Sound Parkway NW, Suite 300, Boca Raton, FL 33487-2742

and by CRC Press
2 Park Square, Milton Park, Abingdon, Oxon, OX14 4RN

© 2022 Robert L. Coffie

CRC Press is an imprint of Taylor & Francis Group, LLC

Library of Congress Control Number: 2021940927

ISBN: 978-0-367-76975-8 (hbk)
ISBN: 978-0-367-76976-5 (pbk)
ISBN: 978-1-003-16918-5 (ebk)

Typeset in Latin font
by KnowledgeWorks Global Ltd.

Access the Support Material: https://www.routledge.com/2D-Electrostatic-Fields-A-Complex-Variable-Approach/Coffie/p/book/9780367769758

*To my parents
and Cathy.*

Contents

Preface

In the fields of engineering and physics, analytic solutions are often sought as they allow for the identification of key variables and their mathematical relationships. This typically leads to a deeper understanding of the physical situation than can be obtained by other means such as numerical simulations. Based on this philosophy, students are often assigned 1D physical problems that have analytic solutions. Due to the mathematical simplification, this is helpful to the beginner, but there are some drawbacks. First, the number of 1D problems that accurately model real world situations is extremely limited. Second, some physics is lost in 1D. For example, the curl of a vector does not exist in 1D. On the other hand, attempting to solve 3D problems analytically can often lead to mathematical complexity that is beyond the scope of most students and engineers. This leads to abandoning analytic techniques in favor of numerical simulations for 3D problems. Two-dimensional analytic solutions are a suitable compromise to these two extremes. A large number of physical problems can be treated as 2D in nature and with functions of a complex variable, the mathematics is greatly simplified compared to 3D analysis.

The purpose of this book is to demonstrate how to use functions of a complex variable to solve engineering problems that obey the 2D Laplace equation (and in some cases the 2D Poisson equation). The book was written with the engineer/physicist in mind and the majority of book focuses on electrostatics. A key benefit of the complex variable approach to electrostatics is the visualization of field lines through the use of field maps (another concept missing with 1D analysis). With todays' powerful computers and mathematical software programs, field maps are easily generated once the complex potential has been determined. Additionally, problems that would have been considered out of scope previously are now easily solved with these mathematical software programs. For example, solutions requiring the use of non-elementary functions such as elliptic and hypergeometric functions would have been viewed as not practical in the past due to the tedious use of look up tables for evaluation. Now, elliptic and hypergeometric functions are built-in functions for most mathematical software programs making their evaluation as easy as a trigonometric function.

Although the text is primarily self-contained, the reader is assumed to have taken differential and integral calculus and introductory courses in complex variables and electromagnetics. Chapter 1 consists of the mathematics required for subsequent chapters. This includes complex variable theory, formulation of vectors and vector operators in terms of complex variable notation, and mathematical theorems. In Chapter 2, the theory of electrostatics is developed. After establishing the requirements for 2D analysis to be valid, all subsequent theorems and analysis are performed in 2D. The 2D treatment allows formulating electrostatics completely in terms of complex variables. Chapter 3 focuses primarily on potential functions that can be described by the superposition of individual line charges. Both indirect and direct methods are developed. Indirect methods describes the approach where potential functions are stated and the physical situation it describes needs to be found. Direct methods describes the approach where the physical problem and boundary conditions are stated and the potential function needs to be found. Green's functions are the primary direct method for obtaining the potential function in Chapter 3. In Chapter 4, full plane conformal mappings are developed and used to demonstrate how the potential function can be obtained with conformal mapping. In Chapter 5, the Schwarz-Christoffel transformation is developed and used to obtain the potential function for various geometries. Due to the importance of the Schwarz-Christoffel transformation for solving problems, more than the usual number of problems have been worked. Chapter 6 combines all the tools developed in Chapters 1–5 to solve several engineering examples in depth. These examples are used to demonstrate not only the utility of conformal mapping and the method of Green's functions, but also the limitations and potential pitfalls often encountered when applying these methods. Although the focus through Chapter 6 is 2D electrostatics, the solving problem techniques developed are valid for other engineering analysis that obey Laplace's equation. This versatility is demonstrated in Chapter 7, by applying them to problems in other engineering disciplines.

The problem sets at the end of each chapter are designed to test the reader's grasp of key points developed in each chapter. The reader is encourage to work the problem sets and can check their work with the solutions posted at the books website. A short conformal mapping dictionary can also be found at the books website.

Despite the desire to eliminate all mistakes from the texts, the author is sure some still persist and will be grateful to anyone pointing them out. The author wishes to thank Cathy Lee for her continuous support throughout the course of writing the book. Without it, the book would never have been completed. He also wishes to thank Dr. Chang-Soo Suh for requesting a better field plate model from the author that launched the deep dive into conformal mapping.

Author

Robert Coffie is the Founder and President of RLC Solutions, a semiconductor/microelectronics consulting company. Robert has over 20 years of experience in compound semiconductor transistor design, processing, electrical characterization, and reliability. He received his B.S. in Engineering Physics from the University of Oklahoma in 1997 and his PhD in Electrical and Computer engineering from the University of California, Santa Barbara in 2003. He has designed, developed, and matured AlGaN/GaN HEMT technologies for RF applications from L-band to Q-band at Northrop Grumman and TriQuint Semiconductor (now Qorvo). Robert also developed the first JEDEC qualified AlGaN/GaN HEMTs for 600 V power switching applications at Transphorm where he served as Director of Device Engineering. Robert has authored numerous journal articles related to AlGaN/GaN HEMTs and a book chapter on new materials for high power RF applications. He is a senior member of IEEE, and an adjunct professor at the Ohio State University.

Symbols

SYMBOL DESCRIPTION

$\arg[\ ]$ argument or phase of a complex number

a_s location of vertex s of a polygon

β_i exterior angle ϕ_i of a polygon divided by $-\pi$

$\text{cn}[\ ]$ Jacobi's elliptic function cn

$\underline{C}$ capacitance per unit length

C_I circle of inversion

χ_e electric susceptibility

$[\![\chi_e]\!]$ electric susceptibility tensor

$\boldsymbol{D}, \mathcal{D}$ electric flux density vector

da infinitesimal area

$ds, |dz|$ infinitesimal length

dv infinitesimal volume

$\delta[\ ]$ Dirac delta function

ϵ_0 free-space dielectric constant

ϵ_r relative dielectric constant of a linear material

ϵ dielectric constant of a linear material, $\epsilon_r \epsilon_0$

$[\![\epsilon]\!]$ permittivity tensor

$\boldsymbol{E}, \mathcal{E}$ electric field intensity vector

E energy

$E[\phi, m]$ incomplete elliptic integral of the second kind

$E[m]$ complete elliptic integral of the second kind

$F[\phi, m]$ incomplete elliptic integral of the first kind

$\boldsymbol{F}$ force vector

$\underline{\boldsymbol{F}}$ force per unit length vector

G Green's function

G_0 portion of Green's function that satisfies Poisson's equation

G_L portion of Green's function that satisfies Laplace's equation

G_D Green's function for Dirichlet boundary conditions

G_N Green's function for Neumann boundary conditions

Γ circulation

$\Gamma[\ ]$ Gamma function

$\text{Im}[\ ]$ imaginary part of a complex value

j $\sqrt{-1}$

κ_{th} thermal conductivity

$K[m]$ complete elliptic integral of the first kind

$K'[m]$ complementary complete elliptic integral of the first kind

λ charge density per unit length

μ_0 permeability of free space

$\underline{M}, M$ dipole moment per unit length

n_p number of curvilinear squares around a conductor

n_s number of curvilinear squares between two conductors

$\hat{n}$ unit normal vector

ψ_{flux} dielectric flux

$\underline{P}, P$ polarization vector

$\underline{P}_{other}$ polarization vector that does not depend on electric field

$\underline{P}_E$ polarization vector that depends on electric field

Φ complex potential function $V + j\Psi$

ϕ_i exterior angle of a polygon

$\phi_{i,\infty}$ exterior angle of a polygon for a vertex at infinity

$\underline{\psi}_{flux}$ flux per unit length

Ψ imaginary part of the complex potential

$\Pi[n, \phi, m]$ incomplete elliptic integral of the third kind

$\Pi[n, m]$ complete elliptic integral of the third kind

q electronic charge

q charge per unit length

q_{enc} total charge per unit length inside a closed surface

q_∞ net charge per unit length at infinity

r polar magnitude

r_I radius of the circle of inversion

$Re[\ ]$ real part of a complex value

$Res[\ ]$ residue

ρ charge density per unit volume

ρ_f free charge density per unit volume

ρ_b bound charge density per unit volume

R, R_{ij} reflection coefficient for a ray of flux density vector and a planar boundary

σ charge density per unit area

σ_f free charge density per unit area

σ_b bound charge density per unit area

$sn[\ ]$ Jacobi's elliptic function sn

θ_s angle of line formed by vertices a_s and a_{s+1} of a polygon and the real positive axis

θ polar angle

$\hat{t}$ unit tangential vector

T, T_{ij} transmission coefficient for a ray flux density vector and a planar boundary

V	electrostatic potential	$(\boldsymbol{a} \times \boldsymbol{b}) \cdot \hat{\boldsymbol{k}}$	2D cross product of $\boldsymbol{a}$ and $\boldsymbol{b}$
W	work		
$\underline{W}$	work per unit length	$\mathrm{Im}[a^*b]$	2D cross product of $\boldsymbol{a}$ and $\boldsymbol{b}$
z	complex number $x + jy$		
z^*	complex conjugate of z	$\boldsymbol{\nabla}$	gradient vector operator
z_c	center of the circle of inversion	$2\frac{\partial}{\partial z^*}$	gradient vector operator
$\|z\|$	magnitude of z	∇^2	Laplace operator
Z	Cartesian axis perpendicular to the x-y plane	$4\frac{\partial^2}{\partial z \partial z^*}$	Laplace operator
$_2F_1[\,]$	Gauss's hypergeometric function	$\oint$	integral around a closed contour in the positive direction (counterclockwise)
$\boldsymbol{a} \cdot \boldsymbol{b}$	scalar product of $\boldsymbol{a}$ and $\boldsymbol{b}$	$\oint$	integral around a closed contour C in the negative direction (clockwise)
$\mathrm{Re}[a^*b]$	scalar product of $\boldsymbol{a}$ and $\boldsymbol{b}$		

CHAPTER 1

Functions of a Complex Variable

Although originally misunderstood and dismissed [4], complex numbers are now commonly used to simplify many areas of analysis. Examples include, using contour integration and the residue theorem to evaluate integrals, using complex notation for sinusoidal functions (also known as phasor notation) to obtain the AC steady state response circuits from algebraic equations instead of solving differential equations, and using conformal mapping to transform the geometry of a problem into a geometric form that can be solved or has already been solved. We use the complex representation of 2D vectors and exploit many features of analytic functions to simplify analysis in later chapters. In this chapter, we review complex numbers, complex variables and functions of a complex variable. Fundamental concepts that are required to analyze and solve 2D electrostatic problems is the focus.

1.1 COMPLEX NUMBERS AND VARIABLES

A general complex number can be written as

$$z = x + jy \tag{1.1}$$

where x and y are real numbers and $j = \sqrt{-1}$. In this notation, x is referred to as the real part of z or $\text{Re}[z]$ and y is the imaginary part of z or $\text{Im}[z]$. For two complex numbers to be equal, their real and imaginary parts must be equal.

Geometrically, real numbers can be viewed as points on a one-dimensional number line and complex numbers can be viewed as points in a plane (commonly referred to as the z-plane, x-y plane, or complex plane). The horizontal axis represents the real part of the complex number and the vertical axis represents the imaginary part (see Fig. 1.1). With this geometric view of complex numbers, $z = x + jy$ represents a number located at (x, y) in the z-plane. Therefore, real numbers are complex numbers confined to the Real or x-axis and imaginary numbers are complex numbers confined to the Imaginary or y-axis of the z-plane.

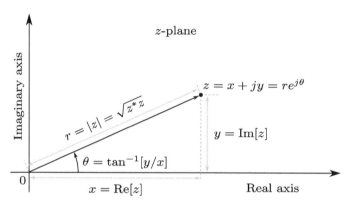

FIGURE 1.1   Geometric representation of a complex number z in the complex plane.

Armed with the geometric view of complex numbers, addition and subtraction of complex numbers coincides with addition and subtraction of (x, y) points

$$(x_1, y_1) \pm (x_2, y_2) = (x_1 \pm x_2, y_1 \pm y_2),$$
$$z_1 \pm z_2 = x_1 \pm x_2 + j(y_1 \pm y_2) \tag{1.2}$$

Addition can be performed geometrically by using the parallelogram rule as shown in Fig. 1.2(a). Lines from the origin to the two points to be added are drawn and form half of the parallelogram. The parallelogram is completed by drawing the two remaining lines (dashed lines in Fig. 1.2(a)). The sum of the two complex numbers is located at the corner of the parallelogram opposite to the origin.

The polar representation of complex numbers is also easily understood geometrically. To transform z into polar representation, the same formulas to transform Cartesian coordinates x and y into polar coordinates r and θ are used

$$x = r \cos[\theta], \quad y = r \sin[\theta] \tag{1.3}$$

and

$$r = \sqrt{x^2 + y^2}, \quad \theta = \arg[z] = \tan^{-1}[y/x] \tag{1.4}$$

This allows expressing z as

$$z = x + jy = r \left(\cos[\theta] + j \sin[\theta]\right) \tag{1.5}$$

The argument (or phase) of the complex number is given by θ. Using Euler's formula

$$e^{j\theta} = \cos[\theta] + j \sin[\theta] \tag{1.6}$$

polar representation of complex numbers can also be written as

$$z = re^{j\theta} \tag{1.7}$$

The mechanics of multiplying complex numbers is easiest seen in polar representation as

$$z_1 z_2 = r_1 r_2 e^{j(\theta_1 + \theta_1)} = x_1 x_2 - y_1 y_2 + j\left(x_1 y_2 + y_1 x_2\right) \tag{1.8}$$

Thus, the distance from the origin of the two points are multiplied and their angles are added. Since division is just multiplication by the reciprocal of a number, we have

$$\frac{z_1}{z_2} = \left(\frac{r_1}{r_2}\right) e^{j(\theta_1 - \theta_2)} = \frac{x_1 x_2 + y_1 y_2}{x_2^2 + y_2^2} + j\frac{y_1 x_2 + x_1 y_2}{x_2^2 + y_2^2} \tag{1.9}$$

To produce a geometric diagram of complex multiplication, the two points to be multiplied (z_1 and z_2) are plotted on the plane and lines from the origin to the points are drawn (see Fig. 1.2(b)). Then a triangle is created with vertices located at the origin, $z = 1$, and z_1. This triangle is then rotated about the origin so the side that coincided with the x-axis (side A) now lies along the line connecting z_2 to the origin. The rotated triangle is then scaled until the length of side A equals the length of z_2. The new location of z_1 (labeled z_3 in Fig. 1.2(b)) on the rotated and scaled triangle is the value of z_1 multiplied by z_2.

The distance of the point z from the origin is the magnitude (or modulus) of z and coincides with the definition of r in polar representation. The magnitude of z can be calculated with (1.3) or as

$$|z| = \sqrt{(z^*)\, z} = r \tag{1.10}$$

where $z^* = x - jy$ is the complex conjugate of z. When taking the

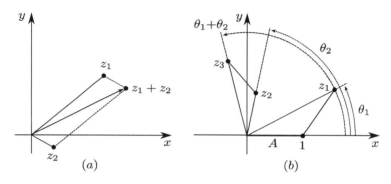

FIGURE 1.2 Geometric representation of (a) addition and (b) multiplication of two complex numbers in the z-plane.

complex conjugate of an expression, j is replaced by $-j$ everywhere in the expression. In polar representation, $z^* = re^{-j\theta}$ showing that the magnitude of z is not changed by taking the complex conjugate, but the angle is now negative. Geometrically, the point z^* is located at the mirror image with respect to the x-axis of point z as shown in Fig. 1.3(a). The real and imaginary parts of a complex expression f can be written in terms of the expression itself and its complex conjugate f^* as

$$\text{Re}[f] = \frac{1}{2}(f + f^*), \quad \text{Im}[f] = \frac{1}{2i}(f - f^*) \tag{1.11}$$

Several properties of $|z|$ are summarized in (1.12). We only prove the generalized triangle inequality (1.12.4) as the other properties are easily verified.

$$
\begin{aligned}
|z| &= |z^*| = |-z| \quad &(1.12.1) \\
|z|^2 &= |z^2| \quad &(1.12.2) \\
|z_1 z_2| &= |z_1|\,|z_2| \quad &(1.12.3) \\
\left|\sum_{i=1}^{n} z_i\right| &\leq \sum_{i=1}^{n} |z_i| \quad &(1.12.4)
\end{aligned}
\tag{1.12}
$$

Before proving the generalized triangle inequality, we first prove the triangle inequality. Let z_1 and z_2 represent two complex numbers. Expanding $|z_1 + z_2|^2$ gives

$$|z_1 + z_2|^2 = (z_1 + z_2)(z_1 + z_2)^* = |z_1|^2 + |z_2|^2 + 2\text{Re}[z_1 z_2^*] \tag{1.13}$$

Using the easily verified inequality

$$\text{Re}[z_1 z_2^*] \leq |z_1 z_2^*| = |z_1|\,|z_2| \tag{1.14}$$

gives

$$|z_1 + z_2|^2 \leq |z_1|^2 + |z_2|^2 + 2\,|z_1|\,|z_2| = (|z_1| + |z_2|)^2 \qquad (1.15)$$

Taking the square root of both sides gives the triangle inequality

$$|z_1 + z_2| \leq |z_1| + |z_2| \qquad (1.16)$$

Extending the triangle inequality to more than two complex points is easily done by replacing z_2 with a sum of complex points. For example, replacing z_2 with $z_2 + z_3$ in (1.16) gives

$$|z_1 + z_2 + z_3| \leq |z_1| + |z_2 + z_3| \leq |z_1| + |z_2| + |z_3| \qquad (1.17)$$

where (1.16) was used on $|z_2 + z_3|$ to produce the final inequality. This approach can be extended to any number of points giving the generalized triangle inequality (1.12.4).

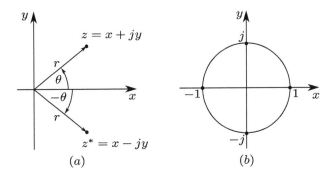

(a)   (b)

FIGURE 1.3 Geometric representation of (a) the complex conjugate of a complex number and (b) the unit circle in the complex plane.

Multiplying a complex number z_0 by $e^{j\theta_1}$ results in a pure rotation of z_0 by θ_1. For example, let

$$z_1 = x_1 = x_1 e^{j0} \qquad (1.18)$$

Multiply z_1 by $e^{j\pi/2}$ gives

$$z_1 e^{j\pi/2} = jx_1 \qquad (1.19)$$

Therefore, an imaginary number is a real number rotated by $\pi/2$ in the complex plane. The computation of j^n is easily performed now as $j^n = e^{jn\pi/2}$ with the values of

$$j^1 = j, \quad j^2 = -1, \quad j^3 = -j, \quad j^4 = 1 \qquad (1.20)$$

These values are easily remembered as the locations where the unit circles intersects the real and imaginary axes as shown in Fig. 1.3(b).

1.2 CONJUGATE COORDINATES

Consider a general function

$$f[x, y] = u[x, y] + jv[x, y] \tag{1.21}$$

where u and v are real functions of x and y with partial derivatives that exist and are continuous. To express f in terms of complex variables, we use what are known as complex conjugate coordinates (or just conjugate coordinates) by replacing x and y with

$$x = \tfrac{1}{2}\left(z + z^*\right), \quad y = -\tfrac{1}{2}j\left(z - z^*\right) \tag{1.22}$$

In general, the function f depends on both z and z^*. Later, mathematical operators are developed in terms of conjugate coordinates which require z and z^* to be treated as independent variables. Thus, general functions of a complex variable in some books use notation that reflects this dependence such as $f[z, z^*]$ to distinguish them from functions that depend only on z or only on z^*. Other books use the simplified notation $f[z]$ to represent a general function of a complex variable even if it depends on z^*. This simplified notation is used throughout this book.

1.3 ANALYTIC FUNCTIONS

Again consider a general function of a complex variable

$$f[z] = f[x, y] = u[x, y] + jv[x, y] \tag{1.23}$$

where u and v are real functions of x and y with partial derivatives that exist and are continuous. Under these conditions, the differentials of u and v are

$$du = \frac{\partial u}{\partial x}dx + \frac{\partial u}{\partial y}dy, \quad dv = \frac{\partial v}{\partial x}dx + \frac{\partial v}{\partial y}dy \tag{1.24}$$

Now we can write the differential of f as

$$df = du + jdv = \left(\frac{\partial u}{\partial x} + j\frac{\partial v}{\partial x}\right)dx + \left(\frac{\partial u}{\partial y} + j\frac{\partial v}{\partial y}\right)dy \tag{1.25}$$

The partial derivative terms in (1.25) can be re-written as

$$\frac{\partial f}{\partial x} = \frac{\partial u}{\partial x} + j\frac{\partial v}{\partial x}, \quad \frac{\partial f}{\partial y} = \frac{\partial u}{\partial y} + j\frac{\partial v}{\partial y} \tag{1.26}$$

allowing the differential of f to be written as

$$df = \frac{\partial f}{\partial x} dx + \frac{\partial f}{\partial y} dy \qquad (1.27)$$

analogous to a real function. Our goal is to express any restrictions for f to be analytic in terms of z. To accomplish this, we use conjugate coordinates. Re-writing (1.27) in terms of dz and dz^* by replacing dx and dy with

$$dx = \tfrac{1}{2}(dz + dz^*), \quad dy = \tfrac{1}{2}(dz - dz^*)/j \qquad (1.28)$$

gives

$$df = \frac{1}{2}\left(\left(\frac{\partial f}{\partial x} - j\frac{\partial f}{\partial y}\right)dz + \left(\frac{\partial f}{\partial x} + j\frac{\partial f}{\partial y}\right)dz^*\right) \qquad (1.29)$$

Defining the operators [1]

$$\frac{\partial}{\partial z} = \frac{1}{2}\left(\frac{\partial}{\partial x} - j\frac{\partial}{\partial y}\right), \quad \frac{\partial}{\partial z^*} = \frac{1}{2}\left(\frac{\partial}{\partial x} + j\frac{\partial}{\partial y}\right) \qquad (1.30)$$

allows writing (1.29) in a form similar to (1.27) as

$$df = \frac{\partial f}{\partial z} dz + \frac{\partial f}{\partial z^*} dz^* \qquad (1.31)$$

In terms of conjugate coordinates, a necessary condition for an analytic function is that it only depend on z [3]. In order for f to be independent of z^*, we have the condition

$$\frac{\partial f}{\partial z^*} = \frac{1}{2}\left(\frac{\partial f}{\partial x} + j\frac{\partial f}{\partial y}\right) = 0 \qquad (1.32)$$

Substituting (1.26) into (1.32) and setting the real and imaginary parts equal to zero leads to the requirements

$$\frac{\partial u}{\partial x} = \frac{\partial v}{\partial y}, \quad \frac{\partial u}{\partial y} = -\frac{\partial v}{\partial x} \qquad (1.33)$$

The requirements of (1.33) are known as the Cauchy-Riemann (C-R) equations. Any function of a complex variable $f = u + jv$ (where u and v are real functions of x and y with partial derivatives that exist and are continuous) that obeys the C-R equations at z_0 is analytic at z_0 and has a complex derivative that exist at z_0.

It should be clear that the sum of two analytic functions is also an analytic function as each term obeys the C-R equations. Now consider a function $g = r + js$ which is an analytic function of $w = u + jv$. The C-R equations in terms of u and v are

$$\frac{\partial r}{\partial u} = \frac{\partial s}{\partial v}, \quad \frac{\partial r}{\partial v} = -\frac{\partial s}{\partial u} \tag{1.34}$$

If w is an analytic function of $z = x + jy$, then we can multiply (1.34) by the first equation of (1.33) to obtain

$$\frac{\partial r}{\partial x} = \frac{\partial s}{\partial y}, \quad \frac{\partial r}{\partial y} = -\frac{\partial s}{\partial x} \tag{1.35}$$

Thus g is also an analytic function of $z = x + jy$.

1.4   REAL AND IMAGINARY PARTS OF ANALYTIC FUNCTIONS

Analytic functions have special properties. One of these special properties is the ability to determine the imaginary part of an analytic function from its real part and vice versa. There are several methods for determining the imaginary part from the real part of an analytic function, but we use the algebraic method given in [6]. Let $u[x, y]$ represent the real part of an analytic function $f[z]$. Then the full analytic function (to within an arbitrary constant) is

$$f[z] = 2u\left[\frac{z + z_1^*}{2}, \frac{z - z_1^*}{2j}\right] - (f[z_1])^* \tag{1.36}$$

where $u[x, y]$ is defined at $z_1 = x_1 + jy_1$. As an example, let $u[x, y] = x$ represent the real part of a complex function $g[z]$. Since $u[x, y]$ is defined at $z = 0$, we choose $z_1 = 0$. Using (1.36), we have

$$g[z] = z - (g[0])^* \tag{1.37}$$

Since $u[0, 0] = 0$, then $(g[0])^* = -jC_0$ where C_0 is an arbitrary real constant. The final form for $g[z]$ is

$$g[z] = z + jC_0 \tag{1.38}$$

The constant C_0 would need to be determined from other information. We use this method in future chapters to determine the complex potential function when the electrostatic potential is known.

1.5 TAYLOR SERIES

The need to represent a function as an infinite series or to approximate a function as a finite polynomial often occurs. The infinite series representation and the finite polynomial approximation of a function are known as Taylor series and Taylor polynomials, respectively. In order to determine the Taylor series representation of an analytic function $f[z]$, we first assume $f[z]$ has derivatives of all orders at $z = z_0$ (to be proven later) and can be represented by an infinite series of the form

$$f[z] = c_0 + c_1 (z - z_0) + c_2(z - z_0)^2 + \cdots = \sum_{n=0}^{\infty} c_n(z - z_0)^n \quad (1.39)$$

where c_n are in general complex constants. Successive differentiation of (1.39) and evaluating at $z = z_0$ gives

$$f[z_0] = c_0, \quad f'[z_0] = c_1, \quad f''[z_0] = (2 \cdot 1)\, c_2, \quad f^{(n)}[z_0] = n!c_n \quad (1.40)$$

where $f^{(n)}[z_0]$ represents the nth derivative of f with respect to z evaluated at $z = z_0$. With the coefficients determined, the infinite series can be written as

$$f[z] = \sum_{n=0}^{\infty} \frac{f^{(n)}[z_0]}{n!}(z - z_0)^n \quad (1.41)$$

and is called the Taylor series for $f[z]$ at z_0. The mth order Taylor polynomial approximation of f is a polynomial made up of the first m terms of the Taylor series. When $z_0 = 0$, the Taylor series and polynomial are also called the Maclaurin series and polynomial for f.

Convergence of the infinite series for values near z_0 (radius of convergence greater than zero) is required in order for (1.39) to be valid. In the complex plane, the radius of convergence is simply the distance from the center of the Taylor series expansion (z_0) to the nearest singularity (location where $f[z]$ does not exist) or branch point [5]. Branch points are encountered when dealing with multi-valued functions in the complex plane.

1.6 MULTI-VALUED FUNCTIONS

The concept that functions of a complex variable can be viewed as mapping points in one complex plane (z-plane) into points in another complex plane (w-plane) is helpful when discussing single valued functions and functions that are multi-valued (also known as multifunctions).

Whether a function is single valued or multi-valued is then dependent on how the function maps the points of one complex plane into the other complex plane. Single valued functions are functions that have only one w value for each value of z. Multifunctions are functions that have more than one w value for each value of z.

An example of a single valued functions is $w = z^2$. To test if $w = z^2$ is single valued or multi-valued, we specify the same point in the z-plane two different ways

$$z = re^{j\theta} \quad \text{and} \quad z = re^{j(\theta+2\pi)} \tag{1.42}$$

Evaluating z^2 for the two representations in (1.42) gives

$$w = r^2 e^{j2\theta} \quad \text{and} \quad w = r^2 e^{j2(\theta+2\pi)} = r^2 e^{j2\theta} \tag{1.43}$$

which are the same points in the w-plane. An example of a multifunction is $w = \sqrt{z}$. Evaluating $w = \sqrt{z}$ for the two representations in (1.42) gives

$$w = \sqrt{r}e^{j\theta/2} \quad \text{and} \quad w = \sqrt{r}e^{j(\theta+2\pi)/2} = -\sqrt{r}e^{j\theta/2} \tag{1.44}$$

which are clearly different points in the w-plane.

A multi-valued function can be viewed as a collection of single valued functions with each member called a branch of the function. Therefore, restricting a multifunction to a single branch results in a single valued function. The line along which the multifunction becomes discontinuous is called a branch line (or a branch cut). For the multifunction $w = \sqrt{z}$, restricting z to a θ range of $\theta_0 \le \theta < \theta_0 + 2\pi$ results in only one w value for every z value and a branch line exist along the $\theta = \theta_0$ line.

In addition to a branch line, there is a branch point defined as a point that when encircled once by a closed loop in the z-plane results in a contour with different start and end points in the w-plane. Using the multifunction example $w = \sqrt{z}$, we draw two small closed loops (only want to encircle one branch point with the loop) in the z-plane and a determine the mapping of these closed loops in the w-plane (see Fig. 1.4). The first loop C_1 does not encircle the origin in the z-plane. The mapping of C_1 in the w-plane is also a closed loop and therefore none of the points enclosed by C_1 in the z-plane are branch points. The second loop C_2 encircles the origin in the z-plane. The mapping of C_2 in the w-plane has different start and end points. Therefore, a point enclosed by C_2 in the z-plane is a branch point. To determine which point is the branch point, a couple of properties of branch points are now given

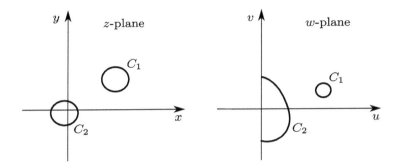

FIGURE 1.4 Mapping of small closed loops in the z-plane into the w-plane for $w = \sqrt{z}$. Loop C_1 does not enclose a branch point since it has the same start and end points in the w-plane. Loop C_2 has different start and end points in the w-plane since it encircles the origin (a branch point for $w = \sqrt{z}$) of the z-plane.

without proof. First, the value of $f[z]$ at the branch point is common to all branches of $f[z]$. Second, branch points exist at the start and end of branch cuts. The only point contained in loop C_2 that gives the same value for all branches is $z = 0$. Thus, $z = 0$ is a branch point. The second branch point for $w = \sqrt{z}$ is $z = \infty$. This can be confirmed by making a change in variable of $z = 1/z_1$, so that $w = 1/\sqrt{z_1}$. Performing the same loop analysis between the z_1-plane and w-plane shows the origin of the z_1-plane is a branch point confirming $z = \infty$ as a branch point for $w = \sqrt{z}$. It is important to note that branch points are a property of the function and independent of the branch lines used. Branch lines are not unique and do not even need to be lines.

The complex logarithm function

$$w = \ln[z] \tag{1.45}$$

is a multifunction with an infinite number of branches. Each branch is separated from the others by multiples of $2\pi j$ with branch points of $z = 0$ and $z = \infty$. To make $\ln[z]$ single valued, the restriction $0 \leq \theta < 2\pi$ (branch line along the positive real axis) or $-\pi < \theta \leq \pi$ (branch line along negative real axis) is commonly used. Unless stated otherwise, we place the branch line along the positive real axis for the $\ln[z]$.

In general, branch points and lines for an arbitrary multifunction can be complicated. Often, knowing a function is a multifunction and the location of the branch lines are enough to perform proper analysis. With mathematical software, this can be accomplished with a surface or contour plot of the imaginary part of the function. A discontinuity in the surface or contour indicates the location of a branch line. Since branch lines are not unique, plotting the function should be performed even if the branch lines have been chosen without the use of software. This allows confirming that the branch lines used by the software are the same as the chosen branch lines. Surface plots of the imaginary parts of $w = \sqrt{z}$ and $w = \ln[z]$ are shown in Fig. 1.5. The discontinuity in the surface plot along the negative real axis shows that the software has placed a branch line along the negative real axis for both of these functions. Since branch lines start and end on branch points, the branch points are $z = 0$ and $z = \infty$ as previously determined. Let

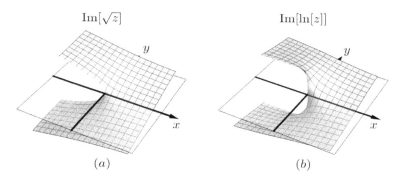

$$\text{Im}[\sqrt{z}] \qquad\qquad \text{Im}[\ln[z]]$$

$$(a) \qquad\qquad\qquad (b)$$

FIGURE 1.5 Surface plots of (a) $\text{Im}\left[\sqrt{z}\right]$ and (b) $\text{Im}[\ln[z]]$. The discontinuities in the surface plots correspond to branch lines for each function.

$$w = \sin^{-1}[z] = -j\ln\left[jz + \sqrt{1 - z^2}\right] \qquad (1.46)$$

A surface plot of the imaginary plot of (1.46) is shown in Fig. 1.6 (a). Two discontinuities corresponding to two branch lines are observed in the surface plot. One branch line occurs along the real axis for $x > 1$ and the other along the real axis for $x < -1$. Thus, the two finite branch points are $z = \pm 1$. For the next multifunction, let

$$w = \ln[z + 1] - \ln[z - 1] \qquad (1.47)$$

A surface plot of the imaginary plot of (1.47) is shown in Fig. 1.6 (b). A single discontinuity along the real axis for $-1 < x < 1$ corresponding to a branch line is observed. Thus, branch lines do not always extend to infinity and in this case, the two branch points are $z = \pm 1$.

1.7 2D VECTORS AND VECTOR OPERATORS

In physics and engineering, quantities that are described by both magnitude and direction are called vector quantities. Quantities that are

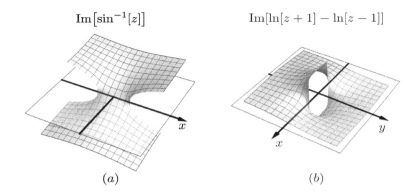

$\mathrm{Im}\left[\sin^{-1}[z]\right]$ $\mathrm{Im}[\ln[z+1]-\ln[z-1]]$

(a)   (b)

FIGURE 1.6 Surface plots of (a) $\mathrm{Im}\left[\sin^{-1}[z]\right]$ and (b) $\mathrm{Im}[\ln[z+1]-\ln[z-1]]$. The discontinuities in the surface plots correspond to branch lines for each function.

independent of direction are called scalar quantities. Examples of scalar quantities include temperature, speed, distance, and mass. Examples of vector quantities include force, electric field, velocity, and acceleration. From these two lists, we see speed is a scalar, but velocity is a vector. Both describe how fast an object is moving, but velocity would also provide the direction the object is moving. Vector equations and mathematical operations are used throughout the field of electromagnetics. In this section we learn how 2D vectors and vector operators are represented in the complex plane.

Let $\boldsymbol{g}$ represent any 2D vector. In Cartesian coordinates, $\boldsymbol{g}$ can be written as

$$\boldsymbol{g} = g_x\hat{\boldsymbol{x}} + g_y\hat{\boldsymbol{y}} \quad \text{or} \quad \boldsymbol{g} = (g_x, g_y) \tag{1.48}$$

where $\hat{\boldsymbol{x}}$ is a vector with a magnitude of one (a unit vector) in the x-direction and $\hat{\boldsymbol{y}}$ is a unit vector in the y-direction. From Section 1.1, complex numbers can also be used to represent (g_x, g_y) and are therefore another way of representing a 2D vector. In complex notation, $\boldsymbol{g}$ is written as

$$g[z] = g_x + jg_y \tag{1.49}$$

This leads to the rule that a 2D vector in Cartesian coordinates can be converted to complex notation by dropping $\hat{\boldsymbol{x}}$ and replacing $\hat{\boldsymbol{y}}$ with j. Additionally, complex notation does not use bold letters for vector quantities.

Vectors can also be written in terms of their magnitude and direction. In this notation, vector $\boldsymbol{g}$ is

$$\boldsymbol{g} = |g|\hat{\boldsymbol{g}} \tag{1.50}$$

where $\hat{g}$ is a unit vector parallel to $\boldsymbol{g}$. Converting magnitude and direction notation into complex notation is easiest done in polar representation of complex numbers as

$$g = |g|e^{j\theta_g} \tag{1.51}$$

where θ_g is the angle between the positive x-axis and vector $\boldsymbol{g}$

$$\theta_g = \tan^{-1}[g_y/g_x] \tag{1.52}$$

In general, the $e^{j\theta}$ term gives the vector its direction in complex notation and can represent any unit vector in 2D with the appropriate choice of θ.

In addition to complex notation for vectors, we also need to develop complex notation for vector operators. We start with the gradient operator $\boldsymbol{\nabla}$ that acts on a scalar function V to produce a vector

$$\boldsymbol{g} = \boldsymbol{\nabla}V \tag{1.53}$$

where in Cartesian coordinates

$$\boldsymbol{\nabla} = \frac{\partial}{\partial x}\hat{\boldsymbol{x}} + \frac{\partial}{\partial y}\hat{\boldsymbol{y}}. \tag{1.54}$$

In complex notation, the gradient and the complex conjugate of the gradient are written as

$$\boldsymbol{\nabla} = \nabla = \frac{\partial}{\partial x} + j\frac{\partial}{\partial y}, \quad \nabla^* = \frac{\partial}{\partial x} - j\frac{\partial}{\partial y} \tag{1.55}$$

Comparing (1.55) to (1.30) allows obtaining the expressions for ∇ and ∇^* in terms of conjugate coordinates as

$$2\frac{\partial}{\partial z^*} = \nabla, \quad 2\frac{\partial}{\partial z} = \nabla^* \tag{1.56}$$

The 2D dot product and cross product of two vectors can also be represented in complex notation. Let the complex functions f and g represent vectors $\boldsymbol{f}$ and $\boldsymbol{g}$ given by

$$\begin{aligned}\boldsymbol{f} = u\hat{\boldsymbol{x}} + v\hat{\boldsymbol{y}} &\quad \text{or} \quad f = u + jv \\ \boldsymbol{g} = s\hat{\boldsymbol{x}} + t\hat{\boldsymbol{y}} &\quad \text{or} \quad g = s + jt\end{aligned} \tag{1.57}$$

In Cartesian coordinates, the 2D dot product and cross product of $\boldsymbol{f}$ and $\boldsymbol{g}$ are defined as

$$\boldsymbol{f} \cdot \boldsymbol{g} = us + vt, \quad \boldsymbol{f} \times \boldsymbol{g} = ut - vs \tag{1.58}$$

Performing the complex multiplication

$$f^*g = (us + vt) + j(ut - vs) \tag{1.59}$$

and comparing to (1.58) gives

$$f^*g = \boldsymbol{f} \cdot \boldsymbol{g} + j(\boldsymbol{f} \times \boldsymbol{g})$$
$$\text{or} \tag{1.60}$$
$$\boldsymbol{f} \cdot \boldsymbol{g} = \text{Re}[f^*g], \quad \boldsymbol{f} \times \boldsymbol{g} = \text{Im}[f^*g]$$

In both 2D and 3D, the dot product of two vectors is a scalar. In 3D, the cross product of two vectors produces another vector that is perpendicular to the plane defined by the two vectors. In 2D, both vectors lie in the complex plane and therefore, the cross product should produce a vector perpendicular to the complex plane. Since this third dimension does not exist in 2D, the definition given by (1.58) for the 2D cross product is used instead and the 2D cross-product is a scalar. In 3D notation, the 2D cross product is

$$(\boldsymbol{f} \times \boldsymbol{g})_{2D} = \hat{\boldsymbol{k}} \cdot (\boldsymbol{f} \times \boldsymbol{g})_{3D} \tag{1.61}$$

where $\hat{\boldsymbol{k}}$ is a unit vector that points in the positive Z direction as shown in Fig. 1.7 (a). The 2D cross product is also the signed area of a parallelogram defined by the two vectors (Fig. 1.7 (b)). In other words, the area of the parallelogram defined by the two vectors is the magnitude of the 2D cross product, but the area can be positive or negative. Positive area corresponds to a 3D cross product that is parallel to $\hat{\boldsymbol{k}}$ and negative area corresponds to a 3D cross product that is anti-parallel to $\hat{\boldsymbol{k}}$.

Graphical representation of the dot product and 2D cross product for two vectors $\boldsymbol{f}$ and $\boldsymbol{g}$ in the complex plane is shown in Fig. 1.8 (a). Rotating vector f^*g by $-\pi/2$ as shown in Figure 1.8 (b) also rotates the triangle in Fig. 1.8 (a) by $-\pi/2$. From Fig. 1.8 (b) it is easily seen that

$$\text{Im}[f^*g] = \text{Re}[f^*(-jg)] \tag{1.62}$$

but $-jg$ is $\boldsymbol{g}$ rotated by $-\pi/2$. This allows the 2D cross product to be recast in dot product form as

$$\text{Im}[f^*g] = (\boldsymbol{f} \times \boldsymbol{g})_{2D} = (\boldsymbol{f} \cdot \hat{\boldsymbol{n}}_g |g|)_{2D} \tag{1.63}$$

where $\hat{\boldsymbol{n}}_g$ is a unit vector normal to $\boldsymbol{g}$ pointing in the direction of $\boldsymbol{g}$ rotated by $-\pi/2$.

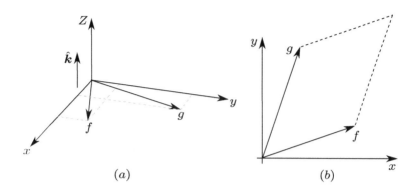

FIGURE 1.7 (a) 3D view of f and g with the direction $\hat{k}$ defined. (b) Parallelogram defined by f and g.

The dot product and cross product of two vectors can also be expressed in terms of their magnitudes and the angle between them. First we write the polar representation of vectors f and g as

$$f = |f|\exp[j\theta_f], \quad g = |g|\exp[j\theta_g] \tag{1.64}$$

Performing the complex multiplication f^*g and using (1.60) gives the alternative dot and cross product formulas

$$f \cdot g = |f||g|\cos[\theta_g - \theta_f], \quad f \times g = |f||g|\sin[\theta_g - \theta_f] \tag{1.65}$$

From (1.65) it is seen that the dot product of two vectors is zero if the vectors are perpendicular to each other ($\theta_f = \theta_g \pm \pi/2$) and the cross product of two vectors is zero if the two vectors are parallel ($\theta_f = \theta_g$) or anti-parallel ($\theta_f = \theta_g + \pi$) to each other.

The dot product of a vector with length one (a unit vector) and the gradient of a scalar function provides the rate of change of the scalar function in the direction of the unit vector. Let $\hat{s}$ be the unit vector and f be the scalar function. We can obtain the derivative of f in the direction of $\hat{s}$ as

$$\frac{\partial f}{\partial s} = \boldsymbol{\nabla} f \cdot \hat{s} = \mathrm{Re}\left[2\frac{\partial f}{\partial z}e^{j\theta_s}\right] \tag{1.66}$$

where θ_s defines the direction of $\hat{s}$.

The dot product of the gradient operator and a vector is called the divergence of the vector. The cross product of the gradient operator and a vector is called the curl of the vector. Letting $\nabla = f$ in (1.60) allows

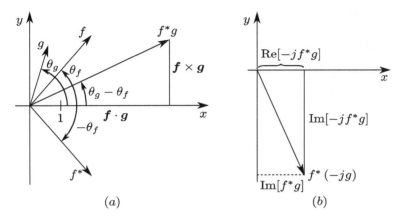

FIGURE 1.8 (a) Graphical representation showing the dot product and 2D cross product of vector $\boldsymbol{f}$ and $\boldsymbol{g}$ in the complex plane. (b) Geometric justification for the dot product representation of the 2D cross product (1.63).

computing the divergence and curl of $\boldsymbol{g}$ as the complex multiplication

$$\nabla^* g = 2\frac{\partial g}{\partial z} = \nabla \cdot \boldsymbol{g} + j\left(\nabla \times \boldsymbol{g}\right)$$

or   (1.67)

$$\nabla \cdot \boldsymbol{g} = \mathrm{Re}[\nabla^* g], \quad \nabla \times \boldsymbol{g} = \mathrm{Im}[\nabla^* g]$$

Another operator that is frequently seen in electrostatics is the dot product of the gradient operator with itself known as the Laplacian operator ∇^2. In Cartesian coordinates the Laplacian operator is

$$\nabla^2 = \nabla \cdot \nabla = \frac{\partial^2}{\partial x^2} + \frac{\partial^2}{\partial y^2}$$ (1.68)

In terms of conjugate coordinates, the Laplacian operator is

$$\nabla^2 = \nabla^* \nabla = 4\frac{\partial^2}{\partial z \partial z^*}$$ (1.69)

The mathematics of complex variable functions is equivalent to 2D vector operations and manipulations. Translating from complex notation to vector notation is typically performed with the use of (1.60). One source of possible confusion when translating from complex notation to vector notation is a scalar quantity vs. a vector that is parallel to the real axis. Since scalar quantities and vectors have different symbols in vector notation, no confusion occurs. In complex notation, this is not the case. As an example, let g be a real function $(\text{Im}[g] = 0)$ and consider the expression

$$2\frac{\partial g}{\partial z^*} \tag{1.70}$$

Assuming g is a vector that is parallel to the real axis and using (1.60) to translate complex notation to vector notation gives

$$2\frac{\partial g}{\partial z^*} = \nabla \cdot \boldsymbol{g} + j\left(\nabla \times \boldsymbol{g}\right) = \frac{\partial g}{\partial x} + j\frac{\partial g}{\partial y} \tag{1.71}$$

Assuming g is a scalar gives

$$2\frac{\partial g}{\partial z^*} = \nabla g = \frac{\partial g}{\partial x} + j\frac{\partial g}{\partial y} \tag{1.72}$$

The final answer is independent of the assumption on what type of quantity (scalar or vector) g is, but the translation to vector notation is very different. Unfortunately, the only way to tell the difference between a scalar and a vector parallel to the real axis in complex notation is through the physics.

1.8 LINE INTEGRALS

Line integrals are integrals where a function is evaluated along a defined path. In the complex plane, line integrals are typically called contour integrals. The line integral of a real function f along a contour C_1 is defined as

$$\int_{C_1} f[x, y]\, ds, \quad ds = \sqrt{dx^2 + dy^2} \tag{1.73}$$

Converting a line integral into a definite integral is done by determining the parametric forms of x, y, and ds. Let t be the parametric variable. Then we have $x[t]$, $y[t]$, and

$$ds = dt\sqrt{\left(dx/dt\right)^2 + \left(dy/dt\right)^2} \tag{1.74}$$

In parametric form, (1.73) is

$$\int_{t_1}^{t_2} f[x[t], y[t]]\left(ds/dt\right) dt \tag{1.75}$$

The line integral depends on the direction. If we let A and B represent the end points of the contour C_1, then

$$\int_{A\ C_1}^{B} f[x, y]\, ds = -\int_{B\ C_1}^{A} f[x, y]\, ds \tag{1.76}$$

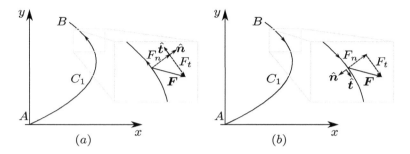

FIGURE 1.9 Definition of different vectors in the line integral of a vector. (a) Direction along contour C is from A to B and (b) direction along contour C is from B to A.

In electromagnetics, the line integral of the normal or parallel component of a vector along a contour is often needed. The line integral of the normal component of a vector F along a contour C_1 has the form

$$\int_{C_1} F \cdot \hat{n} \, ds \qquad (1.77)$$

where $\hat{n}$ is a unit vector normal to the path. If the component of F parallel to C_1 needs to be evaluated, then the line integral has the form

$$\int_{C_1} F \cdot \hat{t} \, ds \qquad (1.78)$$

where $\hat{t}$ is a unit vector parallel to the path. The definition of the unit vector normal ($\hat{n}$) and tangential ($\hat{t}$) to C_1 for the two possible directions along C_1 are shown in Fig. 1.9. In complex notation, ds, $\hat{t}$, and $\hat{n}$ are

$$ds = |dz|, \quad \hat{t} = dz/|dz|, \quad \hat{n} = -jdz/|dz| \qquad (1.79)$$

Using (1.60) and (1.63), vector line integrals for the normal and parallel components along a path in the complex plane have the compact notation

$$\int_{C_1} F^* dz = \int_{C_1} F \cdot \hat{t} \, ds + j \int_{C_1} F \cdot \hat{n} \, ds \qquad (1.80)$$

An upper bound on the magnitude of a line integral can be obtained for a finite contour if the integrand $f[z]$ exist over the entire contour C. Contour integration in summation form is given by

$$\int_C f[z] \, dz = \lim_{n \to \infty} \sum_{i=1}^{n} f[\xi_i] \, \Delta z_i \qquad (1.81)$$

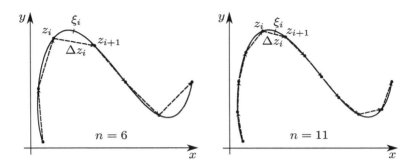

FIGURE 1.10 Definition of Δz_i and ξ_i for the summation approximation of a contour integral.

where ξ_i is a point on the contour between z_i and z_{i+1} (see Fig. 1.10). An accurate estimate of the contour integral can be obtained without passing to the limit if ξ_i are the midpoints between z_i and z_{i+1} [5]. Evaluating the magnitude of the contour integral in summation form gives

$$\left| \int_C f[z]\, dz \right| = \left| \lim_{n \to \infty} \sum_{i=1}^{n} f[\xi_i]\, \Delta z_i \right|$$

$$\leq \lim_{n \to \infty} \sum_{i=1}^{n} |f[\xi_i]|\, |\Delta z_i| = \int_C |f[z]|\, |dz| \tag{1.82}$$

The inequality used in (1.82) is the generalized triangle inequality. Letting f_M equal the maximum value of $|f[z]|$ on C provides another inequality that may be easier to evaluate

$$\lim_{n \to \infty} \sum_{i=1}^{n} |f[\xi_i]|\, |\Delta z_i| \leq f_M \lim_{n \to \infty} \sum_{i=1}^{n} |\Delta z_i| = f_M L \tag{1.83}$$

where L is the length of the contour

$$L = \lim_{n \to \infty} \sum_{i=1}^{n} |\Delta z_i| = \int_C |dz| \tag{1.84}$$

Combining the inequalities gives

$$\left| \int_C f[z]\, dz \right| \leq \int_C |f[z]|\, |dz| \leq f_M L \tag{1.85}$$

Thus the magnitude of the line integral must be less than the length of the contour times the maximum magnitude of $f[z]$ on the contour.

1.9 DIVERGENCE THEOREM IN 2D

Consider a vector $\boldsymbol{F}$ existing everywhere in space and an arbitrary closed contour C. The flux per unit length of $\boldsymbol{F}$ through C is defined as

$$\underline{\psi}_{flux} = \oint_C \boldsymbol{F} \cdot \hat{\boldsymbol{n}} \, ds \tag{1.86}$$

where $\hat{\boldsymbol{n}}$ is a unit vector normal to the contour pointing away from the enclosed area of interest. When moving along the boundary of the enclosed area in a counterclockwise direction, the enclosed area is to the left. The symbol $\oint$ represents a closed integral with the contour integrated in the counterclockwise direction.

If we break the contour up into two pieces C_1 and C_2 as shown in Fig. 1.11, the two new contours have a common section between points A and B that is interior to the original contour C which we label B_1. Note for C_1, B_1 is in the B to A direction and for C_2, B_1 is in the A to B direction.

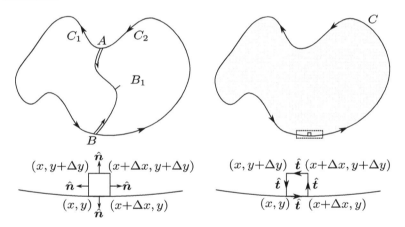

FIGURE 1.11 Figure for the derivation of the Divergence and Curl Theorems.

First we assume $\boldsymbol{F}$ is continuous throughout the enclosed region C. Due to (1.76), the flux through B_1 for C_1 and C_2 cancel each other. Now we allow B_1 to represent the boundary of discontinuity in $\boldsymbol{F}$. We enclose this boundary of discontinuity in a third contour C_3. The left boundary of C_3 along the discontinuity cancels the portion of C_1 along the discontinuity and the right boundary of C_3 cancels the portion of C_2 along the discontinuity. Therefore, the flux through interior boundaries created by breaking the original contour C up into smaller contours can always be made to cancel even in the presence of a discontinuity. Thus,

$$\underline{\psi}_{flux} = \oint_C \boldsymbol{F} \cdot \hat{\boldsymbol{n}} \, ds = \sum_{i=1}^{N} \oint_{C_i} \boldsymbol{F} \cdot \hat{\boldsymbol{n}} \, ds \qquad (1.87)$$

where the sum includes C_i that completely enclose discontinuities in $\boldsymbol{F}$ if they exist. Now divide the interior into small rectangles with sides Δx and Δy. The flux through the ith contour C_i can then be approximated as the value of $\boldsymbol{F}$ at the center of each side of the rectangle times the length of each side with the dot product of $\hat{\boldsymbol{n}}$ and $\boldsymbol{F}$ taken into account

$$\begin{aligned}\underline{\psi}_{flux,C_i} &= F_x\left[x, y + \tfrac{1}{2}\Delta y\right](-\Delta y) + F_x\left[x + \Delta x, y + \tfrac{1}{2}\Delta y\right](\Delta y) \\ &\quad + F_y\left[x + \tfrac{1}{2}\Delta x, y\right](-\Delta x) + F_y\left[x + \tfrac{1}{2}\Delta x, y + \Delta y\right](\Delta x)\end{aligned} \qquad (1.88)$$

Since Δx and Δy are small, we can use first order Taylor polynomial approximations of

$$\begin{aligned}F_x\left[x + \Delta x, y + \tfrac{1}{2}\Delta y\right]\Delta y &\approx F_x\left[x, y + \tfrac{1}{2}\Delta y\right]\Delta y + \frac{\partial F_x}{\partial x}\Delta x \Delta y \\ F_y\left[x + \tfrac{1}{2}\Delta x, y + \Delta y\right]\Delta x &\approx F_y\left[x + \tfrac{1}{2}\Delta x, y\right]\Delta x + \frac{\partial F_y}{\partial y}\Delta y \Delta x\end{aligned} \qquad (1.89)$$

Plugging (1.89) into (1.88), taking the limit as $\Delta x, \Delta y \to 0$, and summing up all the infinitesimal contours that make up the original contour C, we arrive at the total flux given as

$$\begin{aligned}\underline{\psi}_{flux} &= \lim_{\Delta x_i, \Delta y_i \to 0} \sum_{i=1}^{N} \left(\frac{\partial F_x}{\partial x} + \frac{\partial F_y}{\partial y}\right)\Delta x_i \Delta y_i \\ &= \int_{area} \left(\frac{\partial F_x}{\partial x} + \frac{\partial F_y}{\partial y}\right) dx dy\end{aligned} \qquad (1.90)$$

But the divergence of a vector in 2D in Cartesian coordinates is defined as

$$\boldsymbol{\nabla} \cdot \boldsymbol{F} = \frac{\partial F_x}{\partial x} + \frac{\partial F_y}{\partial y} \qquad (1.91)$$

Equating the total flux equations of (1.86) and (1.90) with $da = dx dy$ gives

$$\int_{area} (\boldsymbol{\nabla} \cdot \boldsymbol{F}) \, da = \oint_C (\boldsymbol{F} \cdot \hat{\boldsymbol{n}}) \, ds = \sum_{i=1}^{N} \oint_{C_i} (\boldsymbol{F} \cdot \hat{\boldsymbol{n}}) \, ds \qquad (1.92)$$

where the integral over the area is the area enclosed by the contour C. This is the divergence theorem in 2D.

Note from (1.91), the divergence of a vector $\boldsymbol{F}$ is nonzero only when there is change in $\boldsymbol{F}$ along the direction parallel to at least one of the component directions of $\boldsymbol{F}$. This leads to the geometrical interpretation of the divergence of a vector $\boldsymbol{F}$ as a measure of how much $\boldsymbol{F}$ spreads out (diverges) from the point in question [2]. To help form a mental picture of the divergence, it is often useful to associate $\boldsymbol{F}$ with the surface velocity of a body of water. Imagine sprinkling particles that float on a small portion of the surface of the water. If particles spread out, then this portion of the surface has positive divergence. If particles collect together, then it is a point of negative divergence. If the distance between particles does not change, then the divergence is zero. A point of positive divergence is known as a source and a point of negative divergence is known as a sink. Thus, it is natural for the divergence of $\boldsymbol{F}$ to be associated with the flux of $\boldsymbol{F}$ through a contour C.

1.10   CURL THEOREM IN 2D

Consider a vector $\boldsymbol{F}$ existing everywhere in space and an arbitrary closed contour C. The circulation (Γ) of $\boldsymbol{F}$ around C is defined as

$$\Gamma = \oint_C \boldsymbol{F} \cdot \hat{\boldsymbol{t}} \, ds \tag{1.93}$$

where the vector $\hat{\boldsymbol{t}}$ is a unit vector parallel to the contour C for any point on C. If we break the contour up into two pieces C_1 and C_2 as shown in Fig. 1.11, the common boundary B_1 interior to the original contour C has $\hat{\boldsymbol{t}}$ vectors that are in opposite directions for contours C_1 and C_2. Therefore, the circulation through boundaries that are interior to the original contour C cancel each other and we are left with

$$\Gamma = \oint_C \boldsymbol{F} \cdot \hat{\boldsymbol{t}} \, ds = \sum_{i=1}^{N} \oint_{C_i} \boldsymbol{F} \cdot \hat{\boldsymbol{t}} \, ds \tag{1.94}$$

which has now been generalized to the contour broken up into N pieces. Now divide the interior into small rectangles with sides Δx and Δy. The circulation through the ith contour C_i can then be approximated as the value of $\boldsymbol{F}$ at the center of each side of the rectangle times the length of each side with the dot product of $\hat{t}$ and F taken into account

$$\Gamma_{C_i} = F_x\left[x + \tfrac{1}{2}\Delta x, y\right](\Delta x) + F_x\left[x + \tfrac{1}{2}\Delta x, y + \Delta y\right](-\Delta x)$$
$$+ F_y\left[x + \Delta x, y + \tfrac{1}{2}\Delta y\right](\Delta y) + F_y\left[x, y + \tfrac{1}{2}\Delta y\right](-\Delta y) \tag{1.95}$$

Since Δx and Δy are small, we can use first order Taylor polynomial approximations of

$$F_x\left[x + \tfrac{1}{2}\Delta x, y + \Delta y\right] \approx F_x\left[x + \tfrac{1}{2}\Delta x, y\right] + \frac{\partial F_x}{\partial y}\Delta y$$
$$\tag{1.96}$$
$$F_y\left[x + \Delta x, y + \tfrac{1}{2}\Delta y\right] \approx F_y\left[x, y + \tfrac{1}{2}\Delta y\right] + \frac{\partial F_y}{\partial x}\Delta x$$

Plugging in (1.96) into (1.95), taking the limit as $\Delta x, \Delta y \to 0$, and summing up all the infinitesimal contours that make up the original contour C, we arrival the total circulation

$$\Gamma = \lim_{\Delta x_i, \Delta y_i \to 0} \sum_{i=1}^{N} \left(\frac{\partial F_y}{\partial x} - \frac{\partial F_x}{\partial y}\right)\Delta x_i \Delta y_i$$
$$= \int_{area} \left(\frac{\partial F_y}{\partial x} - \frac{\partial F_x}{\partial y}\right) dx dy \tag{1.97}$$

But the curl of a 2D vector is defined as

$$(\boldsymbol{\nabla} \times \boldsymbol{F})_{2D} = (\boldsymbol{\nabla} \times \boldsymbol{F})_{3D} \cdot \hat{\boldsymbol{k}} = \frac{\partial F_y}{\partial x} - \frac{\partial F_x}{\partial y} \tag{1.98}$$

Equating the total circulation equations of (1.93) and (1.97) with $da = dx dy$ gives

$$\int_{area} (\boldsymbol{\nabla} \times \boldsymbol{F})_{2D}\, da = \oint_C \boldsymbol{F} \cdot \hat{\boldsymbol{t}}\, ds \tag{1.99}$$

where the integral over the area is the area enclosed by the contour C. This is the curl theorem (also known as Stokes' theorem) in 2D. Moving forward, the curl of a vector is understood as the 2D definition and the 2D subscript is dropped.

Note from (1.98), the curl of a vector F is nonzero only when there is change in F along the direction perpendicular to at least one of the component directions of F. This leads to the geometrical interpretation of the curl of a vector F as a measure of how much F swirls around the point in question [2]. To help visualize the curl of F, it is often useful to associate F with the surface velocity of a body of water. Imagine placing a small paddlewheel at the point of interest. If the paddlewheel rotates, then the curl is nonzero there. Thus, it is natural for the curl of F to be associated with the circulation of F around a contour C.

1.11 DIVERGENCE AND CURL THEOREM IN CONJUGATE COORDINATES

The divergence and curl theorems can be expressed as a single equation in terms of conjugate coordinates often labeled Green's theorem. Again, consider a vector F that exists everywhere in space with continuous derivatives and an arbitrary closed contour C. Using (1.60), we have

$$j(\nabla^* F)^* = (\nabla \times F) + j\nabla \cdot F \tag{1.100}$$

Using (1.60), (1.63), and (1.79), we have

$$F^* dz = F \cdot \hat{t}\, ds + j F \cdot \hat{n}\, ds \tag{1.101}$$

With the use of (1.100) and (1.101), the divergence and curl theorems can be expressed as

$$\int_{area} j\nabla F^* da = 2j \int_{area} \left(\frac{\partial F}{\partial z}\right)^* da = \oint_C F^* dz \tag{1.102}$$

where the integral over the area is the area enclosed by the contour C. The divergence theorem is given by the imaginary part of (1.102) and the curl theorem is given by the real part of (1.102). Note if F has a singularity, then (1.102) is still valid if the singularity is excluded from the region enclosed by C. In many books, (1.102) is written as

$$2j \int_{area} \frac{\partial g}{\partial z^*} da = \oint_C g\, dz \tag{1.103}$$

which is easily obtained from (1.102) by letting $F^* = g$.

1.12 CAUCHY'S FIRST INTEGRAL THEOREM

Let F be analytic with derivative $dF/dz = f$. The complex line integral of f is

$$\int_C f\, dz = \int_{t_1}^{t_2} \frac{dF}{dz}\frac{dz}{dt} dt = F[z[t_2]] - F[z[t_1]] \tag{1.104}$$

which states that the line integral of an analytic function depends only on the start and end points of the path and not on the path used for integration. Thus we are free to choose the path that joins the start and end points. If the start and end points are the same, we have

$$\oint_C f\,dz = 0 \tag{1.105}$$

which is known as Cauchy's first integral theorem. Note that Cauchy's first integral theorem is easily proven with (1.103) by letting g be analytic. Since g is analytic, it does not depend on z^* and the left side of (1.103) is zero.

1.13 CAUCHY'S SECOND INTEGRAL THEOREM

Consider the function

$$g = \frac{f[z]}{z - z_0} \tag{1.106}$$

where $f[z]$ is analytic. Thus, g has a single singularity at $z = z_0$. To evaluate the integral

$$\oint_C g\,dz = \oint_C \frac{f[z]}{z - z_0}\,dz \tag{1.107}$$

when the closed contour contains z_0, we use the contour $C_a = C + C_1 + C_2 + C_3$ as shown in Fig. 1.12. Since the region enclosed by C_a does not contain z_0, Cauchy's first integral theorem requires

$$\oint_{C_a} g\,dz = \oint_C g\,dz + \int_{C_1} g\,dz + \oint_{C_2} g\,dz + \int_{C_3} g\,dz = 0 \tag{1.108}$$

Note that the symbol $\oint$ represents that the closed integral of C_2 is in the clockwise direction. Using the identity

$$\oint_{C_2} g\,dz = -\oint_{C_2} g\,dz \tag{1.109}$$

and since C_1 and C_3 cancel each other (they are the same contour but in opposite directions), we have

$$\oint_C g\,dz = \oint_{C_2} g\,dz \tag{1.110}$$

On C_2, $z = z_0 + re^{j\theta}$ and $dz = re^{j\theta}jd\theta$ resulting in

$$\oint_C g\,dz = \oint_{C_2} \frac{f[z]}{z - z_0}dz = j\int_0^{2\pi} f\left[z_0 + re^{j\theta}\right]d\theta \tag{1.111}$$

The reason for using C_2 was to exclude z_0 from our closed contour which allowed using Cauchy's first integral theorem. In order to exclude only z_0, we take the limit of (1.111) as $r \to 0$ giving

$$\oint_C \frac{f[z]}{z - z_0}dz = 2\pi jf[z_0] \tag{1.112}$$

which is known as Cauchy's second integral theorem. An important consequence of Cauchy's second integral theorem is the value of an analytic function $f[z]$ at any point inside a closed contour C is completely determined by the value of $f[z]$ on C.

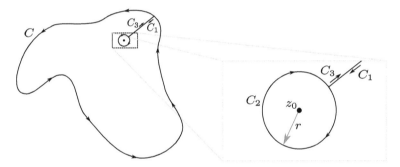

FIGURE 1.12 Figure for the derivation of Cauchy's second integral theorem.

We now extend Cauchy's second integral theorem to the case when z_0 is on the boundary. For the most general case, consider a closed contour C that has a corner located at z_0 as shown in Fig. 1.13. The contour C_a is composed of C_1 that overlaps the original contour C but stops a distance r from z_0 and C_2 which connects the start and end points of C_1 with a small circular arc of radius r centered at z_0 (see Fig. 1.13). In the limit that the radius of the circular arc goes to zero, C_1 becomes the original contour C if the rays that define the start and end of C_2 are parallel to the lines of C that create the corner. Since the region enclosed by C_a does not contain z_0 and $f[z]$ is analytic, we can use Cauchy's first integral theorem to obtain

$$\oint_{C_a} \frac{f[z]}{z - z_0}dz = \int_{C_1} \frac{f[z]}{z - z_0}dz + \int_{C_2} \frac{f[z]}{z - z_0}dz = 0 \tag{1.113}$$

On C_2, $z = z_0 + re^{j\theta}$ and $dz = re^{j\theta}jd\theta$ which allows the integral over C_2 to become

$$\int_{C_2} \frac{f[z]}{z - z_0} dz = j \int_{\theta_2}^{\theta_1} f\left[z_0 + re^{j\theta}\right] d\theta \tag{1.114}$$

Substituting (1.114) into (1.113) and taking the limit as $r \to 0$ gives

$$\oint_C \frac{f[z]}{z - z_0} dz = j \left(\theta_2 - \theta_1\right) f[z_0] \tag{1.115}$$

where $\theta_2 - \theta_1$ is the central angle of the arc. When z_0 is located on a smooth boundary (not a corner), $\theta_2 - \theta_1 = \pi$.

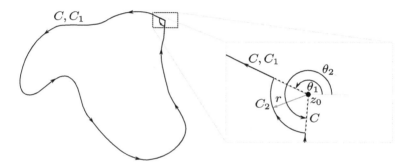

FIGURE 1.13 Figure for extending the derivation of Cauchy's second integral theorem to singularities on the boundary.

Cauchy's second integral theorem can also be used to prove analytic functions are infinitely differentiable. Since the integrand is continuously differentiable (the singularity has been excluded from the region of integration), we can use Leibniz's rule for differentiating an integral (see Appendix A). First we re-write (1.112) as

$$f[z] = \frac{1}{2\pi j} \oint_C \frac{f[w]}{w - z} dw \tag{1.116}$$

Taking the derivative of both sides with respect to z gives

$$\frac{\partial f[z]}{\partial z} = \frac{1}{2\pi j} \oint_C \frac{f[w]}{(w - z)^2} dw \tag{1.117}$$

Since $f[w]$ is analytic in the region enclosed by C, the only singularity that exist is at $w = z$ which has been excluded from the region. Therefore, the derivative is analytic in the region enclosed by C. This process can be repeated to determine higher order derivatives as

$$\frac{\partial^n f[z]}{\partial z^n} = \frac{n!}{2\pi j} \oint_C \frac{f[w]}{(w-z)^{n+1}} dw \qquad (1.118)$$

which are clearly analytic based on the same arguments. Since $(w-z)^{-1}$ is infinitely differentiable, so are analytic functions.

1.14 LAURENT SERIES

Often the series representation of a function in the region around a singularity is desired. Due to limits imposed by the singularity on the radius of convergence, more than one Taylor series would need to be used for the region around a singularity. To overcome this limitation, we now develop what is known as a Laurent series.

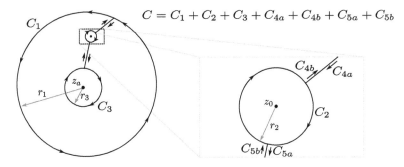

FIGURE 1.14 Figure for the Laurent series derivation.

Let $f[z]$ have a singularity at $z = z_a$. Our goal is to determine the series representation of $f[z]$ inside a circular contour C_1 with center of z_a. We first define a contour C that allows $f[z]/(z-z_0)$ to be analytic within C_1 by excluding the singularities z_a and z_0 enclosed by C_1. As shown in Fig. 1.14, the contour C encloses each singularity with a circle centered at the singularity. Cauchy's first integral theorem requires the closed contour integral of C to be zero. Since the integrals of the lines C_4, C_5 connecting the closed contours C_1, C_2, and C_3 are traced once in each direction ($C_{4a} = -C_{4b}$ and $C_{5a} = -C_{5b}$), they cancel each other and we are left with

$$\oint_C \frac{f[z]}{z-z_0} dz = \oint_{C_1} \frac{f[z]}{z-z_0} dz + \oint_{C_2} \frac{f[z]}{z-z_0} dz + \oint_{C_3} \frac{f[z]}{z-z_0} dz = 0 \qquad (1.119)$$

Note that the contour integrals for C_2 and C_3 are in the clockwise direction. From Cauchy's second integral theorem, we have:

$$2\pi j f[z_0] = \oint_{C_2} \frac{f[z]}{z - z_0} dz \tag{1.120}$$

Converting the integral over C_3 into the standard counterclockwise integral, replacing the integral over C_2 with (1.120), and rearranging (1.119) gives

$$f[z_0] = \frac{1}{2\pi j} \oint_{C_1} \frac{f[z]}{z - z_0} dz - \frac{1}{2\pi j} \oint_{C_3} \frac{f[z]}{z - z_0} dz \tag{1.121}$$

Since z is just a dummy integration variable in (1.121) we can change it to w with no change to the result. Also, z_0 can represent any point contained in the region between C_3 and C_1. Therefore, we can change z_0 to z in (1.121). Making these changes gives

$$f[z] = \frac{1}{2\pi j} \oint_{C_1} \frac{f[w]}{w - z} dw - \frac{1}{2\pi j} \oint_{C_3} \frac{f[w]}{w - z} dw \tag{1.122}$$

The factor $1/(w - z)$ of the first term in (1.122) can be written as

$$\frac{1}{w - z} = \frac{1}{(w - z_a)\left(1 - \frac{z - z_a}{w - z_a}\right)} \tag{1.123}$$

On C_1, $|w - z_a| = r_1$ which means $|(z - z_a)/(w - z_a)| < 1$ for any point z located in the region between C_1 and C_3. We use the geometric series

$$\sum_{n=0}^{\infty} r^n = \frac{1}{1 - r}, \quad |r| < 1 \tag{1.124}$$

to series expand (1.123) as

$$\frac{1}{w - z} = \frac{1}{(w - z_a)} \sum_{n=0}^{\infty} \left(\frac{z - z_a}{w - z_a}\right)^n \tag{1.125}$$

On C_3, $|w - z_a| = r_3$ which requires $|(w - z_a)/(z - z_a)| < 1$ for any point z located in the region between C_1 and C_3. The factor $1/(w - z)$ of the second term in (1.122) can be written as

$$-\frac{1}{w - z} = \frac{1}{(z - z_a)\left(1 - \frac{w - z_a}{z - z_a}\right)} = \frac{1}{(z - z_a)} \sum_{n=0}^{\infty} \left(\frac{w - z_a}{z - z_a}\right)^n \tag{1.126}$$

Using (1.125) and (1.126) and interchanging the summation and integration order allows rewriting (1.122) as

$$f[z] = \frac{1}{2\pi j}\left(\sum_{n=0}^{\infty}\oint_{C_1}\frac{f[w]\,(z-z_a)^n}{(w-z_a)^{n+1}}dw + \sum_{n=0}^{\infty}\oint_{C_3}\frac{f[w]\,(w-z_a)^n}{(z-z_a)^{n+1}}dw\right) \quad (1.127)$$

which is the Laurent series for $f[z]$ in the region between C_3 and C_1. By re-indexing the second series of (1.127), we can write the Laurent series in more traditional form. Letting the second sum start with $n=1$ instead of $n=0$ gives

$$f[z] = \sum_{n=0}^{\infty} a_n\,(z-z_a)^n + \sum_{n=1}^{\infty}\frac{b_n}{(z-z_a)^n} \quad (1.128)$$

where

$$a_n = \frac{1}{2\pi j}\oint_{C_1}\frac{f[w]}{(w-z_a)^{n+1}}dw, \quad b_n = \frac{1}{2\pi j}\oint_{C_3}\frac{f[w]}{(w-z_a)^{-n+1}}dw \quad (1.129)$$

Letting the second sum index over negative integers gives

$$f[z] = \sum_{n=-\infty}^{\infty} c_n\,(z-z_a)^n \quad (1.130)$$

where

$$c_n = \frac{1}{2\pi j}\oint_{\gamma}\frac{f[w]}{(w-z_a)^{n+1}}dw \quad (1.131)$$

and γ is any path that lies in the region between C_3 and C_1 and encircles z_a once in the counterclockwise direction. Note that if z_a is not a singularity but instead an analytic point for $f[z]$, then the integrand for the b_n terms in (1.129) is analytic. The b_n terms must then be zero and the series becomes a Taylor series. The infinite series with a_n coefficients is known as the analytic part of the Laurent series. The infinite series with b_n coefficients is known at the principle part of the Laurent series.

1.15 CLASSIFICATION OF SINGULARITIES

We have previously defined a singularity as a point where $f[z]$ does not exist. Now that the Laurent series has been developed, we can classify different types of singularities based on their Laurent series. There are three types of singularities: Poles, removable singularities, and essential singularities. Their definitions are [7]:

Poles. If the Laurent series of form (1.128) centered at z_a has a finite number of b_n terms, then z_a is called a pole of order m where for $n > m$, $b_n = 0$. A simple pole is when $m = 1$.   (P1)

Removable singularities. If $f[z]$ is not defined at z_a but the $\lim_{z \to z_a} [f[z]]$ exist, then z_a is called a removable singularity. In such cases, all the b_n terms are zero for the Laurent series of form (1.128) centered at z_a.   (P2)

Essential singularities. Any singularity that is not a pole or a removable singularity is called an essential singularity. In such cases, the Laurent series of form (1.128) centered at z_a have an infinite number of b_n terms.   (P3)

Singularities can also occur at infinity. To classify the singularity of $f[z]$ at infinity, let $z = 1/w$ to generate a new function $f[1/w] = F[w]$. The classification of the singularity at infinity is the same classification as $F[w]$ at $w = 0$.

1.16   THE RESIDUE THEOREM

The Laurent series allows evaluating the closed contour integral of functions with singularities within the enclosed region. Let $f[z]$ have a singularity at $z = z_a$ and a Laurent series centered at z_a given in the form of (1.128). Let the region of interest be enclosed by contour C with z_a inside of C. As with previous analysis, we use our standard technique of generating a closed contour C_a that encloses the singularity with a closed circle C_2 centered at z_a that is connected to C with a single path that is traced in both directions (same as Fig. 1.12 with z_0 replace with z_a). In the region between C_2 and C, we can use the Laurent series representation of $f[z]$ and perform a term by term closed contour integration of $f[z]$. By Cauchy's first integral theorem, the closed contour integration over the analytic part of the Laurent series is zero. The integration of the b_n series is

$$\oint_C f[z]\,dz = \oint_{C_2} \left(\frac{b_1}{z - z_a} + \sum_{n=2}^{\infty} \frac{b_n}{(z - z_a)^n} \right) dz \qquad (1.132)$$

On C_2, $z = re^{j\theta} + z_a$ and $dz = jre^{j\theta}d\theta$ allowing (1.132) to be written as

$$\oint_C f[z]\,dz = j\int_0^{2\pi} b_1 d\theta + j\sum_{n=2}^{\infty}\int_0^{2\pi}\frac{b_n}{re^{j(n-1)\theta}}d\theta \qquad (1.133)$$

Performing the integration gives

$$\oint_C f[z]\,dz = 2\pi jb_1 + j\sum_{n=2}^{\infty}\frac{b_n\left(e^{-j(n-1)2\pi} - 1\right)}{-j(n-1)r} \qquad (1.134)$$

Since

$$e^{\pm j2\pi n} = 1 \qquad (1.135)$$

for n any integer, the summation term in (1.134) is zero. We are left with

$$\oint_C f[z]\,dz = 2\pi jb_1 \qquad (1.136)$$

where b_1 is known as the residue. If there are multiple singularities in the enclosed region, then the above method can be applied to each of the singularities resulting in

$$\oint_C f[z]\,dz = 2\pi j\,(\text{the sum of the residues of } f[z] \text{ inside } C_1) \qquad (1.137)$$

which is known as the residue theorem.

Although computing the Laurent series for each singularity provides the value of b_1, another method exists for computing b_1 when the singularity is a pole. Let z_a be the location of the singularity and let the singularity be a pole of order m. Multiplying the Laurent series by $(z - z_a)^m$ gives

$$(z - z_a)^m f[z] = \sum_{n=0}^{\infty} a_n(z - z_a)^{n+m} + \sum_{n=1}^{m} b_n(z - z_a)^{m-n} \qquad (1.138)$$

Differentiating both sides $m - 1$ times with respect to z, taking the limit as $z \to z_a$ and solving for b_1 gives

$$\operatorname*{Res}_{z=z_a}[f[z]] = b_1 = \frac{1}{(m-1)!}\lim_{z\to z_a}\left[\frac{\partial^{m-1}}{\partial z^{m-1}}(z - z_a)^m f[z]\right] \qquad (1.139)$$

For the case of a simple pole ($m = 1$), (1.139) reduces to

$$\operatorname*{Res}_{z=z_a}[f[z]] = \lim_{z\to z_a}[(z - z_a)f[z]] \qquad (1.140)$$

If $f[z]$ can be written as $g[z]/h[z]$ where $g[z]$ is analytic and not zero at z_a and $h[z_a] = 0$, then we can use L'Hopital's rule to determine the limit as $z \to z_a$ of (1.137)

$$\operatorname*{Res}_{z=z_a}[f[z]] = \left(\frac{\partial \left((z - z_a) \, g[z] \right)}{\partial z} \Big/ \frac{\partial h[z]}{\partial z} \right)\Bigg|_{z=z_a} = \frac{g[z_a]}{h'[z_a]} \qquad (1.141)$$

Unfortunately, essential singularities often require the Laurent series expansion in order to obtain the residue.

To define the residue of $f[z]$ at infinity, let $f[z]$ be analytic except for a finite number (n) of singularities. Now enclose all the singularities of $f[z]$ with a contour C. From (1.137), the closed loop integral of C in the counterclockwise direction is

$$\oint_C f[z] \, dz = -\oint_C f[z] \, dz = 2\pi j \sum_{i=1}^{n} \operatorname*{Res}_{z=z_i}[f[z]] \qquad (1.142)$$

But tracing C in the clockwise direction can be viewed as enclosing the region that contains the point at infinity but no singularities at any finite z values. In order for the clockwise integral of (1.142) to be consistent with the residue theorem, we must define the residue at infinity as

$$\operatorname*{Res}_{z=\infty}[f[z]] = -\frac{1}{2\pi j} \oint_C f[z] \, dz \qquad (1.143)$$

Note that if $f[z]$ has n singularities at finite z and all the residues are known, the residue at infinity can be determined as

$$\operatorname*{Res}_{z=\infty}[f[z]] = -\sum_{i=1}^{n} \operatorname*{Res}_{z=z_i}[f[z]] \qquad (1.144)$$

The residue at infinity can also be calculated by using the same change in variable $z = 1/w$ used to classify the singularity at infinity. This change in variable gives

$$f[z] \, dz = -\left(f[1/w] \, / w^2 \right) dw \qquad (1.145)$$

and the residue of $f[z]$ at infinity is

$$\operatorname*{Res}_{z=\infty}[f[z]] = \operatorname*{Res}_{z=0}\left[-\left(f[1/w] \, / w^2 \right) \right] \qquad (1.146)$$

Now the techniques previously described for finding the residue can be used to find the residue at $w = 0$.

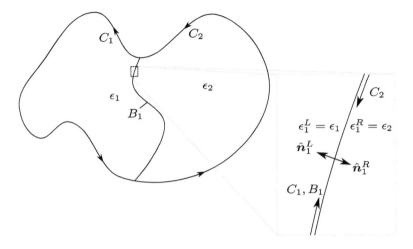

FIGURE 1.15 Figure defining different parameters in Green's identities.

1.17 GREEN'S IDENTITIES IN 2D

In Chapter 3, it is shown how Green's functions can be used to determine the electrostatic potential given a set of boundary conditions. The foundation for Green's functions are two identities known as Green's Identities. Green's Identities can be derived using the divergence theorem by letting $\boldsymbol{F} = G(\epsilon\boldsymbol{\nabla}V)$. G and V are scalar quantities which are finite continuous in the region of integration and can be differentiated twice. ϵ is a scalar quantity which may be differentiated once and may be discontinuous along certain boundaries within the enclosed region. Since we are allowing ϵ to be discontinuous, we break up the integral along the boundary into integrals along boundaries of regions with continuous ϵ and integrals along boundaries that completely enclose the regions of discontinuity as described in the divergence theorem derivation. Also, let the regions enclosing the discontinuity fit so closely to the discontinuity boundary that the normal vector to the boundary on one side of the discontinuity is equal to the negative of the normal vector on the other side of the boundary. Let there be m regions with continuous ϵ and p regions of discontinuities in ϵ, then we have

$$\underline{\psi}_{flux} = \sum_{i=1}^{m} \oint_{C_i} \boldsymbol{F} \cdot \hat{\boldsymbol{n}} \, ds + \sum_{i=1}^{p} \int_{B_i} \left(\boldsymbol{F}_i^R - \boldsymbol{F}_i^L \right) \cdot \hat{\boldsymbol{n}}_i^R \, ds \qquad (1.147)$$

where B_i is the ith discontinuity boundary and the integral is taken from the start and end of B_i in the direction such that the L region is on the left side when tracing the boundary (see Fig. 1.15), F_i^L is F on the L region side, F_i^R is F on the R region side and $\hat{n}_i^R$ points from the L region to the R region. Plugging $F = G(\epsilon \nabla V)$ into (1.92) and substituting $\nabla A \cdot \hat{n} = \partial A / \partial n$ for $A = G$ or V gives

$$\sum_{i=1}^{m} \oint_{C_i} G \epsilon \frac{\partial V}{\partial n} ds + \sum_{i=1}^{p} \int_{B_i} \left(G_i^R \epsilon_i^R \frac{\partial V_i^R}{\partial n_i^R} - G_i^L \epsilon_i^L \frac{\partial V_i^L}{\partial n_i^R} \right) ds$$
$$= \int_{area} \left(\epsilon \nabla V \cdot \nabla G + G \nabla \cdot (\epsilon \nabla V) \right) da \tag{1.148}$$

which is known as Green's first identity. Interchanging G and V we can write a similar equation as

$$\sum_{i=1}^{m} \oint_{C_i} V \epsilon \frac{\partial G}{\partial n} ds + \sum_{i=1}^{p} \int_{B_i} \left(V_i^R \epsilon_i^R \frac{\partial G_i^R}{\partial n_i^R} - V_i^L \epsilon_i^L \frac{\partial G_i^L}{\partial n_i^R} \right) ds$$
$$= \int_{area} \left(\epsilon \nabla G \cdot \nabla V + V \nabla \cdot (\epsilon \nabla G) \right) da \tag{1.149}$$

Subtracting (1.149) from (1.148), we arrive at Green's second identity

$$\sum_{i=1}^{p} \int_{B_i} \left(\epsilon_i^R \left(G_i^R \frac{\partial V_i^R}{\partial n_i^R} - V_i^R \frac{\partial G_i^R}{\partial n_i^R} \right) + \epsilon_i^L \left(V_i^L \frac{\partial G_i^L}{\partial n_i^R} - G_i^L \frac{\partial V_i^L}{\partial n_i^R} \right) \right) ds$$
$$+ \sum_{i=1}^{m} \oint_{C_i} \epsilon \left(G \frac{\partial V}{\partial n} - V \frac{\partial G}{\partial n} \right) ds = \int_{area} (G \nabla \cdot (\epsilon \nabla V) - V \nabla \cdot (\epsilon \nabla G)) da \tag{1.150}$$

When ϵ is a constant without discontinuities, Green's Identities reduce to

$$\int_{area} \left(\nabla G \cdot \nabla V + V \nabla^2 G \right) da = \oint_C V \frac{\partial G}{\partial n} ds$$
$$\int_{area} \left(G \nabla^2 V - V \nabla^2 G \right) da = \oint_C \left(G \frac{\partial V}{\partial n} - V \frac{\partial G}{\partial n} \right) ds \tag{1.151}$$

EXERCISES

1.1 Prove

    a. $|z|^2 = |z^2|$

b. $|z_1 z_2| = |z_1||z_2|$

1.2 Using the complex exponential forms for $\sin[\theta]$ and $\cos[\theta]$ show

a. $\sin[jz] = j\sinh[z]$

b. $\cos[jz] = \cosh[z]$

1.3 Let $f[z]$ be an analytic function with $\mathrm{Re}[f] = u[x,y] = 1+x^3-3xy^2$ and $f[0] = 1$. Using (1.36) show

a. $f[z] = z^3 + 1$

b. $\mathrm{Im}[f] = 3x^2y - y^3$

1.4 Show
$$f[x,y] = 4 - x + x^2 - y^2 - jy + j2xy$$
is analytic for all finite points in the complex plane.

1.5 Show the area of the parallelogram with vertices (in terms of (x,y)) (0,0), (2,1), (3,5) and (1,4) is 7.

1.6 Show the rate of change at $z = 1$ for $f[z, z^*] = z^3 + (z^*)^3$

a. in the vertical direction is 0.

b. in the horizontal direction is 6.

1.7 Contour integration in the complex plane is often used to evaluate improper integrals along the real axis where the integrand is finite but at least one integration limit is infinite. When the integration limits are $\pm\infty$ and the integrand $f[x]$ is a rational function, the contour in the z-plane chosen for integration is often a semicircle in the upper half plane and the real axis connecting the two end points of the semicircle. The integrand is then replaced with $f[z]$, contour integration is performed, and the limit $R \to \infty$ evaluated where R is the radius of the semicircle. Show the integral over the semicircle portion of the contour is zero if $|f[z]| \leq M/R^k$ where $k > 1$ and M is a constant.

1.8 If the integrand in the previous problem is now $e^{jmz} f[z]$ with $m > 0$ and $|f[z]| < M/R^k$, show the integral along the semicircle in the limit $R \to \infty$ is still zero, but the restriction on k is now $k > 0$. A useful inequality for this problem is $\sin[\theta] \geq 2\theta/\pi$ for $0 \leq \theta \leq \pi/2$.

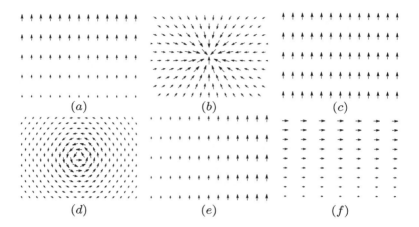

FIGURE 1.16   Vector field plots of six different fields.

1.9   A vector field plot represents the magnitude and direction of the field by the length and direction of an arrow at the location of interest. Vector field plots of six different fields are shown in Figure 1.16. Based on the definition of the divergence and curl of a vector, which plots have a divergence only, curl only, both divergence and curl, and neither divergence or curl? Explain your answers.

1.10   Using Green's theorem, show the area A enclosed by a contour C is given by

$$A = \frac{1}{2j} \oint_C z^* dz$$

then use this result to show the area of a circle of radius r is $A = \pi r^2$.

1.11   Let

$$f[z] = \frac{1}{(z+a)(z+b)}$$

with $0 < a < b$. Find the Laurent series valid for

a. $a < |z| < b$

b. $|z| > b$

c. $|z| < a$

1.12   Let

$$f[z] = \frac{1}{(z+a)(z+b)}$$

with $0 < a < b$. Determine the sum of all the residues contained in the closed contour $|z| = b + 1$ using

a. the Laurent series expansion

b. the residue formula for functions with poles singularities

REFERENCES

[1] Louis Brand. Non-analytic functions of a complex variable. *The American MathematicalMonthly*, 40(5):260–265, May 1933.

[2] David J. Griffiths. *Introduction to Electrodynamics*. Pearson, 4th edition, 2013.

[3] Peter Henrici. *Applied and Computational Complex Analysis, Vol 1*. John Wiley & Sons, Inc., 1974.

[4] Paul J. Nahin. *An Imaginary Tale The Story of $\sqrt{-1}$*. Princeton University Press, 1998.

[5] Tristan Needham. *Visual Complex Analysis*. Oxford University Press, 1998.

[6] William T. Shaw. Recovering holomorphic functions from their real or imaginary parts without the Cauchy–Riemann equations. *SIAM Review*, 46(4):717–728, January 2004.

[7] Murray R. Spiegel. *Complex Variables*. McGraw Hill, 2nd edition, 2009.

CHAPTER 2

Electrostatics

In this chapter, the electromagnetic equations that govern 2D electrostatics are developed in both vector and complex variable representations. This development uses a mix of 3D and 2D equations. Differences exist between some 2D and 3D equations. Equations valid in 3D but not 2D are labeled 3D. Any equation using complex notation is clearly a 2D equation. Additionally, Z is used for the Cartesian axis perpendicular to the x-y plane to reduce the possibility of confusing it with the complex variable z.

2.1 COULOMB'S LAW

Electrostatics is governed by Coulomb's law and the principle of superposition. Coulomb experimentally determined that the force between two small charged bodies is proportional to the product of the charges, inversely proportional to the square of the distance between them, and directed along the line joining them. In linear, homogeneous, isotropic, nonconducting medium, Coulomb's law in 3D is [5]

$$\boldsymbol{F} = \frac{q_1 q_2}{4\pi\epsilon r_{12}^2}\hat{\boldsymbol{r}}_{12}, \quad (3\text{D}) \tag{2.1}$$

where $\boldsymbol{F}$ (unit: N) is the force on charge q_2 (unit: C) due to charge q_1, r_{12} (unit: m) is the distance between q_2 and q_1, $\hat{\boldsymbol{r}}_{12}$ is a unit vector along the line that joins the locations of q_2 and q_1 pointing from q_1 to q_2, and ϵ (unit: F/m) the permittivity which is a constant characteristic of the medium (see Fig. 2.1 (a)). Further details regarding the permittivity are given in Section 2.6.

In order for a 3D problem to be effectively a 2D problem, no variation in one direction is required. Figure 2.1 (a) shows a diagram of Coulomb's law for two point charges in 3D. In order to remove the variation in the Z direction, the point charges must become line charges with uniform line charge density $dq/dZ - \underline{dq}$ (unit: C/m) and axis aligned with the Z-axis as shown in Fig. 2.1 (b). The 2D representation of Fig. 2.1 (b) is shown in Fig. 2.1 (c) where $\underline{q}_i$ now replaces q_i of Fig. 2.1 (a). Additionally, Coulomb's inverse square law of distance in 3D must be modified to the inverse of distance between charged bodies in 2D. In linear, homogeneous, isotropic, nonconducting medium, Coulomb's law in 2D is

$$\underline{F} = \frac{\underline{q}_1 \underline{q}_2}{2\pi\epsilon r_{12}}\hat{r}_{12} = \frac{\underline{q}_1 \underline{q}_2}{2\pi\epsilon\left(z_2^* - z_1^*\right)} \tag{2.2}$$

where $\underline{F}$ (unit: N/m) is now the force per unit length on line charge $\underline{q}_2$ located at z_2 due to line charge $\underline{q}_1$ located at z_1, r_{12} is the distance between z_2 and z_1, $\hat{r}_{12}$ is a unit vector along the line joining z_2 and z_1 pointing from z_1 to z_2, and ϵ is still the permittivity of the medium.

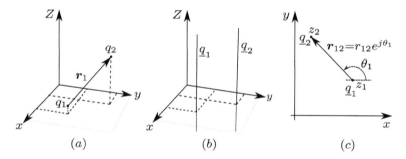

(a) (b) (c)

FIGURE 2.1 (a) Diagram for the force between two point charges in 3D. (b) 3D view of two line charges that are parallel to the Z direction. This is effectively a 2D problem since there is no variation in the Z direction. (c) 2D diagram for the force between two line charges $\underline{q}_1$ located at z_1 and $\underline{q}_2$ located at z_2.

2.2 ELECTRIC FIELD INTENSITY

If there are n charges in a region and we introduce an infinitesimal test charge (so that no redistribution of charge occurs) into the region at point z, the net force on the test charge $\underline{q}$ is given by the vector sum of the forces from all n charges

$$\underline{F} = \frac{\underline{q}}{2\pi\epsilon}\sum_{i=1}^{n}\frac{\underline{q}_i}{z^* - z_i^*} \tag{2.3}$$

This is known as the principle of superposition. The quantity defined by

$$\pmb{\mathcal{E}} = \pmb{F}\big/\underline{q} = \frac{1}{2\pi\epsilon} \sum_{i=1}^{n} \frac{q_i}{z^* - z_i^*} \qquad (2.4)$$

is called the electric field intensity (or electric field). The electric field has unit of V/m in 2D and 3D.

To visualize the electric field for a line charge in 3D, a vector field plot can be generated by plotting arrows at select points. There are different conventions on where to place the arrow. For the vector plots in this book, the center of the arrow corresponds to the location of the electric field represented by the arrow. The length of the arrow is related to the magnitude of the electric field and the arrow points in the direction of the electric field. A vector field plot of a positive line charge in 3D is shown in Fig. 2.2 (a) along with the 2D representation in Fig. 2.2 (b). Although the vector field is only shown for one plane in Fig. 2.2 (a), since there is no variation in the Z direction, the vector field is the same in any plane parallel to the x-y plane. Note that electric field points away from the positive line charge as indicated by the arrows and decreases as the location of interest moves away from the line charge as indicated by the smaller arrow size. There is no electric field due to the line charge in the Z direction which allows this problem to be equivalent to a 2D problem.

Based on the geometric interpretation of the divergence and curl of a vector (Sections 1.9 and 1.10), the vector field plot of a positive line charge shows a positive divergence (the electric field is spreading out as the distance from the line charge increases) and the curl is zero (there is not circulation). These are important observations that come up again when Gauss's law and the electrostatic potential are discussed.

2.3   ELECTRIC FIELDS OF DIPOLES AND MULTIPOLES

The electric field of n line charges can be computed directly from (2.4), but a useful approximation is possible when the net charge of all the line charges is zero and the line charges are separated by small distances compared to the distance where knowledge of the electric field is desired. The simplest configuration consists of two line charges separated by a small distance and is called a dipole. To calculate the electric field of a dipole, first consider the electric field at z due to a line charge $\underline{q}$ located at z_0 which we label $\mathcal{E}_+[z, z_0]$. If $\underline{q}$ is displaced by $\frac{1}{2}\Delta z_0$, the electric field at z becomes $\mathcal{E}_+[z, z_0 + \frac{1}{2}\Delta z_0]$. We want to compute the electric field

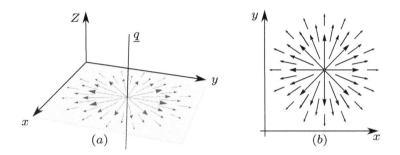

FIGURE 2.2 (a) Vector plot of $\boldsymbol{\mathcal{E}}$ in (a) 3D and (b) 2D for a positive line charge. Note the radial dependence is depicted by the arrows pointing away from the charge and the reduction of $\mathcal{E}$ with distance from the charge is represented by the reduced arrow size further from the location of the charge.

under the condition $|\Delta z_0| \ll |z - z_0|$. Since $\mathcal{E}^*$ is analytic at z, we can expand $\mathcal{E}_+^*\left[z, z_0 + \frac{1}{2}\Delta z_0\right]$ as a first order Taylor polynomial and then take the complex conjugate to obtain

$$\mathcal{E}_+\left[z, z_0 + \tfrac{1}{2}\Delta z_0\right] \approx \mathcal{E}_+[z, z_0] + \frac{\partial \mathcal{E}_+}{\partial z_0^*}\left(\tfrac{1}{2}\Delta z_0\right)^* \tag{2.5}$$

Similar arguments lead to the electric field at z due to a line charge $-\underline{q}$ located at $z_0 - \frac{1}{2}\Delta z_0$ of

$$\mathcal{E}_-\left[z, z_0 - \tfrac{1}{2}\Delta z_0\right] \approx \mathcal{E}_-[z, z_0] + \frac{\partial \mathcal{E}_-}{\partial z_0^*}\left(-\tfrac{1}{2}\Delta z_0\right)^* \tag{2.6}$$

The total electric field at z is the sum of these two fields

$$\mathcal{E}_{dip} = \mathcal{E}_+\left[z, z_0 + \tfrac{1}{2}\Delta z_0\right] + \mathcal{E}_-\left[z, z_0 - \tfrac{1}{2}\Delta z_0\right] \tag{2.7}$$

Since $\mathcal{E}_+[z, z_0] = -\mathcal{E}_-[z, z_0]$, we have

$$\mathcal{E}_{dip} \approx \frac{\partial \mathcal{E}_+}{\partial z_0^*}\left(\Delta z_0\right)^* = \frac{\underline{q}}{2\pi\epsilon}\frac{\left(\Delta z_0\right)^*}{\left(z^* - z_0^*\right)^2} \tag{2.8}$$

The operation $(\Delta z_0)^* \partial/\partial z_0^*$ can be viewed as an electric field dipole operator that acts on the electric field of a single line charge to produce the electric field of a dipole. When the electric field dipole operator acts on the electric field of a positive line charge, the dipole produced has a dipole moment per unit length (unit: C) defined as

$$\underline{\boldsymbol{M}} = \underline{M} = \underline{q}\Delta z_0 \tag{2.9}$$

where $\underline{q}$ is the value of the positive line charge of the dipole and Δz_0 is a vector with magnitude equal to the distance between the line charges and points from negative to positive charge. In general, the dipole moment per unit length of a group of line charges with a net charge of zero is independent of the origin and defined as

$$\underline{M} = \sum_i \underline{q}_i z_i = \underline{q}_+ (\bar{z}_+ - \bar{z}_-) \qquad (2.10)$$

where $\underline{q}_+$ is the total amount of positive charge, $\bar{z}_+$ is the center of positive charge, and $\bar{z}_-$ is the center of negative charge defined as

$$\bar{z}_+ = \sum_i \underline{q}_{+,i} z_{+,i} \Big/ \sum_i \underline{q}_{+,i}, \quad \bar{z}_- = \sum_i \underline{q}_{-,i} z_{-,i} \Big/ \sum_i \underline{q}_{-,i} \qquad (2.11)$$

The electric field dipole operator can be defined in terms of the dipole moment per unit length it produces when it acts on the electric field of a positive line charge $\underline{q}$ as

$$\text{Electric field dipole operator} = \left(\frac{\underline{M}}{\underline{q}}\right)^* \frac{\partial}{\partial z_0^*} \qquad (2.12)$$

If the electric field dipole operator acts on the electric field of a negative line charge $-\underline{q}$, we have

$$\mathcal{E}_{dip} = \left(\frac{\underline{M}}{\underline{q}}\right)^* \frac{\partial \mathcal{E}_-}{\partial z_0^*} = \left(\frac{\underline{M}}{\underline{q}}\right)^* \frac{\partial (-\mathcal{E}_+)}{\partial z_0^*} = \left(\frac{-\underline{M}}{\underline{q}}\right)^* \frac{\partial \mathcal{E}_+}{\partial z_0^*} \qquad (2.13)$$

which is the same dipole produced by operating on the electric field of a positive line charge $\underline{q}$, but with the opposite polarity dipole moment per unit length $(-\underline{M})$. The electric field dipole operator acting on a positive and negative line charge is shown in Fig. 2.3.

Although (2.8) is an approximation for the electric field of a dipole with charges separated by a distance $|\Delta z_0|$, the electric field of a point dipole is exactly described by (2.8). A point dipole is obtained by taking the limit of (2.7) as $\Delta z_0 \to 0$ while $q \to \infty$ in such a way that $\underline{M} = \underline{q}\Delta z_0$ remains a constant. All the higher terms of the series expansion are exactly zero in this limit. The vector plot of the electric field for a point dipole compared to line charges with a finite separation are shown in Fig. 2.4. Note that along the axis for both dipole types, the electric field and the dipole moment point in the same direction when the distance is large from the dipole's center location.

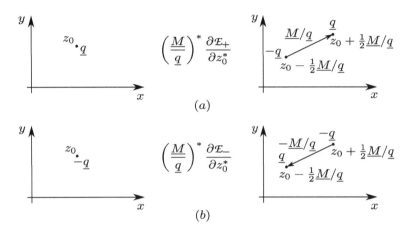

FIGURE 2.3 The creation of a dipole through the electric field dipole operator acting on the electric field of (a) a positive and (b) a negative line charge. The dipole moment per unit length $\underline{M}$ is the same for both diagrams.

If the electric field dipole operator acts on the electric field of a dipole, two dipoles are produced and we obtain the electric field of a quadrupole. The quadrupole configuration depends on the orientation of the dipole moment of the electric field dipole operator relative to the original dipole moment. For example, if the electric field dipole operator has the same dipole moment as the original dipole, we have a linear quadrupole (see Fig. 2.5 (a)) with electric field

$$\mathcal{E}_{quad,lin}[z] = \left(\frac{M}{\underline{q}}\right)^* \frac{\partial}{\partial z_0^*}\left(\left(\frac{M}{\underline{q}}\right)^* \frac{\partial \mathcal{E}_+}{\partial z_0^*}\right) \qquad (2.14)$$

If the electric field dipole operator has a dipole moment with the same magnitude but perpendicular to the original dipole moment, we have a square quadrupole (see Fig. 2.5 (b)) with electric field

$$\mathcal{E}_{quad,sq}[z] = \left(\frac{j\underline{M}}{\underline{q}}\right)^* \frac{\partial}{\partial z_0^*}\left(\left(\frac{M}{\underline{q}}\right)^* \frac{\partial \mathcal{E}_+}{\partial z_0^*}\right) \qquad (2.15)$$

Other quadrupole configurations can be obtained by changing the dipole moment of the electric field dipole operator. The electric fields of more complex multipole configurations where the total charge is zero can be obtained by using the electric field dipole operator on the electric field of more complex charge configurations.

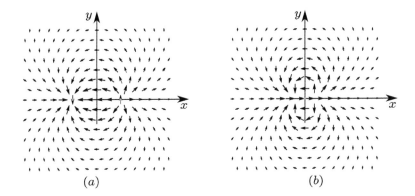

FIGURE 2.4 (a) Vector plot of a finite dipole. (b) Vector plot of a point dipole. The distance in both plots has been normalized to the distance between charges of the finite dipole. The finite dipole moment and point dipole moment are equal.

2.4 CONTINUOUS CHARGE DISTRIBUTIONS

In many cases, the number of charges in a region is large and the spacing between charges is small enough so that a continuous charge distribution can be used to describe the charge configuration. If we designate dq as the charge located at (x, y, Z) , then three different continuous charge distributions can be defined to describe dq. When the charge distribution varies within a volume, the volume charge density ρ (unit: C/m^3) is defined as

$$dq = \rho \, dv, \quad (3D) \qquad\qquad (2.16)$$

where dv (unit: m^3) is an infinitesimal volume. When the charge distribution varies over a surface, the surface charge density σ (unit: C/m^2) is defined as

$$dq = \sigma \, da, \quad (3D) \qquad\qquad (2.17)$$

where da (unit: m^2) is an infinitesimal area. When the charge distribution varies along a line, the line charge density λ (unit: C/m) is defined as

$$dq = \lambda \, ds, \quad (3D) \qquad\qquad (2.18)$$

where ds (unit: m) is an infinitesimal length. Converting 3D continuous charge distributions into 2D continuous charge distributions is done in the same way point charges in 3D were transformed into 2D line charges. Again we require the 3D continuous charge distribution to be constant along the Z direction. Then in 2D, a charge distribution that varies along a contour is a sheet of charge that is perpendicular to the z-plane

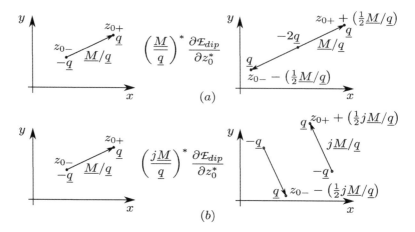

FIGURE 2.5 (a) Creation of a linear quadrupole through the electric field dipole operator acting on the electric field of a dipole. (b) Creation of a square quadrupole through the electric field dipole operator acting on the electric field of a dipole. The value of $\underline{M}$ is the same in all plots.

as shown in Fig. 2.6. Let the sheet of charge have a surface charge density σ that depends only on its z position. If we take a thin slice ds of the sheet as shown in Fig. 2.6, then we have a line of charge with value $\sigma[z]ds$. In other words, the sheet can be viewed as being made up of closely spaced line charges. This leads to the 2D relationship

$$d\underline{q} = \sigma[x, y]\, ds = \sigma[z]\, |dz| \tag{2.19}$$

A 2D charge distribution that varies over an area is a volume charge density ρ in 3D that depends only on the z position

$$d\underline{q} = \rho[x, y]\, da = \rho[z]\, da \tag{2.20}$$

The electric field at position z due to $d\underline{q}$ at z_0 is

$$d\boldsymbol{E} = \frac{1}{2\pi\epsilon}\left(\frac{d\underline{q}[z_0]}{z^* - z_0^*}\right) \tag{2.21}$$

To obtain the total electric field due to a charge distribution that varies over an area in the z-plane, we substitute (2.20) into (2.21) and integrate over the area that contains charge

$$\boldsymbol{E} = \frac{1}{2\pi\epsilon}\int\limits_{area}\frac{\rho[z_0]\, da_0}{z^* - z_0^*} \tag{2.22}$$

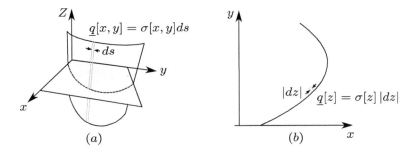

FIGURE 2.6 (a) 3D view of surface charge that does not vary in the Z direction. It can be viewed as being composed of different line charges. (b) 2D representation of the surface charge shown in (a).

To obtain the total electric field due to a charge distribution that varies along a contour C in the z-plane, we substitute (2.19) into (2.21) and integrate over the contour

$$\mathbf{E} = \frac{1}{2\pi\epsilon} \int_C \frac{\sigma[z_0]\,|dz_0|}{z^* - z_0^*} \tag{2.23}$$

2.5 GAUSS'S LAW IN 2D

Consider a line charge q_i located at z_i and a point z_A on a closed contour C (see Fig. 2.7) in a linear, homogeneous, isotropic, nonconducting medium. Using (2.3), the electric field at z_A due to q_i is

$$\mathcal{E}[z_A] = \frac{q_i}{2\pi\epsilon\left(z_A^* - z_i^*\right)} = \frac{q_i e^{j\theta_1}}{2\pi\epsilon r_1} \tag{2.24}$$

Now add a small angle $d\theta$ to $z_A - z_i$ and label this point z_C. Label the location where C and the line that goes through z_i and z_C intersect as z_B. The flux per unit length, $d\psi_{flux}$ (unit: C/m), of $\epsilon\mathcal{E}$ through the small section of the closed contour defined by $dz = z_B - z_A$ is

$$d\psi_{flux} = \epsilon\mathbf{E}\cdot\hat{\mathbf{n}}\,ds = \text{Im}[\epsilon\mathcal{E}^*dz] \tag{2.25}$$

where (1.63) was used to obtain the last equality. With the angles α and β defined in Fig. 2.7, $dz = |dz|\exp[j\beta]$ and (2.25) can be evaluated as

$$d\psi_{flux} = \text{Im}\left[\frac{q_i e^{j(\beta-\theta_1)}\,|dz|}{2\pi r_1}\right] = \frac{q_i \cos[\alpha]\,|dz|}{2\pi r_1} \tag{2.26}$$

The infinitesimal triangle formed by points z_A, z_B, and z_C gives

$$\cos[\alpha] = r_1 d\theta/|dz| \tag{2.27}$$

allowing (2.26) to be written as

$$2\pi d\underline{\psi}_{flux} = \underline{q}_i d\theta \tag{2.28}$$

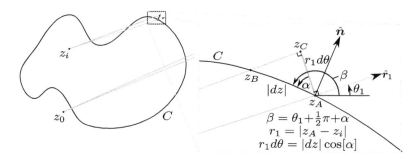

FIGURE 2.7 Incremental flux through an incremental portion of a closed contour used to prove Gauss's law.

Tracing C in the direction such that the bounded region is on the left, leads to a positive $d\theta$ if dz is in same direction and a negative $d\theta$ if dz is in the opposite direction. When the point charge is located inside the closed contour (z_i for example), the cone cuts the contour n times where n is an odd integer and $d\theta$ is positive $\frac{1}{2}(n+1)$ times and negative $\frac{1}{2}(n-1)$ times. The net flux through the cone is $\underline{q}_i d\theta/(2\pi)$. When the point charge is located outside the closed contour (z_0 for example), the cone cuts the contour an even number of times with equal positive and negative $d\theta$ values so that the net flux is zero. To obtain the total flux through the contour surrounding the charge, we integrate the normal component over the entire contour and obtain

$$2\pi \oint_C d\underline{\psi}_{flux} = \underline{q}_i \int_0^{2\pi} d\theta \tag{2.29}$$

$$\underline{\psi}_{flux} = \underline{q}_i$$

By adding up the flux due to all the charges inside the contour, we arrive at Gauss's law

If any closed contour is taken in an electric field and if $\boldsymbol{\mathcal{E}}$ is the electric field intensity at any point on the contour and $\hat{n}$ the unit outward normal vector to the surface, then

$$\oint_C \epsilon \boldsymbol{E} \cdot \hat{\boldsymbol{n}} \, ds = \sum_i \underline{q}_i = \underline{q}_{enc} \tag{2.30}$$

where the integration extends over the whole contour and $\underline{q}_{enc}$
is the total net charge enclosed by the contour.

Cauchy's second integral theorem provides a much simpler derivation of Gauss's law. For a line charge $\underline{q}_i$ located at z_i, the conjugate of electric field $\boldsymbol{E}^*$ is a function of z only and has a simple pole at $z = z_i$. Letting

$$f[z] = (z - z_i) \, \epsilon \boldsymbol{E}^* = \frac{\underline{q}_i}{2\pi} \tag{2.31}$$

in (1.112) immediately gives

$$\oint_C \epsilon \boldsymbol{E}^* dz = j \underline{q}_i \tag{2.32}$$

The principle of superposition leads to

$$\oint_C \epsilon \boldsymbol{E}^* dz = \sum_i j \underline{q}_i = j \underline{q}_{enc} \tag{2.33}$$

where $\underline{q}_i$ is the ith line charge located at the ith location z_i inside C. Since

$$\oint_C \epsilon \boldsymbol{E}^* dz = \oint_C \epsilon \boldsymbol{E} \cdot \hat{\boldsymbol{t}} \, ds + j \oint_C \epsilon \boldsymbol{E} \cdot \hat{\boldsymbol{n}} \, ds \tag{2.34}$$

the imaginary part of (2.34) is Gauss's law. Note that the real part of (2.33) is always zero which requires

$$\oint_C \epsilon \boldsymbol{E} \cdot \hat{\boldsymbol{t}} \, ds = 0 \tag{2.35}$$

Physically, this is a consequence of the static electric field being a conservative field.

Using the extended version of Cauchy's second integral theorem, Gauss's law for charge on the boundary can be obtained. Assuming a smooth boundary and letting

$$\epsilon \boldsymbol{E}^* = \frac{f[z]}{z - z_0} \tag{2.36}$$

the imaginary part of (1.115) gives

$$\oint_C \epsilon \boldsymbol{\mathcal{E}} \cdot \hat{n}\, ds = \tfrac{1}{2} \underline{q}_{boundary} \tag{2.37}$$

Physically this means that only half of the flux for a charge on a smooth boundary is considered to originate from within the enclosed region.

One example of the power of Gauss's law is to prove that all 2D systems are charge neutral if the point at infinity is included in the system. The same arguments made back in Section 1.16 regarding the residue at infinity can be made with Gauss's law leading to

$$\oint_C \epsilon \boldsymbol{\mathcal{E}}^* dz = -\oint_C \epsilon \boldsymbol{\mathcal{E}}^* dz = j\underline{q}_{enc} \tag{2.38}$$

With the view that tracing C in a clockwise direction encloses the area that would normally be considered exterior to C means that an equal magnitude but opposite polarity of total charge enclosed by C must exist external to C. If all the charge at finite z has been enclosed by C and the net charge enclosed is not zero, then an equal magnitude but opposite polarity of the total charge enclosed by C is located at infinity.

2.6 POLARIZATION

In matter, dipole moments can be created from different stimuli. For example, an electric field produces a force on the positive nucleus of an atom in the direction of the electric field and electrons of an atom experience a force in the opposite direction. These forces produce a small separation between the positive and negative charges in the atom creating a dipole. When the dipoles line up, the material has become polarized. The measure of this effect is the polarization defined in 3D as the dipole moment per unit volume. In 2D the polarization vector $\boldsymbol{\mathcal{P}}$ (unit: C/m^2) at z_0 is defined as the dipole moment per unit length per unit area

$$\boldsymbol{\mathcal{P}} = \mathcal{P}[z_0] = d\underline{M}[z_0]/da_0 \tag{2.39}$$

The electric field due to a polarized region is the sum of the electric fields from all the dipoles in that region

$$\mathcal{E} = \int_{area} \frac{1}{2\pi\epsilon} \frac{(d\underline{M}[z_0])^*}{(z^* - z_0^*)^2} = \int_{area} \frac{1}{q} \frac{\partial \mathcal{E}_+}{\partial z_0^*} \mathcal{P}[z_0]^* \, da_0 \tag{2.40}$$

We can recast (2.40) into an alternative form by using

$$\frac{\partial\left(\mathcal{E}_+\mathcal{P}^*\right)}{\partial z_0^*} = \mathcal{E}_+\frac{\partial\mathcal{P}^*}{\partial z_0^*} + \frac{\partial\mathcal{E}_+}{\partial z_0^*}\mathcal{P}[z_0]^*, \quad \frac{\partial\left(\mathcal{E}_+\mathcal{P}\right)}{\partial z_0} - \mathcal{E}_+\left(\frac{\partial\mathcal{P}}{\partial z_0}\right) = 0 \quad (2.41)$$

which allows writing (2.40) as

$$q\underline{E} = \int_{area}\frac{\partial\left(\mathcal{E}_+\mathcal{P}^*\right)}{\partial z_0^*}da_0 + \int_{area}\frac{\partial\left(\mathcal{E}_+\mathcal{P}\right)}{\partial z_0}da_0 - \int_{area}\mathcal{E}_+\left(\frac{\partial\mathcal{P}}{\partial z_0} + \frac{\partial\mathcal{P}^*}{\partial z_0^*}\right)da_0 \quad (2.42)$$

The first two integrals of (2.42) can be converted to closed contour integrals with the aid of (1.103) which leads to

$$q\underline{E} = \oint_C \mathcal{E}_+\left(\frac{\mathcal{P}^*(-jdz_0) + \mathcal{P}(-jdz_0)^*}{2}\right) - \int_{area}\mathcal{E}_+\left(\frac{\partial\mathcal{P}^*}{\partial z_0^*} + \frac{\partial\mathcal{P}}{\partial z_0}\right)da_0 \quad (2.43)$$

Recognizing

$$\frac{\mathcal{P}^*\left(-jdz_0\right) + \mathcal{P}(-jdz_0)^*}{2} = \mathrm{Re}\left[\mathcal{P}^*\left(\frac{-jdz_0}{|dz_0|}\right)\right]|dz_0| = \boldsymbol{\mathcal{P}}\cdot\hat{\boldsymbol{n}}_0\,|dz_0| \quad (2.44)$$

where $\hat{\boldsymbol{n}}_0$ is a unit vector normal to contour C and

$$\frac{\partial\mathcal{P}^*}{\partial z_0^*} + \frac{\partial\mathcal{P}}{\partial z_0} = \mathrm{Re}\left[2\frac{\partial\mathcal{P}}{\partial z_0}\right] = \boldsymbol{\nabla}_0\cdot\boldsymbol{\mathcal{P}} \quad (2.45)$$

gives

$$\mathcal{E} = \oint_C\frac{\mathcal{E}_+}{\underline{q}}\left(\boldsymbol{\mathcal{P}}\cdot\hat{\boldsymbol{n}}_0\right)|dz_0| + \int_{area}\frac{\mathcal{E}_+}{\underline{q}}\left(-\boldsymbol{\nabla}_0\cdot\boldsymbol{\mathcal{P}}\right)da_0 \quad (2.46)$$

Comparing the terms of (2.46) with (2.22) and (2.23) indicates that the electric field of a polarized region is the same as the electric field produced by a surface charge density

$$\sigma_b = \boldsymbol{\mathcal{P}}\cdot\hat{\boldsymbol{n}}_0 \quad (2.47)$$

and a volume charge density

$$\rho_b = -\boldsymbol{\nabla}_0\cdot\boldsymbol{\mathcal{P}} \quad (2.48)$$

The charges associated with polarization are called bound charges and all other charges are called free charges. The total charge in a given region is the sum of the free and bound charges. Knowledge of the total charge allows calculating the electric field.

With the polarization vector now defined, we can define the electric flux density vector $\boldsymbol{D}$ (unit: C/m²) as

$$\boldsymbol{D} = \epsilon_0 \boldsymbol{E} + \boldsymbol{P} \tag{2.49}$$

where ϵ_0 is a constant with an approximate value of 8.85×10^{-12} F/m. To define a linear medium, the polarization vector is broken up into a component dependent on the electric field $\boldsymbol{P}_E$ and a component independent of electric field $\boldsymbol{P}_{other}$

$$\boldsymbol{P} = \boldsymbol{P}_E + \boldsymbol{P}_{other} \tag{2.50}$$

A linear medium is defined as having polarization that is linearly related to the electric field,

$$\boldsymbol{P}_E[\boldsymbol{E}] = \epsilon_0 \, [\![\chi_e]\!] \, \boldsymbol{E} \tag{2.51}$$

where

$$[\![\chi_e]\!] = \begin{bmatrix} \chi_{11} & \chi_{12} \\ \chi_{12} & \chi_{22} \end{bmatrix} \tag{2.52}$$

is the electric susceptibility tensor of the medium that quantifies the amount of polarization an applied electric field can produce in the medium. Although the medium is linear, the polarization may be different in different directions depending on the χ_e tensor values.

When $\boldsymbol{P}_{other} = 0$, $\boldsymbol{D}$ and $\boldsymbol{E}$ are related through a symmetrical permittivity tensor

$$\boldsymbol{D} = [\![\epsilon]\!] \, \boldsymbol{E} \tag{2.53}$$

where

$$[\![\epsilon]\!] = \begin{bmatrix} \epsilon_{11} & \epsilon_{12} \\ \epsilon_{12} & \epsilon_{22} \end{bmatrix} \tag{2.54}$$

If the linear medium is also isotropic, the polarization is independent of direction. The susceptibility and permittivity tensors reduce to single quantities and we have the simplified relationship for linear isotropic mediums of

$$\boldsymbol{P} = \boldsymbol{P}_E = \epsilon_0 \chi_e \boldsymbol{E} \tag{2.55}$$

and

$$\boldsymbol{D} = \epsilon \boldsymbol{E} \tag{2.56}$$

The permittivity ϵ of a linear, isotropic medium is related to the electric susceptibility as

$$\epsilon = \epsilon_0 \left(1 + \chi_e\right) \tag{2.57}$$

In vacuum, the electric susceptibility is zero (nothing to polarize) and the permittivity is ϵ_0. The unitless quantity $\epsilon_r = \epsilon/\epsilon_0$ is called the relative permittivity or dielectric constant of the material and measures the ability of a material to become polarized by an electric field. The larger ϵ_r is, the more easily the material is polarized with an electric field.

2.7 MAXWELL'S EQUATIONS

Electrostatics can also be described in terms of Maxwell's equations. Since there are no time varying fields or current flow, the magnetic field is zero and Maxwell's equations consists of only two vector equations. The first vector equation can be obtained by combining the divergence theorem (1.92) with Gauss's law (2.30) giving

$$\int_{area} (\boldsymbol{\nabla} \cdot \epsilon \boldsymbol{E})\, da = \oint_C \epsilon \boldsymbol{E} \cdot \hat{\boldsymbol{n}}\, ds = \underline{q}_{enc} \qquad (2.58)$$

When deriving (2.30), we assumed a linear, homogeneous, isotropic, nonconducting medium, but did not specify what type of charge (free or bound) we were considering. Since the permittivity ϵ in (2.24) already takes into account the polarization charge created by the electric field, $\underline{q}_{enc}$ must be the sum of bound charge not due to the electric field $(\underline{q}_{b,other})$ and free charge $(\underline{q}_f)$. We can re-write $\underline{q}_{enc}$ as

$$\underline{q}_{enc} = \int_{area} (\rho_f + \rho_{b,other})\, da = \int_{area} (\rho_f - \boldsymbol{\nabla} \cdot \boldsymbol{P}_{other})\, da \qquad (2.59)$$

and $\epsilon \boldsymbol{E}$ as

$$\epsilon \boldsymbol{E} = \epsilon_0 \boldsymbol{E} + \boldsymbol{P}_{\mathcal{E}} \qquad (2.60)$$

Substituting (2.59) and (2.60) into (2.58) gives

$$\int_{area} (\boldsymbol{\nabla} \cdot (\epsilon_0 \boldsymbol{E} + \boldsymbol{P}_{\mathcal{E}} + \boldsymbol{P}_{other}))\, da = \int_{area} (\boldsymbol{\nabla} \cdot \boldsymbol{D})\, da = \int_{area} \rho_f da \qquad (2.61)$$

or in differential form

$$\boldsymbol{\nabla} \cdot \boldsymbol{D} = \rho_f, \quad \text{ME}\,(I) \qquad (2.62)$$

which is the first Maxwell's equation. An alternative version of ME(I) in terms of the electric field is

$$\boldsymbol{\nabla} \cdot \boldsymbol{E} = \rho/\epsilon_0, \quad \text{ME}\,(Ia) \qquad (2.63)$$

where $\rho = \rho_f + \rho_b$.

The second vector equation for electrostatics can be obtained by the fact that the static electric field is a conservative field. The amount of work done in a conservative field is independent of the path. For example, the work done in moving a test charge q from A to B in an electric field $\boldsymbol{\mathcal{E}}$ is not dependent on the path taken to get from A to B. In 2D, we have a force per unit length $\underline{\boldsymbol{F}}$ which leads to work per unit length $\underline{W}$ (unit: N) defined as

$$\underline{W} = \int_A^B \underline{\boldsymbol{F}} \cdot \hat{\boldsymbol{t}} \, ds = \int_A^B -\underline{q}\boldsymbol{\mathcal{E}} \cdot \hat{\boldsymbol{t}} \, ds \qquad (2.64)$$

The minus sign in (2.64) results from the fact the electric field is exerting a force of $\underline{q}\boldsymbol{\mathcal{E}}$ on $\underline{q}$ requiring a force of $-\underline{q}\boldsymbol{\mathcal{E}}$ in opposition. Using the curl theorem and the property that work is zero for a closed loop in a conservative field gives

$$-\int_{area} \left(\boldsymbol{\nabla} \times \underline{q}\boldsymbol{\mathcal{E}}\right) da = -\oint_C \underline{q}\boldsymbol{\mathcal{E}} \cdot \hat{\boldsymbol{t}} \, ds = 0 \qquad (2.65)$$

which was already proven in Section 2.5. We can re-state (2.65) in differential form as

$$\boldsymbol{\nabla} \times \boldsymbol{\mathcal{E}} = 0, \quad \text{ME}\,(II) \qquad (2.66)$$

which is the second Maxwell's equation. The 2D electrostatic Maxwell's equations ME (Ia) and ME (II) can be expressed as a single equation in the complex plane as

$$\nabla^* \boldsymbol{\mathcal{E}} = \rho/\epsilon_0 \qquad (2.67)$$

2.8 BOUNDARY CONDITIONS

In general, discontinuities in $\boldsymbol{\mathcal{D}}$ and $\boldsymbol{\mathcal{E}}$ occur at the boundary between two different mediums or at a boundary with charge. The boundary conditions that determine the discontinuities are now discussed.

Consider a boundary in the z plane as shown in Fig. 2.8. Let the boundary have a 3D free surface charge density perpendicular to the z-plane of σ_f. Enclose the boundary with a loop composed of sides parallel (C_1, C_2) and perpendicular (C_3, C_4) to the boundary as shown in Fig. 2.8 (a). Integrating the flux of $\boldsymbol{\mathcal{D}}$ over the closed loop and using Gauss's law gives

$$\oint_C \boldsymbol{\mathcal{D}} \cdot \hat{\boldsymbol{n}} \, ds = \underline{q}_{f,enc} \qquad (2.68)$$

Letting the sides of the loop that are perpendicular to the enclosed boundary go to zero $(C_3, C_4 \rightarrow 0)$ leaves only the charge along the boundary enclosed by the loop and the integrals along C_1 and C_2. The integral of (2.68) reduces to

$$\int_{C_2} \boldsymbol{D}_2 \cdot \hat{\boldsymbol{n}}_2 \, ds + \int_{C_1} \boldsymbol{D}_1 \cdot \hat{\boldsymbol{n}}_1 \, ds = \int_{C_2} \sigma_f \, ds \qquad (2.69)$$

Since C_1 and C_2 are the same contour in opposite directions, (2.69) can be written as

$$\int_{C_2} \boldsymbol{D}_2 \cdot \hat{\boldsymbol{n}}_2 \, ds - \int_{C_2} \boldsymbol{D}_1 \cdot \hat{\boldsymbol{n}}_2 \, ds = \int_{C_2} \sigma_f \, ds \qquad (2.70)$$

Defining $\hat{\boldsymbol{n}}_{12}$ as a unit vector normal to the boundary with a direction pointing from medium 1 into medium 2 as shown in Fig. 2.8, the discontinuity in $\boldsymbol{D}$ is

$$(\boldsymbol{D}_2 - \boldsymbol{D}_1) \cdot \hat{\boldsymbol{n}}_{12} = \sigma_f \qquad (2.71)$$

Thus, free surface charge at a boundary causes a discontinuity in the normal component of $\boldsymbol{D}$. If the surface charge is zero, the normal component of $\boldsymbol{D}$ is continuous at the boundary.

Again enclose the boundary with a loop containing sides parallel and perpendicular to the boundary as shown in Fig. 2.8. Integrating ME(II) over the area enclosed by the loop and using the curl theorem to convert the integral of $(\nabla \times \boldsymbol{E})$ over this area into a line integral along the loop gives

$$\oint_C \boldsymbol{E} \cdot \hat{\boldsymbol{t}} \, ds = 0 \qquad (2.72)$$

Letting the sides of loop that are perpendicular to the boundary go to zero results in the second boundary condition

$$(\boldsymbol{E}_2 - \boldsymbol{E}_1) \cdot \hat{\boldsymbol{t}} = 0 \qquad (2.73)$$

where $\hat{\boldsymbol{t}}$ is a unit vector tangential to the surface as shown in Fig. 2.8. Thus, the tangential component of the electric field is continuous at boundaries.

2.9   ELECTROSTATIC POTENTIAL

For electrostatics, Maxwell's equations can also be stated in terms of a potential function. Since the curl of the gradient of any scalar function

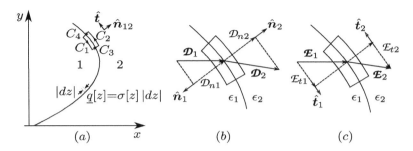

FIGURE 2.8 Unit vector definitions for normal and tangential boundary conditions.

is always zero,

$$\boldsymbol{\nabla} \times (\boldsymbol{\nabla}\phi) = 0, \tag{2.74}$$

we can define the electrostatic potential function for $\boldsymbol{\mathcal{E}}$ as

$$\boldsymbol{\mathcal{E}} = -\boldsymbol{\nabla}V \tag{2.75}$$

The electrostatic potential V (unit: V) is an extremely power tool for solving electrostatic problems. Instead of dealing with vector equations, you only need to determine a scalar quantity. Once V is known, then $\boldsymbol{\mathcal{E}}$ is found using (2.75).

In terms of the electrostatic potential, ME(Ia) (2.63) becomes

$$\boldsymbol{\nabla} \cdot \boldsymbol{\nabla}V = \nabla^2 V = -\rho/\epsilon_0 \tag{2.76}$$

which is known as Poisson's equation. In regions absent of charge ($\rho = 0$), Poisson's equation reduces to

$$\nabla^2 V = 0 \tag{2.77}$$

which is known as Laplace's equation.

To understand the physical significance of the potential function V, we evaluate the rate of change in V along a contour C_1. Let two points (z_1 and z_2) on C_1 be separated by an infinitesimal distance $ds = |z_2 - z_1|$. The rate of change in V between these two points, $\partial V/\partial s$, is the dot product of the gradient of V and the unit vector $\hat{t}$ in the direction from z_1 to z_2

$$\frac{\partial V}{\partial s} = \boldsymbol{\nabla}V \cdot \hat{t} \tag{2.78}$$

With the aid of (2.75) and (2.78), work per unit length can be written in terms of the electrostatic potential as

$$\underline{W} = \int_A^B -\underline{q}\boldsymbol{\mathcal{E}} \cdot \hat{t} \, ds = \int_A^B \underline{q}\frac{\partial V}{\partial s} ds = \underline{q}\left(V[B] - V[A]\right) \tag{2.79}$$

Thus, the work per unit length to move a line charge q from location A to B is just the electrostatic potential difference between A and B times the value of the line charge.

Note that adding an arbitrary constant to V does not change (2.75) or (2.79). This means the difference in potential values has physical significance and the zero of the potential function can be located anywhere V is defined. In 3D, the potential of a point charge is proportional to $1/r$ and the zero of potential for a point charge is typically chosen as infinity. In 2D, the potential for a line charge q located at z_0 can be obtained by integrating (2.79) with $\mathcal{E}$ given by (2.4)

$$V = -\text{Re}\left[\int \mathcal{E}^* dz\right] = -\text{Re}\left[\int \frac{q}{2\pi\epsilon(z - z_0)} dz\right] \qquad (2.80)$$

Performing the integration gives

$$V = -\frac{q}{2\pi\epsilon}\text{Re}[\ln[z - z_0]] + C_0 = -\frac{q}{2\pi\epsilon}\ln[|z - z_0|] + C_0 \qquad (2.81)$$

where C_0 is a real constant. Since V is undefined at $z = \infty$, the zero of potential for a line charge cannot be located at infinity.

2.10 COMPLEX POTENTIAL

In regions that obey Laplace's equation (charge free regions), we can define what is known as the complex potential. Working with the complex potential for 2D electrostatics leads to many mathematical simplifications. In general, the complex potential $\Phi[z]$ (unit: V) is defined as an analytic function (except for a discrete number of singularities) with the real part equal to the electrostatic potential

$$\text{Re}[\Phi[z]] = V[x, y] \qquad (2.82)$$

Since the Imaginary part of an analytic function can be obtained from its Real part using (1.36), the complex potential can be defined in terms of the electrostatic potential (to within an arbitrary constant) as

$$\Phi[z] = V + j\Psi = 2V\left[\frac{z + z_1^*}{2}, \frac{z - z_1^*}{2i}\right] - (\Phi[z_1])^* \qquad (2.83)$$

where $\Phi[z_1]$ has a defined value.

Although the electric field $\mathcal{E}$ can be computed in conjugate coordinates using (1.56) as

$$\mathcal{E}[z] = -\boldsymbol{\nabla}V = -2\frac{\partial V}{\partial z^*} \qquad (2.84)$$

it is often more convenient to compute $\mathbfcal{E}$ directly from Φ in charge free regions. Since Φ is an analytic function, we have

$$\frac{\partial \Phi}{\partial z^*} = 0, \quad \frac{\partial \Phi^*}{\partial z} = 0 \tag{2.85}$$

We can now re-write the gradient of V as

$$\mathbfcal{E}[z] = -\nabla V = -2\frac{\partial}{\partial z^*}\left(\frac{\Phi + \Phi^*}{2}\right) = -\left(\frac{\partial \Phi}{\partial z}\right)^* \tag{2.86}$$

allowing $\mathbfcal{E}$ to be obtained directly from $\Phi[z]$. When $\mathbfcal{E}$ is known, the relationship given by (2.86) can be reversed allowing $\Phi[z]$ to be determined as

$$\Phi[z] = -\int (\mathbfcal{E}[z])^* dz \tag{2.87}$$

From (2.81), the complex potential for a line charge q located at z_0 is

$$\Phi[z] = -\frac{q}{2\pi\epsilon}\ln[z - z_0] + C_0 \tag{2.88}$$

where C_0 is now an arbitrary complex constant. The principle of superposition allows generalizing (2.88) to n charges as

$$\Phi[z] = -\frac{1}{2\pi\epsilon}\sum_{i=1}^{n} q_i \ln[z - z_i] + C_0 \tag{2.89}$$

Similar to the electrostatic potential, it is the difference in complex potential values that has physical significance. The value of C_0 allows setting the reference location for V and Ψ.

For a general charge distribution, the integral of (2.21) can be used for $\mathbfcal{E}$ in (2.87) and the integration over z peformed to obtain

$$\Phi[z] = -\frac{1}{2\pi\epsilon}\int_{charge} d\underline{q}[z_0]\ln[z - z_0] + C_0 \tag{2.90}$$

For a charge distribution that varies over an area, $d\underline{q}$ can be replaced with (2.20) giving

$$\Phi[z] = -\frac{1}{2\pi\epsilon}\int_{area} \rho[z_0]\ln[z - z_0]\, da_0 + C_0 \tag{2.91}$$

For a charge distribution that varies along a contour C, $d\underline{q}$ can be replaced with (2.19) giving

$$\Phi[z] = -\frac{1}{2\pi\epsilon}\int_{C} \sigma[z_0]\ln[z - z_0]\,|dz_0| + C_0 \tag{2.92}$$

Work per unit length can be written in terms of the complex potential as

$$W = \int_A^B -q\boldsymbol{E}\cdot\hat{\boldsymbol{t}}\,ds = \mathrm{Re}\left[\int_{z_A}^{z_B} q\frac{\partial\Phi}{\partial z}dz\right] = q\mathrm{Re}[\Phi[z_B] - \Phi[z_A]] \quad (2.93)$$

which is just a re-statement of (2.79) in terms of the complex potential.

Analytic functions have mixed partial derivatives that are equal requiring Φ to have the following properties

$$\frac{\partial^2\Phi}{\partial z^*\partial z} = \frac{\partial^2\Phi}{\partial z\partial z^*} \quad \text{and} \quad \frac{\partial^2\Phi}{\partial x\partial y} = \frac{\partial^2\Phi}{\partial y\partial x} \quad (2.94)$$

Expanding out the first equation of (2.94) in terms of x and y gives

$$\frac{\partial^2\Phi}{\partial z^*\partial z} = \frac{\partial^2\Phi}{\partial x^2} + \frac{\partial^2\Phi}{\partial y^2} = \nabla^2\Phi \quad (2.95)$$

Since Φ is independent of z^*, (2.95) must equal zero and

$$\nabla^2 V = \nabla^2\Psi = 0 \quad (2.96)$$

The Real and Imaginary parts of Φ (or any analytic function) are both solutions of Laplace's equation (2.77).

2.11 COMPLEX POTENTIAL FOR A DIPOLE

The complex potential for a point dipole located at z_0 with dipole moment per unit length $\underline{M}$ can be obtained by integrating the electric field of the dipole

$$\Phi_{dip}[z] = -\int (\mathcal{E}_{dip}[z])^*dz = -\int \left(\frac{M}{q}\right)\frac{\partial\mathcal{E}_+^*}{\partial z_0}dz \quad (2.97)$$

where

$$\mathcal{E}_+^* = -\frac{\partial\Phi_+}{\partial z} \quad (2.98)$$

and Φ_+ are the electric field and complex potential of a positive line charge q located at z_0, respectively. Replacing $\mathcal{E}_+$ in (2.97) with (2.98) and reversing the order of integration and differentiation gives

$$\Phi_{dip}[z] = \left(\frac{M}{q}\right)\frac{\partial}{\partial z_0}\int\frac{\partial\Phi_+}{\partial z}dz = \left(\frac{M}{q}\right)\frac{\partial\Phi_+}{\partial z_0} = \left(\frac{M}{q}\right)\frac{1}{2\pi\epsilon}\frac{1}{(z - z_0)} \quad (2.99)$$

2.12   COMPLEX POTENTIAL FOR A DOUBLE LAYER

A double layer consists of a layer of positive line charges on one side of a contour and an identical layer of negative line charges on the other side of the contour. The strength of the line charges can vary along the contour which results in a position dependent dipole moment per unit length. Let z_0 correspond to a point on the contour. The direction of the dipole moment per unit length $d\underline{M}$ at z_0 due to a differential length $|dz_0|$ is always normal to the contour

$$d\underline{M} = |d\underline{M}|\,\hat{n}_0 = |d\underline{M}|\left(-j\frac{dz_0}{|dz_0|}\right) \tag{2.100}$$

Note we have taken the convention that the positive charge is to the right of the negative charge when integrating along C. The complex potential $d\Phi$ at z due to $d\underline{M}$ is obtained from (2.99) as

$$d\Phi = \left(\frac{d\underline{M}}{\underline{q}}\right)\frac{\partial\Phi_+}{\partial z_0} \tag{2.101}$$

where Φ_+ is the complex potential due to a positive line charge $\underline{q}$ located at z_0. Integrating over the entire dipole distribution along C gives the total complex potential at z as

$$\Phi = \frac{1}{\underline{q}}\int_C \frac{\partial\Phi_+}{\partial z_0}d\underline{M} = -\frac{j}{\underline{q}}\int_C \frac{|d\underline{M}|}{|dz_0|}\frac{\partial\Phi_+}{\partial z_0}dz_0 \tag{2.102}$$

where (2.100) was used on the right hand side of (2.102).

The electrostatic potential is obtained by taking the real part of (2.102)

$$V = \mathrm{Re}[\Phi] = \mathrm{Im}\left[\frac{1}{\underline{q}}\int_C \frac{|d\underline{M}|}{|dz_0|}\frac{\partial\Phi_+}{\partial z_0}dz_0\right] = \frac{1}{2\pi\epsilon}\int_C \frac{|d\underline{M}|}{|dz_0|}\mathrm{Im}\left[\frac{dz_0}{z-z_0}\right] \tag{2.103}$$

where θ is the angle of vector $(z-z_0)$ as shown in Fig. 2.9 (a). With the definitions of α, γ, and θ in Fig. 2.9,

$$\mathrm{Im}\left[\frac{dz_0}{z-z_0}\right] = \frac{|dz_0|}{|z-z_0|}\sin[\beta-\theta] \tag{2.104}$$

and

$$\beta + \gamma + \pi = \theta, \quad \sin[\beta-\theta] = \sin[\gamma] \tag{2.105}$$

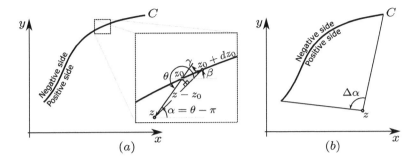

FIGURE 2.9 Distribution of point dipoles along a contour C is called a double layer. The potential integration is performed in the direction such that the positive side of the double layer is on the right (from bottom left to upper right in these two diagrams). (a) Definition of θ and α. (b) Definition of $\Delta\alpha$ with $\Delta\alpha > 0$ in the counterclockwise direction.

For an infinitesimal dz_0, the angle at vertex $z_0 + dz_0$ of the infinitesimal shaded triangle shown in Fig. 2.9 is γ, the length of the hypotenuse is $|dz_0|$, and the length of the shortest side is $|z - z_0|\, d\alpha$. We can now re-write (2.103) as

$$V = \frac{1}{2\pi\epsilon} \int_{\alpha_1}^{\alpha_2} \frac{|d\underline{M}|}{|dz_0|} d\alpha \qquad (2.106)$$

If $|d\underline{M}|\,/|dz_0|$ is constant, then the electrostatic potential at point z is

$$V = \frac{\Delta\alpha}{2\pi\epsilon} \frac{|d\underline{M}|}{|dz_0|} \qquad (2.107)$$

where $\Delta\alpha$ is defined as the angle contained between the line connecting z to the end of contour C and the line connecting z to the start of contour C as shown in Fig. 2.9 (b). Note $\Delta\alpha$ is measured from the line formed by connecting z to the end of C and is positive in the counterclockwise direction.

Crossing the double layer in the normal direction at any point z_1 on C results in a well defined potential difference. To determine the potential drop across the double layer, we divide the double layer contour into three segments and calculate the potential for each segment separately. Then using the principle of superposition, the potentials for all segments are summed to produce the potential due to the original double layer. Let segment 1 begin at the start of the contour and ends at $z_1 - dz_0$. Let segment 2 start at $z_1 + dz_0$ and end where the contour ends. Segment 3 is the remaining contour between $z_1 - dz_0$ and $z_1 + dz_0$. Let the potential due to segments 1 and 2 be V_1. When considering only segments 1 and 2, the crossing point z_1 is free of any dipole requiring V_1 to be continuous at z_1

$$\Delta V_1[z_1] = 0 \tag{2.108}$$

Thus, the dipoles outside of the crossing point do not contribute to the potential drop due to crossing the double layer. Any difference in potential when crossing the double layer at z_1 must be due to the dipole moment per unit area $|d\underline{M}| / |dz_0|$ at z_1. Now let dz_0 be so small that $|d\underline{M}| / |dz_0|$ is a constant over segment 3. Let the point on the positive side and normal to the double layer be

$$z = z_1 - j\frac{dz_0}{|dz_0|}\Delta z \tag{2.109}$$

An expanded view of segment 3 that includes this point is shown in Fig. 2.10. The angle contained between the lines connecting z to the start and ends points of segment 3 is

$$\Delta\alpha = 2\tan^{-1}\left[\frac{|dz_0|}{|\Delta z\,(dz_0/\,|dz_0|)|}\right] = 2\tan^{-1}\left[\frac{|dz_0|}{|\Delta z|}\right] \tag{2.110}$$

In the limit that $|\Delta z| \to 0$, $\Delta\alpha \to \pi$ and the potential on the positive side of the double layer is given by (2.103) as

$$V_3\left[z_1^+\right] = \frac{1}{2\epsilon}\left.\left|\frac{d\underline{M}}{dz_0}\right|\right|_{z_0=z_1} \tag{2.111}$$

Similar analysis on the negative side of the double layer gives $\Delta\alpha = -\pi$ and the potential

$$V_3\left[z_1^-\right] = -\frac{1}{2\epsilon}\left.\left|\frac{d\underline{M}}{dz_0}\right|\right|_{z_0=z_1} \tag{2.112}$$

The potential difference by crossing the double layer (negative side to positive side) is

$$\Delta V = \Delta V_1[z_1] + V_3\left[z_1^+\right] - V_3\left[z_1^-\right] = \frac{1}{\epsilon}\left.\left|\frac{d\underline{M}}{dz_0}\right|\right|_{z_0=z_1} \tag{2.113}$$

2.13 TRANSFORMING POISSON'S EQUATION INTO LAPLACE'S EQUATION

A large portion of this book is devoted to finding the electrostatic potential and electric field in regions that are free of charge (governed by

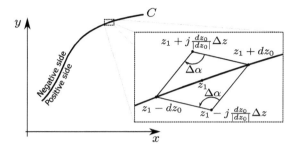

FIGURE 2.10 Calculating $\Delta\alpha$ for the potential drop across a double layer.

Laplace's equation) where the complex potential Φ is valid. In regions where charge exist (governed by Poisson's equation), Φ is no longer valid and the advantages obtained by using analytic functions of a complex variable would seem to be lost. Fortunately, it is possible in some cases to transform Poisson's equation into Laplace's equation by letting [3]

$$V = \Omega + f[x, y] \tag{2.114}$$

and choosing the function f so that

$$\nabla^2 \Omega = 0 \tag{2.115}$$

Once transformed into Laplace's equation, all the methods developed to solve Laplace's equation are valid. The solution to Poisson's equation is then obtained by evaluating (2.114). Thus the key to this method is the ability to obtain a suitable $f[x, y]$.

To demonstrate the procedure, consider the case when $\rho = \rho_0$ is a constant and the material is free space. Poisson's equation is

$$\nabla^2 V = -\rho_0/\epsilon_0 \tag{2.116}$$

If we let

$$f[x, y] = -\rho_0 x^2 / (2\epsilon_0) \quad \text{or} \quad f[x, y] = -\rho_0 y^2 / (2\epsilon_0) \tag{2.117}$$

then plugging (2.114) into (2.116) gives

$$\nabla^2 \Omega = 0 \tag{2.118}$$

Once the solution for Ω is found, V can be computed as

$$V = \Omega - \frac{\rho_0 x^2}{2\epsilon_0} \quad \text{or} \quad V = \Omega - \frac{\rho_0 y^2}{2\epsilon_0} \tag{2.119}$$

depending on which function of f was used.

When boundaries are present, boundary conditions lead to unique solutions of Poisson's equation (see Section 2.19). Thus, any changes to the boundary conditions when transforming from Poisson's to Laplace's equation must be taken into account. In general, the boundary conditions for Ω are

$$\Omega|_C = (V - f)|_C \tag{2.120}$$

where C represents any position on the boundary. Unless f is zero on the boundary, the boundary conditions for V and Ω are different.

When the regions of charge do not extend to the boundaries, the boundary conditions do not change and determining Ω is equivalent to determining Φ in the charge free regions. When the complex potential Φ_1 for a single line of charge of unit value that satisfies the boundary conditions is known, the complex potential Φ in the charge free regions can be obtained by superposition

$$\Phi = \int_{\text{all charges}} \Phi_1[z, z_0]\, \rho[z_0]\, da_0 \tag{2.121}$$

Once Φ is determined, the real part gives

$$\text{Re}[\Phi] = \begin{cases} V, & \text{charge free regions} \\ \Omega, & \text{regions with charge} \end{cases} \tag{2.122}$$

If the transforming function f is known, the potential in the regions with charge is obtained by using (2.114). Assuming the boundaries for the regions of charge are free of double layers, V is continuous across the boundary, but $\text{Re}[\Phi]$ may not be. The discontinuity depends on the transforming function f. For these cases, it is often necessary to add

$$Ax + B \quad \text{or} \quad Ay + B \tag{2.123}$$

(where A and B are constants) to f so that V is continuous across the boundary.

2.14   EQUIPOTENTIAL CONTOURS

With the electrostatic potential defined, the properties of equipotential contours (or boundaries) can be developed. An equipotential contour is a contour where the potential function is a constant. Therefore, no work is performed when a charge moves along an equipotential boundary.

Additionally, $\partial V/\partial s = 0$ on an equipotential contour. From (2.78), this leads to the requirement

$$\frac{\partial V}{\partial s} = \boldsymbol{\nabla} V \cdot \hat{\boldsymbol{t}} = 0 \tag{2.124}$$

for equipotential contours. In order for (2.124) to be true for all cases, the gradient of V must be perpendicular to equipotential contours (the dot product vanishes). Thus the electric field is always perpendicular to equipotential contours. This leads to the alternative expression for the gradient of

$$\boldsymbol{\nabla} V = (\partial V/\partial n)\,\hat{\boldsymbol{n}} \tag{2.125}$$

where dn is evaluated along a path perpendicular to the equipotential contours of V and $\hat{\boldsymbol{n}}$ is a unit vector that points in the direction of increasing potential perpendicular to the equipotential contour. In the limit of infinitesimally spaced equipotential contours, the contours are parallel and the perpendicular distance dn between the equipotential contours is the shortest distance between the contours. In other words, $\partial V/\partial s$ is maximized when $ds = dn$. Thus, the direction of the gradient of V coincides with the direction of greatest rate of increase of V.

2.15 LINES OF FORCE

In Section 2.2, the vector field plot was introduced to visualize the electric field. An alternative method for visualizing the electric field is to plot the equipotential contours along with lines of force contours. A physical description for a line of force contour can be obtained by first considering a region that has an electric field. Now introduce an infinitesimal charge and release it. If the charge is somehow prevented from gaining any significant momentum, the path the infinitesimal charge traces as it moves through the region is a line of force contour. The tangent at any point along a line of force contour is parallel to the electric field vector at that point. Let ds represent an infinitesimal distance between two points on a line of force contour and let $\hat{\boldsymbol{t}}$ represent a vector that is parallel to the line of force contour where ds is located. The line of force contour is then defined as

$$ds\,\hat{\boldsymbol{t}} = \zeta\boldsymbol{\mathcal{E}} \tag{2.126}$$

where ζ is a real constant. Writing out the components in Cartesian coordinates and equating values of ζ produces the differential equation for the lines of force

$$\frac{dx}{E_x} = \frac{dy}{E_y} \quad \text{or} \quad \frac{dy}{dx} = \frac{E_y}{E_x} \tag{2.127}$$

Note that since the lines of force are parallel to the electric field, they must be perpendicular to equipotential contours. Although solving (2.127) is one approach for obtaining the lines of force equations, a better method exist if the complex potential can be determined.

In Section 2.14, we established that the gradient of an analytic function f has a direction that is perpendicular to contours of constant value of f. Therefore, the contours generated by $\text{Im}[\Phi] = \Psi$ constant have the direction given by the vector $\pm j\nabla\Psi$. The direction of $\pm j\nabla\Psi$ is

$$\frac{dy}{dx} = \frac{\text{Im}[\pm j\nabla\Psi]}{\text{Re}[\pm j\nabla\Psi]} = \frac{\partial\Psi/\partial x}{-\partial\Psi/\partial y} \tag{2.128}$$

From the Cauchy-Riemann equations (1.33), we can re-write (2.128) as

$$\frac{dy}{dx} = \frac{-dV/dy}{-dV/dx} = \frac{E_y}{E_x} \tag{2.129}$$

showing that the imaginary part of the complex potential is the solution to the differential equation (2.127). If the complex potential is known, $\text{Im}[\Phi] = \Psi$ equal to a constant are the lines of force equations and $\text{Re}[\Phi] = V$ equal to a constant are the equipotential contour equations.

2.16 FIELD MAPS

Using the complex potential for a line charge (2.88), we can plot the equipotential contours and the lines of force contours creating a field map (see Fig. 2.11). In contrast to the vector field plot, the field map does not use arrows of different sizes to represent the field strength of a region. Instead, the number of lines of force per unit length in the region represents the field strength. For example, the equipotential contours of a line charge are circles with length $2\pi r$ where r is the distance from the line charge. The number of field lines crossing the equipotential contours is constant as we move away from the line charge. Thus the field strength is proportional to $1/r$ which is consistent with (2.4).

When plotting field maps, a constant value between adjacent equipotential curves (ΔV) and a constant value between adjacent lines of force $(\Delta\Psi)$ should be used. When the system has a constant permittivity (ϵ),

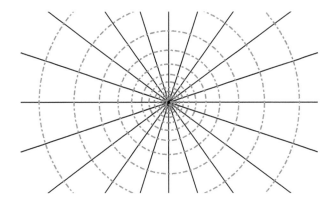

FIGURE 2.11   Field map of a single line charge located at the center of the plot.
$\Delta V = \Delta \Psi = 0.05\underline{q}_0/\epsilon$.

plotting field maps with $|\Delta V| = |\Delta \Psi|$ allows an estimation of the capacitance through the method of curvilinear squares (see Section 2.22).

In systems with constant permittivity and charge existing only on the boundaries, a simple relationship between field lines and the dielectric flux per unit length can be established. The dielectric flux per unit length through a contour between any two points in the field is

$$\underline{\psi}_{flux} = \int_{s_1}^{s_2} \boldsymbol{D} \cdot \hat{n}\, ds = \int_{z_1}^{z_2} \mathrm{Im}[\mathcal{D}^* dz] \qquad (2.130)$$

where the identity given in (1.63) was used. Maxwell's equations for linear media give

$$\boldsymbol{D} = \epsilon \boldsymbol{E} = -\epsilon \left(\frac{\partial \Phi}{\partial z} \right)^*, \quad \mathcal{D}^* = -\epsilon \frac{\partial \Phi}{\partial z} \qquad (2.131)$$

We can rewrite (2.130) as

$$\underline{\psi}_{flux} = \int_{z_1}^{z_2} \mathrm{Im}\left[-\epsilon \frac{\partial \Phi}{\partial z} dz \right] = -\epsilon \int_{\Psi[z_1]}^{\Psi[z_2]} d\Psi = \epsilon \left(\Psi[z_1] - \Psi[z_2] \right) \quad (2.132)$$

The flux through a contour between two points is just the permittivity times the difference between the imaginary part of the complex potential evaluated at the two points. Since the charge on the boundary and the flux are related through Gauss's law, the direction of the integral is such that the charged boundary of interest is to the left when going from z_1 to z_2.

Many field maps are drawn throughout the book. All field maps have the following properties. Since charges are the sources of electric fields, lines of force start on positive charges and terminate on negative charges. The tangent of a point on a line of force is the direction of the electric field at that point. Therefore, lines of force cannot cross. If they did, this would mean the electric field has two different directions.

When only conductors (see Section 2.20) of different potential are present, the conductors at the highest potential must be entirely positive and lines of force are initiated from them, but do not terminate on them. The conductors at the lowest potential must be entirely negative and lines of force terminate on them, but no lines of force are initiated from them. There are no lines of force between conductors of the same potential. Based on the properties discussed, regions of high and low electric field and the direction of the electric field at any point should be obvious when analyzing a field map of conductors at different potential drawn with $\Delta V = \Delta \Psi$. For example, the field map shown in Fig. 2.12 has the center conductor at potential $V_1 > 0$ and the circular outer conductor grounded. The field lines therefore, go from the center conductor to the circular ground. The largest distance between field lines occurs along the horizontal sides of the center conductor. Therefore, the horizontal sides of the center conductor are the regions of lowest electric field and lowest surface charge density. The field lines are closest at the point labeled A of the center conductor. Thus, point A has the highest electric field. Although points A and C are the same distance from the outer conductor, the electric field profile is very different near these points. The spacing between lines of force is much more uniform for the region near point C compared to the region near point A. Thus, the curved region near point C results in a much more uniform electric field compared to the nonuniform electric field near point A. Once we learn the method of curvilinear squares, the capacitance between the conductors as well as the energy stored in different locations can also be determined from field maps.

2.17 GAUSS'S LAW FOR INHOMOGENEOUS MEDIUMS

In Section 2.5, the derivation of Gauss's law assumed the permittivity of the enclosed area was a constant. We now show that the total flux around a closed contour is equal to the total net charge enclosed by the contour even when the permittivity and electric field intensity are functions of position. Let C enclose an area that has a permittivity that

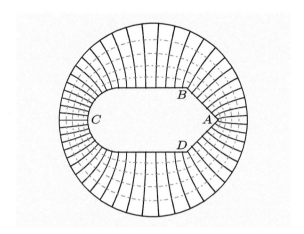

FIGURE 2.12 Field map of a conductor at potential $V_1 > 0$ inside a grounded circular conductor. The field map has been drawn with $\Delta V = \Delta \Psi = 0.2 V_1$.

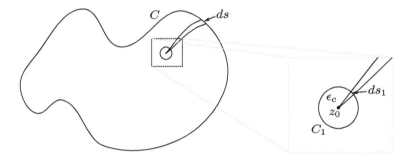

FIGURE 2.13 Definition of different parameters used to prove Gauss's law for inhomogeneous mediums.

is a function of position and contains a line charge $\underline{q}_0$ at $z = z_0$. We can draw a contour C_1 around z_0 so small that the permittivity inside C_1 is a constant ϵ_c (see Fig. 2.13). Now take an element ds of the contour C so small that ϵ is a constant over ds. The constant lines of flux from the edges of ds on C can be traced back to $\underline{q}_0$. A portion ds_1 of C_1 lies between these lines of flux. Now apply Gauss's law to the uncharged dielectric bound by the flux lines, ds, and ds_1. The normal component of $\boldsymbol{D}$ is zero on the sides since they are constant flux contours. The only contribution to the line integral is from ds and ds_1, so that

$$\epsilon_c \boldsymbol{E}_1 \cdot \hat{\boldsymbol{n}} \, ds_1 = \epsilon \boldsymbol{E} \cdot \hat{\boldsymbol{n}} \, ds \qquad (2.133)$$

Integrating over the two contours C and C_1 gives

$$\epsilon_c \oint_{C_1} \boldsymbol{E}_1 \cdot \hat{\boldsymbol{n}}_1 \, ds_1 = \oint_C \epsilon \boldsymbol{E} \cdot \hat{\boldsymbol{n}} \, ds \qquad (2.134)$$

The left side of (2.134) can be evaluated using (2.30) as $\underline{q}_0$ since ϵ_c is a constant and (2.134) becomes

$$\oint_C \epsilon \boldsymbol{E} \cdot \hat{\boldsymbol{n}} \, ds = \underline{q}_0 \qquad (2.135)$$

The results of (2.135) can be extended to multiple charges by the principle of superposition leading to $\underline{q}_0$ being replaced by $\underline{q}_{enc}$.

2.18   DIELECTRIC BOUNDARY CONDITIONS FOR Φ

The dielectric boundary conditions with free boundary charge density σ_f were given in terms of $\boldsymbol{D}$ and $\boldsymbol{E}$ by (2.71) and (2.73). We now want to establish them for Φ. Since a single reference point is used for the electrostatic potential of the entire system, condition (2.73) requires the electrostatic potential be continuous across the boundary

$$V_1 = V_2 \quad \text{on boundary} \qquad (2.136)$$

To obtain the boundary conditions for Ψ, we integrate (2.70) using (2.132) to obtain

$$\epsilon_1 \left(\Psi_1[z] - \Psi_1[z_1] \right) + \epsilon_2 \left(\Psi_2[z_1] - \Psi_2[z] \right) = \int_{C_1} \sigma_f \, ds \qquad (2.137)$$

where C_1 is the portion of the boundary between z_1 and z. If we impose the additional requirement that $\Psi_1[z_1] = \Psi_2[z_1] = 0$, then (2.137) reduces to

$$\epsilon_1 \Psi_1[z] - \epsilon_2 \Psi_2[z] = \int_{C_1} \sigma_f \, ds \quad \text{on boundary} \qquad (2.138)$$

where C_1 is the portion of the boundary from where $\Psi = 0$ to z. When the free boundary charge is zero, we have

$$\epsilon_1 \Psi_1 = \epsilon_2 \Psi_2 \quad \text{on boundary} \qquad (2.139)$$

2.19 UNIQUENESS THEOREM

The uniqueness theorem states there is only one solution to the potential function that conforms to the boundary conditions and Poisson's equation. In order to establish the uniqueness theorem, we assume the permittivity of the region is a constant and use Green's first identity (1.151) considering each of the following boundary conditions:

− Dirichlet boundary conditions where the potential on the boundary is specified.

− Neumann boundary conditions when the charge density on the boundary is specified.

− Mixed boundary conditions where the potential is specified on some portions of the boundary and charge density is specified on the portion of the boundary where the potential has not been specified.

Then it is shown how the proof can be extended to the case where the permittivity is a function of position.

We assume there are two solutions V_1 and V_2 that satisfy the same Poisson equation and boundary conditions. Defining U as the difference between these two solutions

$$U = V_1 - V_2 \qquad (2.140)$$

then leads to the following relationships

$$
\begin{aligned}
\nabla^2 U &= 0, \quad \text{inside boundary} \\
U &= 0, \quad \text{on boundary for Dirichlet B.C.} \\
\frac{\partial U}{\partial n} &= 0, \quad \text{on boundary for Neumann B.C.}
\end{aligned}
\qquad (2.141)
$$

Letting $G = V = U$ in Green's first identity leads to

$$\int_{area} (\nabla U \cdot \nabla U)\, da = 0 \qquad (2.142)$$

for all three types of boundary conditions. Since the integrand of (2.142) can only be a positive number or zero, each term must vanish in order for the integral to be zero. Therefore,

$$\frac{\partial U}{\partial x} = \frac{\partial U}{\partial y} = 0 \qquad (2.143)$$

or U must be a constant. If any portion of the boundary has Dirichlet boundary conditions, $U = V_1 - V_2 = 0$ on those portions of the boundary and the constant is zero (the two solutions are the same). For boundaries completely specified by Neumann boundary conditions, the constant cannot be determined since we only know $\partial U/\partial n = 0$ on the boundary. Therefore, at most, electrostatic potential functions that satisfy the same Poisson equation and boundary conditions can differ by an additive constant.

Clearly these argument can be extended to regions where the permittivity is a function of position by assuming both solutions satisfy $\nabla \cdot (\epsilon \nabla V) = -\rho_f$. The uniqueness theorem tells us that no matter how we determine the potential function to a problem, it is the only solution (to within an additive constant). This is extremely important as many different methods to find the potential function are used throughout the book.

2.20   CONDUCTORS AND INSULATORS

The properties of the surfaces and regions that contain charges and electric fields are often categorized by the ability of charge to flow through the region or on its surface. This property is known as the conductivity of the region or surface. The conductivity of different materials spans many orders of magnitude. Conductors which allow the flow of charge are on one end of the spectrum and insulators which do not allow the flow of charge are on the other end of the spectrum. Although ideal conductors with infinite conductivity do not exist, they are good approximations to many metals. Ideal insulators do not exist as well, but they can be good approximations for many non-metals. Throughout this book, we use ideal conductors to represent metals and ideal insulators to represent dielectric materials. The properties of ideal conductors and insulators are now discussed.

In equilibrium, an ideal conductor (or just conductor) has a single electrostatic potential value for the region it occupies. One consequence of this property is that all the charge of a conductor must reside on its surface. If charge existed inside a conductor, an electric field would be present and a potential drop would occur internal to the conductor violating the definition of the conductor. Thus the charge on the surface arranges itself so that the electric field is zero inside the conductor. Additionally, the requirement that the tangential component of the electric field be continuous (2.73) combined with $\boldsymbol{\mathcal{E}} = 0$ inside the conductor requires that the electric field be perpendicular at the surface of a conductor. Since the surface of a conductor is an equipotential surface, any equipotential contour can be replaced with an infinitesimal thick ideal conductor at the same potential and the electric fields in the system would not change.

The surface charge density on a conductor is related to the surface electric field through (2.71). Since the electric field inside the conductor is zero, at the surface of the conductor we have

$$\sigma_f = \epsilon \boldsymbol{\mathcal{E}} \cdot \hat{n} = \epsilon \operatorname{Im}[\mathcal{E}^* dz / |dz|] \tag{2.144}$$

where $\hat{n}$ is a unit vector normal to and pointing away from the surface of the conductor. Therefore, dz is in the direction such that the conductor is to the left as z is moved along the boundary.

There are two types of electrostatic configurations for a conductor. The first configuration is when the conductor has a fixed electrostatic potential that does not change when the electrostatic environment around the conductor changes. The total charge (and possibly the charge distribution) on the conductor changes instead. The second configuration is when the conductor has a fixed amount of charge that does not change when the electrostatic environment around the conductor changes. The electrostatic potential of the conductor (and possibly the charge distribution) changes instead. A conductor with the second configuration is known as a floating conductor.

An ideal insulator (or just insulator) does not allow the flow of charge through it. Thus charges are allowed on surfaces and inside insulators, but are not allowed to move with time. Additionally, ideal insulators may be polarized with the amount of polarization defined by their permittivity. Insulators are typically used to prevent the flow of charge between conductors at different potentials. Real insulators have a maximum electric field that can be supported before charge begins to flow termed the dielectric breakdown field. Ideal insulators have an infinite dielectric breakdown field.

2.21 CAPACITANCE

Before defining capacitance, we need to define complete and incomplete electrostatic systems. A complete system is defined as a system that has a net charge of zero. In other words, the dielectric flux of a closed contour surrounding the system is zero and all the lines of force start and end within the system. An incomplete system is defined as a system that has net flux through a contour surrounding it. An incomplete system can be transformed into a complete system by including the residual charge that is responsible for the net flux from the original system. All 2D systems can be made complete by including the point at infinity. If the system can be made complete without including the point at infinity, then the system is a finite complete system.

If we take two conductors separated by an insulator and deposit positive charge $\underline{Q}_1$ on one of the conductors and $-\underline{Q}_1$ on the other conductor, we have a complete system. The charge distributes itself over the surface of each conductor to ensure the electric field is zero inside the conductors. An electric field $\mathcal{E}_0[z]$ exists outside of the conductors resulting in a potential difference ΔV_0 between the conductors. The distribution of charge $\sigma_0[z]$ on each conductor may not be known, but we do know

$$\mathcal{E}_0[z] = \frac{1}{2\pi\epsilon} \int \frac{\sigma_0[z_i]|dz_i|}{z^* - z_i^*} \quad \text{and} \quad \Delta V_0 = -\text{Re}\left[\int_{z_1}^{z_2} (\mathcal{E}_0[z])^* dz\right] \quad (2.145)$$

where the first integral for the electric field is over all of the charges in the system. If we now increase the charge density by a factor α everywhere in the system $(\sigma_0[z] \to \alpha\sigma_0[z])$ without changing the geometry in anyway, then

$$\underline{Q}_1 \to \alpha\underline{Q}_1, \quad \mathcal{E}_0[z] \to \alpha\mathcal{E}_0[z], \quad \Delta V_0 \to \alpha\Delta V_0 \quad (2.146)$$

and the conductor boundary conditions are still met. Due to the uniqueness theorem, this is the charge distribution when the potential difference $\alpha\Delta V_0$ is applied to the conductors. This leads to the general conclusion that for a given geometry, the electrostatic potential difference between two conductors is proportional to the total charge (or flux) on one of the conductors. In 2D, the proportionality constant is the capacitance per unit length $\underline{C}$ (unit: F/m) with the definition

$$\underline{Q} = \underline{\psi}_{flux} = \underline{C}\Delta V \quad (2.147)$$

The capacitance is dependent only on the geometry of the problem and the permittivity of the insulator.

One of the simplest two conductor capacitor configurations is two infinite parallel conductors called the parallel plate capacitor. We orient the plates so that the real axis coincides with the bottom plate of the capacitor. Since the plates are infinitely wide, symmetry ensures the charge density is uniform over both conductors. We let the top plate located at $y = \delta$ have a charge density $\sigma > 0$. Since this is a complete system, the bottom plate must have the opposite charge density $-\sigma$. The electric field is a constant value with a direction pointing from the top plate to the bottom plate

$$\mathcal{E}[z] = -j\sigma/\epsilon \qquad (2.148)$$

The complex potential is obtained by plugging (2.148) into (2.87) and evaluating the integral

$$\Phi[z] - \Phi[0] = -\int_0^z (j\sigma/\epsilon)\, dz = -j\sigma z/\epsilon \qquad (2.149)$$

We are free to choose the complex potential reference point $\Phi[0]$ which allows determining σ/ϵ in terms of potentials. If we let

$$\Phi[0] = V_0, \quad \Phi[j\delta] = V_1 \qquad (2.150)$$

we obtain

$$\sigma/\epsilon = (V_1 - V_0)/\delta \qquad (2.151)$$

We can now write (2.149) as

$$\Phi[z] = -j\,(V_1 - V_0)\,(z/\delta) + V_0 \qquad (2.152)$$

The capacitance of the parallel plate capacitor is infinite since the plates are of infinite width, but the capacitance due to a portion of the plate w_{pp} wide can be computed as the charge on this portion of the plate (σw_{pp}) divided by the potential difference between plates $(V_1 - V_0)$

$$\underline{C} = \epsilon w_{pp}/\delta \qquad (2.153)$$

An alternative approach to calculating the capacitance is to use the flux through w_{pp}. Since we know the complex potential, the flux is given by (2.132) and the capacitance is

$$\underline{C} = \epsilon \, \mathrm{Im}[\Phi[x + j\delta] - \Phi[x + w_{pp} + j\delta]]\,/\,(V_1 - V_0) = \epsilon w_{pp}/\delta \quad (2.154)$$

A field map of the parallel plate capacitor with equipotential contour and lines of force spacings of $\Delta V = \Delta \Psi$ is shown in Fig. 2.14. If we look closely we notice that w_{pp} is proportional to the number or horizontal squares it spans and δ is proportional to the number of vertical squares it spans. This implies that w_{pp}/δ is just the number of horizontal squares w_{pp} spans divided by the number of vertical squares that δ spans. This observation indicates there may be an easy way to calculate the capacitance directly from the field map. This is the method of curvilinear squares.

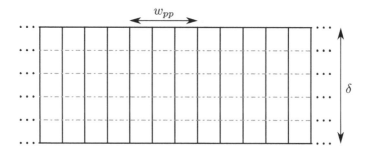

FIGURE 2.14 Field map of an infinitely wide parallel plate capacitor. $\Delta V = \Delta \Psi = 0.2\,(V_1 - V_0)$

2.22 METHOD OF CURVILINEAR SQUARES

A visual method for calculating the capacitance between conductors is called the method of curvilinear squares. In a two conductor system, the capacitance per unit length between conductors is

$$\underline{C} = \left| \underline{\psi}_{flux} / \Delta V \right| \tag{2.155}$$

where $\underline{\psi}_{flux}$ is the total flux per unit length from one of the conductors and ΔV is potential difference between the two conductors. Using Gauss's law, the flux per unit length through a small contour Δl of a conductor is

$$d\underline{\psi}_{flux} = \int_l^{l+\Delta l} (\boldsymbol{D} \cdot \hat{\boldsymbol{n}}) dl_1 \tag{2.156}$$

where $\hat{\boldsymbol{n}}$ is an outward pointing unit vector normal to the contour Δl. Assuming we start on the conductor with positive charge, the potential difference between the conductor and a distance Δn normal to the surface is

$$\int_{V_1}^{V_1 - dV_1} dV = -dV_1 = - \int_n^{n+\Delta n} (\boldsymbol{E} \cdot \hat{\boldsymbol{n}})\, dn_1 \tag{2.157}$$

Thus, the capacitance per unit length of this small area is

$$d\underline{C} = \left(\int_l^{l+\Delta l} (\boldsymbol{D} \cdot \hat{\boldsymbol{n}}) dl_1 \right) \Big/ \int_n^{n+\Delta n} (\boldsymbol{E} \cdot \hat{\boldsymbol{n}})\, dn_1 \tag{2.158}$$

In linear, homogeneous, isotropic media, we have $\boldsymbol{D} = \epsilon \boldsymbol{E}$. For $\Delta n \ll n$ and $\Delta l \ll l$, $\boldsymbol{E} \cdot \hat{\boldsymbol{n}}$ is approximately constant and can be pulled outside the integrals of (2.158) resulting in the capacitance per unit length of

$$d\underline{C} = \epsilon\, dl/dn \approx \epsilon \Delta l / \Delta n \tag{2.159}$$

where Δl is taken along the equipotential line and Δn is along the line of force (see Fig. 2.15). When

$$\int_{l}^{l+\Delta l} (\boldsymbol{E} \cdot \hat{\boldsymbol{n}})\, dl_1 = \int_{n}^{n+\Delta n} (\boldsymbol{E} \cdot \hat{\boldsymbol{n}})\, dn_1 \qquad (2.160)$$

the spacing between equipotential lines and lines of force are equivalent and the smallest region bounded by lines of force and equipotential lines is a curvilinear square. Independent of size, all curvilinear squares have the same capacitance per unit length of $\underline{C} = \epsilon$. When subdivided into smaller squares, curvilinear squares approach true squares, with infinitesimal lengths resulting in true squares. Therefore, each curvilinear square can be viewed as a capacitor of value $\underline{C} = \epsilon$. All the squares around the perimeter of the conductors can be viewed as capacitors connected in parallel and the squares along a line of force from one conductor to the other conductor can be viewed as capacitors connected in series. The total electric flux from a conductor is proportional to the total number of curvilinear squares (n_p) along the perimeter of the conductor times the permittivity. The voltage drop between conductors is proportional to the total number of curvilinear squares (n_s) from one conductor to the other conductor without crossing any line of force. The proportionality constant is the same in both cases allowing the total capacitance to be written as

$$\underline{C}_{total} = \epsilon n_p / n_s \qquad (2.161)$$

In order to use the method of curvilinear squares, a field map with $\Delta V = \Delta \Psi$ for the system is generated. This can be done by hand [4, 2] or with a computer program. The capacitance is calculated by simply counting the number of curvilinear squares around the perimeter of a conductor and between conductors through the use of (2.161). Two other benefits of generating plots with curvilinear squares is (1) regions of high capacitance and low capacitance are quickly identified through the density of curvilinear squares and (2) contributions of different regions to the total capacitance are easily determined by counting the number of curvilinear squares along the region of interest to the total number of curvilinear squares around the entire conductor.

Figure 2.15 is an example of a two conductor system with lines of force and equipotential lines plotted to produce curvilinear squares. The top conductor is at potential V_1 and the bottom conductor at potential V_2. It is assumed that the top and bottom plates extend to $\pm\infty$. This geometry can be viewed as two parallel plate capacitors of different plate spacing joined by a vertical metal transition in the top plates.

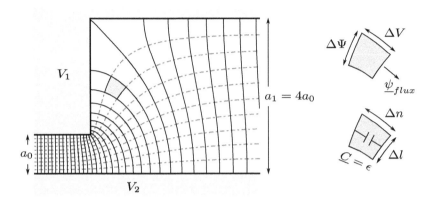

FIGURE 2.15 Example of a two conductor system with field lines and equipotential lines plotted to produce curvilinear squares. $\Delta V = \Delta \Psi = 0.1 \, (V_1 - V_2)$.

The capacitor on the left is labeled $\underline{C}_1$ and has a plate separation of a_0. The capacitor on the right is labeled $\underline{C}_2$ and has a plate separation of $a_1 = 4a_0$. Before performing any calculations, the larger density of curvilinear squares in the $\underline{C}_1$ region compared to the $\underline{C}_2$ region indicates the capacitance associated with $\underline{C}_1$ is much larger than the capacitance associated with $\underline{C}_2$. Using the two capacitor model, we would expect the capacitance far away from the vertical top plate transition to be

$$\underline{C}_1 = \epsilon \Delta l / a_0 = \epsilon n_{p,1} / n_{s,1}$$
$$\underline{C}_2 = \epsilon \Delta l / a_1 = \underline{C}_1 / 4 = \epsilon n_{p,2} / n_{s,2} \tag{2.162}$$

Equation (2.162) states that for the same distance along the capacitor plates, the number of parallel squares along $\underline{C}_1$ is four times larger than that of $\underline{C}_2$ since n_s is the same for both capacitors. This is consistent with Fig. 2.15. Additionally, the plot allows an estimation of the sidewall capacitance $\underline{C}_{sw}$. The vertical metal transition has $n_p \approx 9$ and $n_s = 10$ leading to $\underline{C}_{sw} \approx 0.9\epsilon$. In Chapter 3, a closed form expression for $\underline{C}_{sw}$ is obtained for this geometry as

$$\underline{C}_{sw} = \frac{\epsilon}{\pi \varphi} \ln[a_1 / a_0] \tag{2.163}$$

where $\pi \varphi$ is the transition angle measured from the horizontal in the counterclockwise direction ($\frac{1}{2}\pi$ for this example). Evaluating (2.163) for this examples gives $\underline{C}_{sw} = 0.883\epsilon$, which is in very good agreement with $\underline{C}_{sw}$ estimated by counting curvilinear squares.

2.23 ENERGY IN THE ELECTROSTATIC FIELD

In Section 2.9, it was shown that work is required to move charges in electric fields. Since the electrostatic field is a conservative field, the work put into the system becomes stored energy in the system. Systems with charge at infinity have infinite energy. Therefore, only finite closed systems have finite stored energy.

As discussed previously, one simple finite complete system is a capacitor formed by two finite sized conductors. The energy stored in a capacitor is calculated by determining the amount of work needed to assemble the charges onto the capacitor plates. First let us consider the electric field due to a line charge $\underline{q}_1$. The work performed by bringing a line charge $\underline{q}_2$ from the zero of potential to a distance r from $\underline{q}_1$ is

$$\underline{W} = \frac{\underline{q}_1 \underline{q}_2}{2\pi\epsilon} \ln[r_{12}] = \underline{q}_2 V_{21} \qquad (2.164)$$

where V_{21} is the potential at the location of $\underline{q}_2$ due to $\underline{q}_1$ and $r_{12} = |z_1 - z_2|$ is the distance between the line charges. If instead $\underline{q}_1$ is brought from the zero of potential in an electric field due to $\underline{q}_2$ and the final positions of $\underline{q}_1$ and $\underline{q}_2$ are the same as before, then the work performed can be written as

$$\underline{W} = \underline{q}_1 V_{12} \qquad (2.165)$$

and has the same value as before. The work (or stored energy) can then be expressed as

$$\underline{W} = \tfrac{1}{2} \left(\underline{q}_1 V_{12} + \underline{q}_2 V_{21} \right) \qquad (2.166)$$

For three charges, the same arguments lead to

$$\underline{W} = \tfrac{1}{2} \left(V_{12} + V_{13} \right) \underline{q}_1 + \tfrac{1}{2} \left(V_{21} + V_{23} \right) \underline{q}_2 + \tfrac{1}{2} \left(V_{31} + V_{32} \right) \underline{q}_3 \qquad (2.167)$$

This approach can be continued to give the expression of stored energy for n line charges as

$$\underline{W} = \tfrac{1}{2} \sum_{i=1}^{n} V_i \underline{q}_i \qquad (2.168)$$

where V_i is the potential at the location of q_i due to the remaining $n-1$ charges

$$V_i = \sum_{k \neq i} V_{ik} \qquad (2.169)$$

We assume no charges exist in the insulator between capacitor plates, the plate with positive charge is at a potential V_1, the plate with negative charge is at potential V_2, and the total charge on the plate with positive charge is $\underline{Q}_0$. Since all charges are located on the equipotential capacitor plates we have

$$\underline{W} = \tfrac{1}{2}\sum_i V_i \underline{q}_i = \tfrac{1}{2}(V_1 - V_2)\,\underline{Q}_0 \qquad (2.170)$$

where the sum signifies summing all the charges on plate i. With the aid of (2.147), the stored energy of a capacitor can be written in many equivalent forms

$$\underline{W} = \tfrac{1}{2}\Delta V \underline{Q}_0 = \tfrac{1}{2}\underline{C}\Delta V^2 = \tfrac{1}{2}\underline{Q}_0^2/\underline{C} \qquad (2.171)$$

where $\Delta V = V_1 - V_2$.

For cases where multiple conductors exist in the system, but charge is still located only on conductors, the energy can be calculated with the following method. Take a small area $da = \Delta l \Delta n$ in the system such that the boundaries coincide with equipotential and lines of force contours. The flux per unit length of this small area is given by (2.156) and the potential drop across this area is given by (2.157) leading to the stored energy per unit length for this area of

$$d\underline{W} = \tfrac{1}{2}d\underline{\psi}_{flux}\Delta V = \tfrac{1}{2}\int_l^{l+\Delta l}(\boldsymbol{D}\cdot\hat{n})\,dl_1\int_n^{n+\Delta n}(\boldsymbol{E}\cdot\hat{n})\,dn_1 \qquad (2.172)$$

For small lengths, $\boldsymbol{E}$ and $\hat{n}$ are parallel $(\boldsymbol{E}\cdot\hat{n} = \mathcal{E})$ and both $\boldsymbol{D}\cdot\hat{n}$ and $\boldsymbol{E}\cdot\hat{n}$ are approximately constant resulting in stored energy per until length of

$$d\underline{W} = \tfrac{1}{2}(\boldsymbol{D}\cdot\boldsymbol{E})\,da \qquad (2.173)$$

Stored energy per unit volume is defined as

$$d\underline{W}/da = \tfrac{1}{2}(\boldsymbol{D}\cdot\boldsymbol{E}) \qquad (2.174)$$

With (2.174), the total energy per unit length of any closed system is

$$\underline{W} = \tfrac{1}{2}\int(\boldsymbol{D}\cdot\boldsymbol{E})\,da \qquad (2.175)$$

where the integral is extended over the entire area occupied by the electric field.

For the case when the permittivity is a constant, the boundaries of the area are conductors, and the complex potential is known, an alternative expression for the stored energy per unit length of the system is [1]

$$\underline{W} = \tfrac{1}{2}\int d\underline{q}\,dV = \tfrac{1}{2}\epsilon\int d\Psi\,dV \qquad (2.176)$$

This final integral is just the enclosed area in the complex potential plane when complex potential along the boundary of the complete system is mapped. A complete description regarding the complex potential plane is given in Chapter 3.

Since each curvilinear square can be viewed as a capacitor, field maps drawn with curvilinear squares allow visualizing how the energy is stored in the system. Each curvilinear capacitor stores $\frac{1}{2}\epsilon\Delta V^2$ amount of energy per unit length perpendicular to the z-plane. Thus, regions containing high density of curvilinear squares are also regions of high energy storage.

2.24 GREEN'S RECIPROCATION THEOREM

Green's reciprocation theorem states that if charges $\underline{Q}_1, \underline{Q}_2, \ldots, \underline{Q}_m$ on the conductors of a system give rise to potentials $V_1, V_2, \ldots, V_m$ and if charges $\underline{Q}'_1, \underline{Q}'_2, \ldots, \underline{Q}'_m$ give rise to potentials $V'_1, V'_2, \ldots, V'_m$, then

$$\sum_{i=1}^{m} \underline{Q}_i V'_i = \sum_{i=1}^{m} \underline{Q}'_i V_i \qquad (2.177)$$

We use Green's second identity (1.150) to prove this theorem.

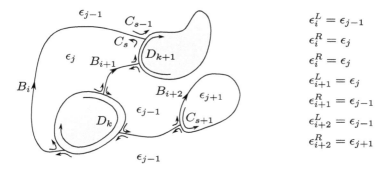

FIGURE 2.16 Definition of different parameters used for deriving Green's reciprocation theorem.

Let G represent V', V remains V, and ϵ represent the permittivity of the each region which is assumed different in (1.150). The B_i then correspond to the boundaries of different permittivity regions (see Fig.

2.16). Making these substitutions gives

$$\sum_{i=1}^{p}\int_{B_i}\left(\epsilon_i^R\left(V_i'^R\frac{\partial V_i^R}{\partial n_i^R}-V_i^R\frac{\partial V_i'^R}{\partial n_i^R}\right)+\epsilon_i^L\left(V_i^L\frac{\partial V_i'^L}{\partial n_i^R}-V_i'^L\frac{\partial V_i^L}{\partial n_i^R}\right)\right)ds$$

$$+\sum_{i=1}^{m}\oint_{C_i}\epsilon\left(V'\frac{\partial V}{\partial n}-V\frac{\partial V'}{\partial n}\right)ds=\int_{area}(V'\boldsymbol{\nabla}\cdot(\epsilon\boldsymbol{\nabla}V)-V\boldsymbol{\nabla}\cdot(\epsilon\boldsymbol{\nabla}V'))\,da$$

$$(2.178)$$

Assuming no free charges at the dielectric interfaces and applying the dielectric boundary condition (2.71) along B_i gives

$$\epsilon_i^L\frac{\partial V_i^L}{\partial n_i^R}=\epsilon_i^R\frac{\partial V_i^R}{\partial n_i^R},\quad \epsilon_i^L\frac{\partial V_i'^L}{\partial n_i'^R}=\epsilon_i^R\frac{\partial V_i'^R}{\partial n_i^R}\qquad(2.179)$$

Applying the boundary condition (2.136) along B_i gives

$$V_i^L=V_i^R,\quad V_i'^L=V_i'^R\qquad(2.180)$$

Thus, the integrals over the different B_i are all zero and so are the portions of C_i that do not lie on conductors. If there are no free charges between the conductors, then the area integral of (2.178) is also zero. This only leaves

$$\sum_{i=1}^{m}\oint_{D_i}\epsilon\left(V'\frac{\partial V}{\partial n}-V\frac{\partial V'}{\partial n}\right)ds=0\qquad(2.181)$$

where D_i corresponds to the boundary of the ith conductor. We can also write (2.144) in terms of surface charges with

$$\sigma_f=-\epsilon\frac{\partial V}{\partial n},\quad \sigma_f'=-\epsilon\frac{\partial V'}{\partial n}\qquad(2.182)$$

Making these changes and rearranging gives

$$\sum_{i=1}^{m}V_i'\oint_{D_i}\sigma_{f,i}\,ds=\sum_{i=1}^{m}V_i\oint_{D_i}\sigma_{f,i}'\,ds\qquad(2.183)$$

Since the integral of the surface charges on the ith conductor is $\underline{Q}_i$ for the unprimed case and $\underline{Q}_i'$ for the primed case, (2.183) is the same as (2.177).

To demonstrate one useful application of this theorem, consider a system of n conductors. For the unprimed case, let conductor 1 have a charge $\underline{Q}_1 = \underline{Q}$ and all other conductors uncharged and floating ($\underline{Q}_i = 0$ for $i \neq 1$). For the primed case, let conductor 2 have a charge $\underline{Q}'_2 = \underline{Q}$ and all other conductors uncharged and floating. Using (2.177), the potential of conductor 1 for the primed case equals the potential of conductor 2 for the unprimed case

$$V'_1 = V_2 \tag{2.184}$$

In other words, the potential to which a floating conductor 1 is raised by putting a charge $\underline{Q}$ on conductor 2 is the same potential conductor 2 is raised when floating if conductor 1 has a charge $\underline{Q}$.

2.25   INDUCED CHARGES ON GROUNDED CONDUCTORS

Another useful application of Green's reciprocation theorem is the calculation of induced charge on grounded conductors when a line charge is introduced near them. Let point P be near a group of grounded conductors. If a line charge $\underline{q}_0$ is placed at P, then each grounded conductor has some charge Q induced on it. Let the unprimed case be line charge present at P and all conductors grounded. Let the primed case be line charge absent, conductor 1 raised to potential V' and all other conductors grounded. From (2.177)

$$\underline{q}_0 V'_P + \underline{Q}_1 V'_1 + \underline{Q}_2 \cdot 0 + \underline{Q}_3 \cdot 0 + \cdots = 0 \cdot V_P + \underline{Q}'_1 \cdot 0 + \underline{Q}'_2 \cdot 0 + \underline{Q}'_3 \cdot 0 + \cdots \tag{2.185}$$

Solving for $\underline{Q}_1$ gives the induced charge on conductor 1 when it is grounded as

$$\underline{Q}_1 = -\underline{q}_0 \frac{V'_P}{V'_1} \tag{2.186}$$

In order to use (2.186), the potential at point P when the line charge is absent and the conductor is raised to a potential V (all others grounded) needs to be known. For example, let a line charge be located between two infinite grounded parallel planes. Let both planes be perpendicular to the complex plane with plane 1 going through the real axis and plane 2 going through the line $y = \delta$. This is the same structure as the infinite parallel plate capacitor except both plates are grounded. Let the line charge be located at $z = jy_0$. The potential when the line charge is absent, plate 2 at potential V'_2, and plate 1 grounded is given by the real part of (2.152) as

$$V = V'_2 y / \delta \tag{2.187}$$

Using (2.186) with $V'_P = V'_2 y_0/\delta$, the induced charge on plate 2 is

$$\underline{Q}_2 = -\underline{q}_o y_0/\delta \tag{2.188}$$

and on plate 1

$$\underline{Q}_1 = -\underline{q}_o - \underline{Q}_2 = \underline{q}_o (y_0 - \delta)/\delta \tag{2.189}$$

EXERCISES

2.1 Calculate the force on a line charge $\underline{q}_1$ located at z_1 due to charge distribution

$$\rho[x, y] = \begin{cases} \rho_0, & |x| < l, |y| < h \\ 0, & \text{otherwise} \end{cases}$$

Assume $|z_1| > l, h$. A useful integral for this problem is

$$\int \ln[aw + z]\, dw = -w + (w + (z/a)) \ln[aw + z] + C_0$$

2.2 A finite dipole is composed of a line charge $\underline{q}_0$ located at x_0 and a line charge $-\underline{q}_0$ located at $-x_0$.

   a. Calculate the electric field.

   b. At what distance from the origin along the real axis does a point dipole located at the origin with the same dipole moment as part (a) differ in electric field by 1 %?

2.3 The charge contained within a finite contour C has a charge density $\rho[z_0]$ and total charge $\underline{q}_{tot}$. Let $|z|$ be much larger than the magnitude of any point on C.

   a. If $\underline{q}_{tot} \neq 0$, show that the electric field at z is

$$\mathcal{E} \approx \frac{1}{2\pi\epsilon} \left(\frac{\underline{q}_{tot}}{z^*} \right)$$

   b. If $\underline{q}_{tot} = 0$, show that the electric field at z is

$$\mathcal{E} \approx \frac{M^*}{2\pi\epsilon (z^*)^2}$$

   where $\underline{M}$ is the dipole moment of the charge distribution.

2.4 A double layer with constant dipole moment per unit area β forms a closed contour with its positive side on the outside of the contour.

a. Show the potential at a point z inside the contour is

$$V[z] = -\beta/\epsilon$$

b. Show the potential at a point z outside the contour is

$$V[z] = 0$$

2.5 If the electric field is

$$\mathcal{E} = \frac{q_0}{2\pi\epsilon} \frac{3z^* - a}{(z^2 - a^2)^*}$$

where $a > 0$, show the flux through a circle centered at the origin with radius a is $3q_0/2$.

2.6 An infinite parallel plate capacitor has a distance between plates of δ and is composed of two dielectric layers. Dielectric layer 1 has a thickness t_1 and permittivity ϵ_1. Dielectric layer 2 has a thickness t_2 and perimittivity ϵ_2. If a potential difference ΔV is placed across the plates, show that the resulting polarization charge at the interface between the two dielectric layers is

$$\sigma_b = \Delta V \epsilon_0 \left(\frac{\epsilon_1 - \epsilon_2}{\epsilon_1 t_2 + \epsilon_2 t_1} \right)$$

2.7 Given

$$V[x, y] = \frac{q_0}{4\pi\epsilon} \ln \left[\frac{x^2 - 2xx_0 + x_0^2 + y^2}{x^2 + 2xx_0 + x_0^2 + y^2} \right]$$

determine the complex potential $\Phi[z]$ and draw the field map. What does this potential correspond to physically?

2.8 An infinite grounded conductor lies along the real axis and the charge density in the upper half plane is

$$\rho = \begin{cases} \rho_0, & |x| \leq l \text{ and } w_1 \leq y \leq w_2 \\ 0, & \text{otherwise} \end{cases}$$

where ρ_0 is a constant. Given the complex potential for a line charge q_0 located at $z = z_0$ above an infinite grounded plate that lies along the real axis is

$$\Phi[z] = \frac{q_0}{2\pi\epsilon} \ln \left[\frac{z - z_0^*}{z - z_0} \right]$$

a. Determine the complex potential in the charge free regions.

b. If $l = 1$, $w_1 = 1$, and $w_2 = 3$, show that the electrostatic potential at the center of the charge distribution is

$$V[0, 2] \approx 1.117\rho_0/\epsilon$$

2.9 An infinite grounded vertical plate lies along the imaginary axis and a second infinite grounded vertical plate lies along the line $x = a$. If a line charge q_0 is located at $z = b$ where $0 < b < a$, the complex potential is

$$\Phi[z] = -\frac{q_0}{2\pi\epsilon} \ln\left[\frac{\sin[\pi (z - b) / (2a)]}{\sin[\pi (z + b) / (2a)]}\right] + C_0$$

a. Determine the induced charge distribution on the plate that is aligned with the imaginary axis.

b. Integrate the charge distribution or use the complex potential directly to determine the total induced charge on the plate that is aligned with the imaginary axis.

c. Use Green's reciprocation theorem to determine the total induced charge on the plate that is aligned with the imaginary axis.

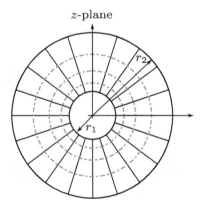

z-plane

FIGURE 2.17   Field map for two concentric cylinders. $\Delta V = \Delta \Psi = (V_2 - V_1)/4$.

2.10 The field map for two concentric cylinders is shown in Fig. 2.17. The smaller cylinder has a radius r_1 and potential V_1. The larger cylinder has a radius r_2 and potential V_2.

a. Using the method of curvilinear squares, determine the capacitance.

b. The complex potential for concentric cylinders with a potential V_2 on the larger cylinder and V_1 on the smaller cylinder is

$$\Phi[z] = (V_2 - V_1)\frac{\ln[z/r_1]}{\ln[r_2/r_1]} + V_1$$

Estimate the ratio of r_2/r_1 from the capacitance calculation in the first part of the problem and from Fig. 2.17 with a ruler to measure r_1 and r_2 directly. How close do they agree?

c. If the voltage difference between cylinders is ΔV, what is the stored energy?

2.11 A circular cylinder with radius r_1 is concentric with a second circular cylinder with radius $r_2 > r_1$. If both cylinders are grounded and a line charge q_0 is placed a distance r_q from the center of the cylinders where $r_1 < r_q < r_2$, show the induced charge on cylinder 1 is

$$\underline{Q}_1 = -\underline{q}_0\frac{\ln[r_2/r_q]}{\ln[r_2/r_1]}$$

The results from problem 2.10 maybe useful.

REFERENCES

[1] LL. G. Chambers. The electrostatic energy of a two-dimensional system. *Mathematical Notes*, 42:1–2, November 1959.

[2] Prescott D. Crout. A flux plotting method for obtaining fields satisfying Maxwell's equations, with applications to the magnetron. *Journal of Applied Physics*, 18(4):348–355, April 1947.

[3] Parry Moon and Domina Eberle Spencer. *Field Theory for Engineers.* D. van Nostrand Company, Inc., 1961.

[4] Arthur Dearth Moore. *Fundamentals of Electrical Design.* McGraw-Hill Book Company, Inc., 1927.

[5] William R. Smythe. *Static and Dynamic Electricity.* McGraw-Hill Book Company, 3rd edition, 1968.

CHAPTER 3

Line Charges

In this chapter, we study complex potentials due to different line charge configurations. A significant number of problems can be transformed into problems composed only of line charges. The complex potential is easily obtained for these types of electrostatic problems by the principle of superposition as

$$\Phi[z] = -\frac{1}{2\pi\epsilon} \sum_{i=1}^{n} \underline{q}_i \ln[z - z_i] + C_0 \qquad (3.1)$$

where $\underline{q}_i$ is the line charge value of the ith line charge located at z_i and C_0 is an arbitrary complex constant. Combining the solutions of this chapter with conformal mapping (developed later) produces the complex potential for a large number of electrostatic configurations with a modest amount of work. Before analyzing different charge configurations, we first develop a better understanding of the complex potential plane.

3.1 THE COMPLEX POTENTIAL PLANE

An understanding of the complex potential plane (or Φ-plane) allows further insight into the complex potential function. The Φ-plane has the value of the electrostatic potential V along the real axis and the value of Ψ along the imaginary axis. Similar to the z-plane where a point in the plane can be described by its (x, y) coordinates, points in the Φ-plane can be described by their (V, Ψ) coordinates. Typically we want to know the resulting curve in the Φ-plane as we trace out a curve in the z-plane. This procedure is called mapping. A point in one plane (the z-plane) maps into a point of another plane (the Φ-plane) which leads to a curve in one plane mapping into a curve in the other plane. We are normally interested in the curve produced in the Φ-plane for the values of the complex potential along the boundary of the system in the z-plane. To illustrate the mapping procedure, we consider the complex potential of a positive charge q_0 located at the origin.

The complex potential for a positive line charge q_0 located at the origin is given by (2.88)

$$\Phi[z] = -\frac{q_0}{2\pi\epsilon} \ln[z] = -\frac{q_0}{2\pi\epsilon} (\ln[r] + j\theta) \qquad (3.2)$$

where we have set $C_0 = 0$. The field map of (3.2) is shown in Fig. 2.11. Now consider only the upper half of the z-plane as shown in Fig. 3.1 (a). The boundary is then the real axis in the z-plane. In order to map this boundary into the Φ-plane, key points along the boundary are labeled by letters. The points to label are typically points where large changes in Φ occur. For example, a jump in Ψ occurs between $z = 0^-$ and $z = 0^+$ for (3.2) and these points should be labeled with letters. When points at $\pm\infty$ are part of the system, they are typically labeled as well and the letters are given the subscript ∞ to indicated they are associated with points at infinity. It is also advisable to generate a table of corresponding points in both planes.

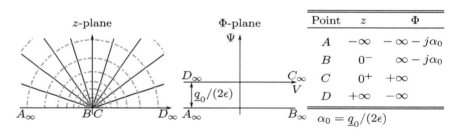

Point	z	Φ
A	$-\infty$	$-\infty - j\alpha_0$
B	0^-	$\infty - j\alpha_0$
C	0^+	$+\infty$
D	$+\infty$	$-\infty$

$\alpha_0 = q_0/(2\epsilon)$

FIGURE 3.1 The complex potential along boundary AD in the z-plane mapped into the complex potential plane (Φ-plane).

In this example, we have four points to label ($z = \pm\infty, 0^\pm$). The point to label first is arbitrary, but once the first point has been selected, the remaining points should be labeled in order and in the direction such that the area of interest is on the left as we go from one point to the next. Since we are interested in the upper half of the z-plane, we choose the first point to label as $z = -\infty$ as shown in Fig. 3.1. As the negative real axis is traced out in the z-plane, we are moving along the constant flux line $\Psi = -q_0/(2\epsilon)$. Therefore, this is a horizontal line in the Φ-plane. Near A in the z-plane, r is large (approaching ∞) and V has a large negative value (approaching $-\infty$). Near B, r is small (approaching 0) and V has a large positive value (approaching ∞). Therefore, the negative real axis of the z-plane maps into the horizontal line $\Psi = -q_0/(2/\epsilon)$ in the Φ-plane. As the positive real axis is traced out in the z-plane, we are moving along the constant flux line $\Psi = 0$. Therefore, this is a horizontal line in the Φ-plane coincident with the real axis of the Φ-plane. The locations for C and D are determined by their values of r. The mapping of the real axis in the z-plane into the Φ-plane, along with a table of the four key points in both planes is shown in Fig. 3.1.

Two items should be noticed from Fig. 3.1. First there is a jump in Ψ of $q_0/(2\epsilon)$ at BC, the location of the charge. There is the same jump in Ψ at DA except in the opposite direction indicating a charge of $-q_0$ at infinity. As discussed in Section 2.5, Gauss's law requires charge neutrality for a system that includes the point at infinity. Thus, a charge of $-q_0$ must exist at infinity. Second, labeling the key points by letters in both planes allows visualizing how these points map between the two planes even without the table.

3.2 SINGLE LINE CHARGE

3.2.1 Coaxial Circular Cylinders

We begin our analysis of complex potentials due to line charges by further analyzing the configuration of a single line charge located at the origin. Analyzing the contours of constant potential (see Fig. 2.11), we observe that they consist of concentric circles. In general, we can replace a constant potential contour by a thin conducting foil at the same potential without disturbing the rest of the field map. Therefore, we can view any two circles of constant potential as conducting cylinders at different potentials. Let cylinder 1 have a radius r_1 and a potential V_1. Let cylinder 2 have a radius r_2 and a potential V_2. Also, let $\Psi = 0$ at $\theta = 0$. We start with (3.1) and set $z_1 = 0$ and $q_1 = q_0$. Next we adjust the values of q_0 and C_0 to match the boundary conditions

$$V_1 = \frac{q_0}{2\pi\epsilon} \ln[r_1] + C_0, \quad V_2 = \frac{q_0}{2\pi\epsilon} \ln[r_2] + C_0 \tag{3.3}$$

Solving for q_0 and C_0 gives

$$q_0 = \frac{(V_2 - V_1)\, 2\pi\epsilon}{\ln[r_2/r_1]}, \quad C_0 = V_1 - \frac{(V_2 - V_1)\ln[r_1]}{\ln[r_2/r_1]} \tag{3.4}$$

The complex potential for two concentric cylinders at different voltages is

$$\Phi[z] = (V_2 - V_1) \frac{\ln[z/r_1]}{\ln[r_2/r_1]} + V_1 \tag{3.5}$$

The capacitance per unit length is easily obtained from (3.4) as

$$\underline{C} = \frac{q_0}{V_2 - V_1} = \frac{2\pi\epsilon}{\ln[r_2/r_1]} \tag{3.6}$$

The complex potential of the concentric boundaries maps into a rectangle in the Φ-plane as shown in Fig. 3.2. As expected, the stored energy of the system is finite. The amount of stored energy per unit length can be calculated with (2.171) or from the area calculation of Fig. 3.2 as

$$\underline{W} = \frac{\pi\epsilon (V_2 - V_1)^2}{\ln[r_2/r_1]} \tag{3.7}$$

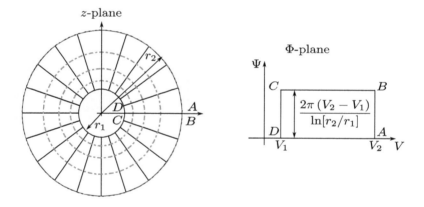

FIGURE 3.2 The complex potential along boundary A to B in the counterclockwise direction and then C to D in the clockwise direction in the z-plane mapped into the complex potential plane.

3.2.2 Two Conductive Plates That Meet at the Origin

For our next problem, we need to recall from Section 2.10 that both the Real and Imaginary parts of Φ are solutions of Laplace's equation. If we let the lines of force defined by (3.2) now represent equipotential contours, the complex potential for a line charge at the origin becomes the solution to semi-infinite conducting plates at different potentials that meet at the origin. Switching back and forth between the electrostatic potential as the Re[Φ] or Im[Φ] can become confusing. To minimize this confusion, the electrostatic potential corresponds to the Re[Φ] in this book unless stated otherwise. It is still expected that when a field map is presented, the reader can visualize constant contours of Ψ as potential contours, recognize them as a solution to a different problem (as was done in this case), and then transform the complex function so that the Re[Φ] corresponds to the electrostatic potential.

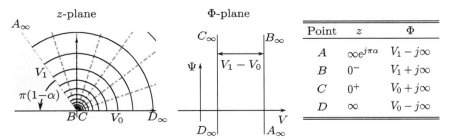

Point	z	Φ
A	$\infty e^{j\pi\alpha}$	$V_1 - j\infty$
B	0^-	$V_1 + j\infty$
C	0^+	$V_0 + j\infty$
D	∞	$V_0 - j\infty$

FIGURE 3.3 The complex potential along boundary $ABCD$ in the z-plane mapped into the complex potential plane. Note that the geometry in the Φ-plane is the same as Fig. 3.1 rotated by $\pi/2$. This rotation transforms constant lines of force into constant potential contours and vice versa.

To illustrate this procedure, we determine the complex potential between a semi-infinite conducting plate (plate CD) with potential V_0 that lies along the positive real axis and a semi-infinite conducting plate (plate AB) with potential $V_1 > V_0$ that lies along the $z = re^{j\pi\alpha}$ line as shown in Fig. 3.3. An infinitesimally thick insulator is located at the origin to prevent the conductors from touching. Referring to the Φ-plane in Fig. 3.1, a rotation of $\pi/2$ transforms lines AB and CD from constant Ψ lines into constant V lines with the potential of line CD at a lower potential than line AB. Therefore, we need to multiply (3.2) by j to obtain the complex potential. Note that the $\pi/2$ rotation in the complex plane converts infinite potentials at the locations of charges in the z-plane into infinite flux per unit length at these locations. This is a consequence of having an infinitesimal insulator between two conductors with different potentials. Later, a method for including finite gaps between conductors is developed. To match the boundary conditions, we add a constant C_0 and adjust the values of q_0 and C_0. The complex potential is

$$\Phi[z] = -j\frac{q_0}{2\pi\epsilon} \ln[z] + C_0 \qquad (3.8)$$

Specifying the boundary conditions requires us to first assign the location of the zero flux line ($\Psi = 0$). As discussed in Section 2.10, we are free to choose the reference location of Ψ. To simplify the final form of Φ, we choose $z = 1$ as the reference point for $\Psi = 0$. The values of q_0 and C_0 can now be determined by the boundary conditions

$$\Phi[1] = V_0, \quad \mathrm{Re}\left[\Phi\left[re^{j\pi\alpha}\right]\right] = V_1 \tag{3.9}$$

Substituting (3.8) into (3.9) and solving for q_0 and C_0 gives

$$q_0 = 2\epsilon\left(V_1 - V_0\right)/\alpha, \quad C_0 = V_0 \tag{3.10}$$

The complex potential then becomes

$$\Phi[z] = -j\frac{(V_1 - V_0)}{\pi\alpha}\ln[z] + V_0 = \frac{(V_1 - V_0)}{\pi\alpha}\left(\theta - j\ln[r]\right) + V_0 \tag{3.11}$$

When one plate occupies the negative real axis and the other plate occupies the positive real axis, $\alpha = 1$ and the complex potential is

$$\Phi[z] = -j\frac{(V_1 - V_0)}{\pi}\ln[z] + V_0 \tag{3.12}$$

The Schwarz-Christoffel transformation is developed in Chapter 5 allowing the complex potential of (3.12) to be mapped onto the boundaries of an arbitrary polygon.

3.3 TWO LINE CHARGES

A system with two line charges allows two more degrees of freedom since we can adjust each charge value and the spacing between charges in addition to the arbitrary constant. We start our two charges configuration discussion by revisiting the dipole. We set up the coordinate system so that both dipole charges lie on the real axis an equal distance ($x_0 > 0$) from the origin and put the positive charge at $z = -x_0$. The complex potential is

$$\Phi[z] = \frac{q_0}{2\pi\epsilon}\ln\left[\frac{z - x_0}{z + x_0}\right] + C_0 \tag{3.13}$$

The field map for (3.13) is shown in Fig. 3.4. This configuration leads to the solutions of four new electrostatic problems: coplanar plates with zero separation, two circular cylinders external to each other, two circular cylinders one contained within the other, and a circular cylinder and grounded plane.

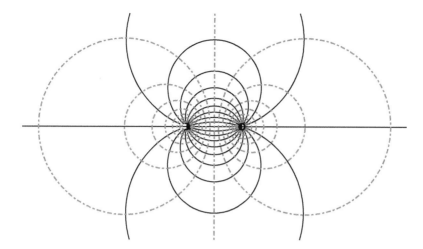

FIGURE 3.4 Field map for two equal but opposite polarity line charges. Note that the plane equidistant from the line charges is an equipotential contour. $\Delta V = \Delta \Psi = 0.05 \underline{q}_0 / \epsilon$.

3.3.1 Three Coplanar Plates

The first solution is similar to the solution just solved with one charge located at the origin. Considering only the upper half of the z-plane, a map of the x-axis into the Φ-plane is shown in Fig. 3.5. We now have six key points to label, $x = \pm\infty, -x_0^\pm, x_0^\pm$. Figure 3.5 shows the contour in the Φ-plane due to mapping the real axis of the z-plane. This contour looks similar to the Φ-plane mapping of Fig. 3.3 for a single charge at the origin except there is a $\Psi = +\underline{q}_0/(2\epsilon)$ line instead of a $\Psi = -\underline{q}_0/(2\epsilon)$. Note that the $\Psi = +\underline{q}_0/(2\epsilon)$ line corresponds to the region between the charges $(-x_0 < z < x_0)$ in the z-plane. A rotation of $-\pi/2$ in the Φ-plane transforms the line segments AB and EF into conductors at zero potential and line segment CD into a conductor at a potential greater than zero which is labeled V_0. The complex potential for this case is obtained by multiplying (3.13) by $-j$

$$\Phi[z] = -j \frac{q_0}{2\pi\epsilon} \ln\left[\frac{z - x_0}{z + x_0}\right] + C_0 \tag{3.14}$$

Setting the zero of the complex potential at $z = \infty$ and applying the boundary conditions

$$\Phi[0] = V_0 = -j \frac{q_0}{2\pi\epsilon} \ln[-1] + C_0, \quad \lim_{z \to \infty}[\Phi[z]] = 0 = C_0 \tag{3.15}$$

gives

$$\underline{q}_0 = 2\epsilon V_0, \quad C_0 = 0 \tag{3.16}$$

The complex potential for three coplanar conductors separated by infinitesimal thick insulators with the center conductor of length $2x_0$ at potential V_0 and the two outer conductors at ground is

$$\Phi[z] = -j\frac{V_0}{\pi}\ln\left[\frac{z - x_0}{z + x_0}\right] \tag{3.17}$$

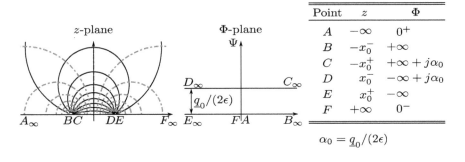

Point	z	Φ
A	$-\infty$	0^+
B	$-x_0^-$	$+\infty$
C	$-x_0^+$	$+\infty + j\alpha_0$
D	x_0^-	$-\infty + j\alpha_0$
E	x_0^+	$-\infty$
F	$+\infty$	0^-

$$\alpha_0 = \underline{q}_0/(2\epsilon)$$

FIGURE 3.5 The complex potential along boundary AF in the z-plane mapped into the complex potential plane. Jumps at BC and DE in Ψ of $\underline{q}_0/(2\epsilon)$ are due to crossing the line charges at those locations.

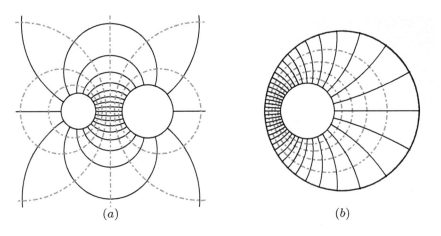

(a)   (b)

FIGURE 3.6 The boundary conditions for two conductive cylinders at different potentials can be produced by the two line charges of a dipole. The equipotential contours of two equal but opposite charge are replaced by conductors to produce two new electrostatic situations. (a) Two conductive cylinders external to each other at different potentials. (b) Two conductive cylinders one contained within the other at different potentials.

3.3.2 Two Noncentric Circular Cylinders

The other three solutions involve replacing equipotential contours of the two point charges with conductors. Field maps for two of the solutions are shown in Fig. 3.6. The first solution (Fig. 3.6 (a)) replaces one equipotential contour to the left and right of the y-axis with conducting cylinders. This solution can be for cylinders of the same diameter or for cylinders of different diameters. The second solution (Fig. 3.6 (b)) replace two equipotential contours on the same side of the y-axis with conductive nonconcentric cylinders. To determine the complex potential for each of these cases, we first separate the real and imaginary parts of (3.13) to show the equipotential contours are circles and the lines of force are also circles. Then the values of q_0 and the constant C_0 are determined by the boundary conditions. Since we are free to choose the location of the $\Psi = 0$ contour, we can let C_0 be a real constant.

Setting the real part of (3.13) equal to V gives

$$V = \frac{q_0}{4\pi\epsilon} \ln\left[\frac{|z - x_0|^2}{|z + x_0|^2}\right] + C_0$$

$$\exp\left[\frac{4\pi\epsilon(V - C_0)}{q_0}\right] = \frac{|z - x_0|^2}{|z + x_0|^2} \tag{3.18}$$

$$\left|z + x_0 \coth\left[2\pi\epsilon(V - C_0)/q_0\right]\right|^2 = x_0^2 \operatorname{csch}^2\left[2\pi\epsilon(V - C_0)/q_0\right]$$

As determined from the field map, the equipotential contours are circles with centers that lie on the x-axis. The radius and location of the center of each equipotential contour depend on the distance of the charges from the origin (x_0) and the ratio of $(V - C_0)/q_0$. For the imaginary part of (3.13) we have

$$\frac{2\pi\epsilon\Psi}{q_0} = \operatorname{Im}\left[\ln\left[\frac{(z - x_0)(z^* + x_0)}{|z + x_0|^2}\right]\right] = \tan^{-1}\left[\frac{-j(z - z^*)x_0}{|z|^2 - x_0^2}\right]$$

$$\tan\left[\frac{2\pi\epsilon\Psi}{q_0}\right] = \frac{-j(z - z^*)x_0}{|z|^2 - x_0^2} \tag{3.19}$$

$$\left|z - jx_0 \cot\left[2\pi\epsilon\Psi/q_0\right]\right|^2 = x_0^2 \csc^2\left[2\pi\epsilon\Psi/q_0\right]$$

which are also circles, but with the centers of the circles on the y-axis. These circles cross the x-axis at the location of the charges $(\pm x_0)$. Both of these features are clearly observable from the field map of Fig. 3.4.

To illustrate how to use (3.18) and (3.19) to solve either situation depicted in Fig. 3.6, let cylinder 1 have a radius r_1 and potential V_1. Let cylinder 2 have a radius r_2 and potential V_2. Let the distance between the center of the two cylinders be D. To determine the complex potential for this problem, we need to determine the location of the two charges ($\pm x_0$), $\underline{q}_0$, and C_0 in terms of r_1, r_2, and D. The analysis is greatly simplified by recognizing there is a circular line of force centered at the origin with radius x_0 when $\Psi = \underline{q}_0/(4\epsilon)$. This circular line of force is shown in Fig. 3.7 with a dashed outline. The $\Psi = \underline{q}_0/(4\epsilon)$ contour must intersect all equipotential contours at right angles. Thus, right triangles can be formed with vertices of the origin, the center of each conductive cylinder, and the intersection of the $\Psi = \underline{q}_0/(4\epsilon)$ contour and each conductive cylinder. These triangles result in the following relationships

$$a^2 = x_0^2 + r_1^2, \quad b^2 = x_0^2 + r_2^2 \tag{3.20}$$

Subtracting a^2 from b^2 gives

$$b^2 - a^2 = (b+a)(b-a) = r_2^2 - r_1^2 \tag{3.21}$$

From Fig. 3.7, we also have

$$D = a \pm b \tag{3.22}$$

where the upper sign corresponds to Fig. 3.7 (a) and the lower sign corresponds to Fig. 3.7 (b) in this analysis. We now have two equations ((3.21) and (3.22)) allowing us to solve for a and b

$$a = (D^2 - r_2^2 + r_1^2)/(2D), \quad b = \pm(D^2 + r_2^2 - r_1^2)/(2D) \tag{3.23}$$

The location of the charges ($\pm x_0$) is then obtained using (3.20) as

$$x_0 = \sqrt{\frac{\left(D^2 - (r_2 - r_1)^2\right)\left(D^2 - (r_2 + r_1)^2\right)}{4D^2}} \tag{3.24}$$

The value of $\underline{q}_0$ is determined by the potential difference between the two conducting cylinders. From (3.18), we have

$$a = \pm x_0 \coth\left[\frac{2\pi\epsilon}{\underline{q}_0}(V_1 - C_0)\right], \quad r_1 = \pm x_0 \operatorname{csch}\left[\frac{2\pi\epsilon}{\underline{q}_0}(V_1 - C_0)\right]$$

$$b = -x_0 \coth\left[\frac{2\pi\epsilon}{\underline{q}_0}(V_2 - C_0)\right], \quad r_2 = -x_0 \operatorname{csch}\left[\frac{2\pi\epsilon}{\underline{q}_0}(V_2 - C_0)\right] \tag{3.25}$$

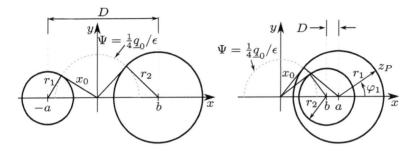

FIGURE 3.7 Definition of the different parameters used to derive the complex potential of two conductive cylinders at different potentials.

Taking the ratios of a/r_1 and b/r_2 gives

$$\frac{a}{r_1} = \cosh\left[\frac{2\pi\epsilon}{\underline{q}_0}(V_1 - C_0)\right], \qquad \frac{b}{r_1} = \cosh\left[\frac{2\pi\epsilon}{\underline{q}_0}(V_2 - C_0)\right] \qquad (3.26)$$

With the help of (3.25) and (3.26), we can write

$$\frac{ab \pm x_0^2}{r_1 r_2} = \cosh[\alpha_1]\cosh[\alpha_2] - \sinh[\alpha_1]\sinh[\alpha_2] \qquad (3.27)$$

where

$$\alpha_i = \frac{2\pi\epsilon\,(V_i - C_0)}{\underline{q}_0} \qquad (3.28)$$

From (3.20), we have

$$x_0^2 = a^2 - r_1^2 = b^2 - r_2^2 = \tfrac{1}{2}\left(a^2 - r_1^2\right) + \tfrac{1}{2}\left(b^2 - r_2^2\right) \qquad (3.29)$$

Inserting the last form of x_0^2 into (3.27) and using the identity

$$\cosh[u - v] = \cosh[u]\cosh[v] - \sinh[u]\sinh[v] \qquad (3.30)$$

gives

$$\cosh\left[\frac{2\pi\epsilon\,(V_1 - V_2)}{\underline{q}_0}\right] = \frac{2ab \pm (a^2 - r_1^2 + b^2 - r_2^2)}{2r_1 r_2}$$
$$= \pm\frac{D^2 - r_1^2 - r_2^2}{2r_1 r_2} \qquad (3.31)$$

Solving for $\underline{q}_0$ gives

$$\underline{q}_0 = 2\pi\epsilon\,(V_1 - V_2)\left(\cosh^{-1}\left[\pm\frac{D^2 - r_1^2 - r_2^2}{2r_1 r_2}\right]\right)^{-1} \qquad (3.32)$$

The constant C_0 can now be determined using (3.25) and (3.32) as

$$C_0 = V_2 + \frac{q_0}{2\pi\epsilon}\coth^{-1}\left[\frac{b}{x_0}\right] \tag{3.33}$$

and the complex potential is now fully determined. Additionally, the capacitance per unit length is now easily obtained from (3.31) as

$$\underline{C} = \underline{q_0}/(V_1 - V_2) = 2\pi\epsilon\left(\cosh^{-1}\left[\pm\frac{D^2 - r_1^2 - r_2^2}{2r_1r_2}\right]\right)^{-1} \tag{3.34}$$

The charge density on the surface of each cylinder is obtained using (2.144). To calculate the charge on cylinder 1, consider the point z_P shown in Fig. 3.7. We define φ_1 as the angle between the real axis and the line that goes through z_P and the center of the cylinder 1 ($z = \mp a$) as shown for cylinder 1 in Fig. 3.7 (b). Any point on cylinder 1 satisfies

$$z_P = \mp a + r_1 e^{j\varphi_1} \tag{3.35}$$

When determining the surface charge on a conductor (see 2.20), dz is in the direction such that the conductor is to the left. Therefore, dz on cylinder 1 is

$$dz = \pm jr_1 e^{j\varphi_1} d\varphi_1 \tag{3.36}$$

The electric field at the surface of cylinder 1 is obtained by differentiating (3.13) and evaluating at z_P as

$$\mathcal{E}[z_p] = -\left(\frac{\partial\Phi}{\partial z}\right)^*\Bigg|_{z=z_P} = \frac{\underline{q_0}x_0}{\pi\epsilon(x_0^2 - z_P^2)^*} = \frac{\underline{q_0}x_0 e^{j\varphi_1}}{2\pi\epsilon r_1\left(\pm a - r_1\cos[\varphi_1]\right)} \tag{3.37}$$

The charge density is

$$\sigma_f = \mathrm{Im}[\epsilon\mathcal{E}^*dz/|dz|] = \frac{\underline{q_0}x_0}{2\pi r_1\left(a \mp r_1\cos[\varphi_1]\right)} \tag{3.38}$$

Performing the same analysis on cylinder 2 gives

$$\mathcal{E} = -\frac{\underline{q_0}x_0 e^{j\varphi_2}}{2\pi\epsilon r_2\left(b + r_2\cos[\varphi_2]\right)}, \quad dz = jr_2 e^{j\varphi_2}d\varphi_2 \tag{3.39}$$

and

$$\sigma_f = -\frac{\underline{q_0}x_0}{2\pi r_2\left(b + r_2\cos[\varphi_2]\right)} \tag{3.40}$$

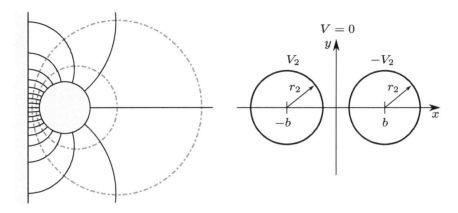

FIGURE 3.8 The boundary conditions for a conductive cylinder and a conductive plane at different potentials can also be produced by two line charges. The equipotential plane equidistant between the charges is replaced by a conductor and an equipotential circular contour is replaced by a conductive cylinder.

3.3.3 Conductive Cylinder and a Conductive Plane

The final solution (Fig. 3.8 (a)) replaces the vertical equipotential contour that is coincident with the y-axis with a conducting plane and an equipotential contour to the right of the y-axis with a conducting cylinder. We still label the cylinder on the positive real axis as cylinder 2 with a radius of r_2. Let the potential difference between cylinder 2 and the conducting plane be V_2. Let the center of cylinder 2 be $z = b$. We can still use all the equations previously developed for two cylinders if we imagine there is a cylinder 1. Let the radius of cylinder 1 be $r_1 = r_2$. From (3.18), the potential along the imaginary axis is zero for any value of x_0. Therefore, the potential on cylinder 2 should be $-V_2$ to comply with the potential difference between cylinder 2 and the conducting plane of V_2. From symmetry, the potential on cylinder 1 is V_2. From (3.20) and (3.32), the values of x_0 and q_0 are

$$x_0 = \sqrt{b^2 - r_2^2}, \quad \underline{q}_0 = 2\pi\epsilon V_2\left(\cosh^{-1}[b/r_2]\right)^{-1} \tag{3.41}$$

where the identity $\cosh^{-1}[2z^2 - 1] = 2\cosh^{-1}[z]$ was used to simplify $\underline{q}_0$. The capacitance per unit length is

$$\underline{C} = \left|\underline{q}_0/V_2\right| = 2\pi\epsilon\left(\cosh^{-1}[b/r_2]\right)^{-1} \tag{3.42}$$

Note that capacitance between the $x = 0$ plane and cylinder 2 is twice the capacitance between cylinders 1 and 2. The charge density on cylinder 2 is still given by (3.38).

3.4 Φ FOR CONDUCTOR BOUNDARY IN PARAMETRIC FORM

Although we could continue exploring potential functions for different charge arrangements and determining the boundary conditions they satisfy (referred to as the indirect method for finding the potential), we typically have the opposite scenario of determining the potential function for a given geometry with boundary conditions (referred to as the direct method for finding the potential). If boundaries did not exist, then (2.89) would be the most direct method for determining the electrostatic potential. When boundaries are present, boundary conditions impose constraints that lead to unique solutions to the electrostatic potential (see Section 2.19). Thus we need to develop general methods for solving what are referred to as boundary value problems.

As an example of a direct method for finding the electrostatic potential, consider the problem of determining the electrostatic potential for a charged conductor at zero potential where the boundary of the conductor can be expressed in terms of a parametric equation

$$x = f_1[t], \quad y = f_2[t] \tag{3.43}$$

The solution to Laplace's equation with the conductor as a boundary at zero potential is

$$z = f_1[bj\Phi] + jf_2[bj\Phi] \tag{3.44}$$

where b is a real number. To prove this, we determine the $V = 0$ constant potential contour by substituting $\Phi = j\Psi$ into (3.44) to obtain

$$z = f_1[-b\Psi] + jf_2[-b\Psi] = x[t] + jy[t] \tag{3.45}$$

Thus, the constant potential contour for $V = 0$ is the boundary of the conductor as desired.

This approach is typically limited to finding Φ when the boundary of the electric field can be defined parametrically. This is the case for confocal conics or for charged conductors with boundaries that can be described parametrically and electric fields that extend to infinity. A much more general method for solving boundary value problems is the method of Green's functions discussed in Section 3.5. As an example, let a grounded conductive cylinder with elliptical cross-section have a charge per unit length of q_0. Let the semi-major axes length be a_1, the semi-minor axes length b_1, and the parametric equation for the ellipse be

$$x = a_1 \cos[t], \quad y = b_1 \sin[t], \quad 0 \le t \le 2\pi \qquad (3.46)$$

From (3.44),

$$z = a_1 \cos[bj\Phi] + jb_1 \sin[bj\Phi] \qquad (3.47)$$

For an ellipse, $a_1^2 - b_1^2 = c^2$ where c is the distance from the center of the ellipse to a focus point. This allows writing

$$a_1 = c \cosh[\alpha], \quad b_1 = c \sinh[\alpha] \qquad (3.48)$$

where $\alpha > 0$. Inserting these values for a_1 and b_1 into (3.47) gives

$$z = c \cosh[\alpha] \cosh[b\Phi] - c \sinh[\alpha] \sinh[b\Phi] = c \cosh[\alpha - b\Phi] \qquad (3.49)$$

Solving for Φ gives

$$\Phi = b^{-1}\left(\alpha - \cosh^{-1}[z/c]\right) \qquad (3.50)$$

On the surface,

$$\Phi = j\Psi, \quad x = a_1 \cos[b\Psi], \quad y = -b_1 \sin[b\Psi], \quad 0 \le b\Psi \le 2\pi \qquad (3.51)$$

The total charge per unit length on the surface is

$$q_0 = \epsilon\left(\Psi_{max} - \Psi_{min}\right) = 2\pi\epsilon/b \qquad (3.52)$$

Replacing b in (3.50) with its value from (3.52) gives the complex potential outside the ellipse as

$$\Phi = \frac{q_0}{2\pi\epsilon}\left(\alpha - \cosh^{-1}[z/c]\right) \qquad (3.53)$$

The field map of a charged ellipse is shown in Fig. 3.9 (a).

As a second example, let a metal sheet be formed into the shape of a cycloid and serve as one boundary of an electric that has uniform field $\mathcal{E}_0$ far from the sheet [4]. The equation of the surface is

$$x[t] = a\left(t - \sin[t]\right), \quad y[t] = a\left(1 - \cos[t]\right) \qquad (3.54)$$

From (3.44),

$$z = a\left(bj\Phi - \sin[bj\Phi]\right) + ja\left(1 - \cos[bj\Phi]\right) \qquad (3.55)$$

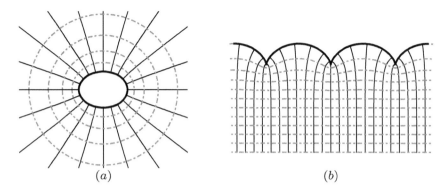

FIGURE 3.9 (a) Field map of a charged ellipse with $c = b_1$ and $\Delta V = \Delta \Psi = 0.05 q_0 / \epsilon$. (b) Field map of an electric field that terminates on a sheet of metal that has the shape of a cycloid. Far from the metal sheet, the field is uniform.

Solving for x and y in terms of Φ gives

$$x = a\left(-b\Psi + e^{bV}\sin[b\Psi]\right), \quad y = a\left(1 + bV - e^{bV}\cos[b\Psi]\right) \qquad (3.56)$$

When V is large and negative, $x \to -ab\Psi$ and $y \to abV$ which corresponds to a uniform electric field of

$$\mathcal{E} = -j\partial V/\partial y = -j/(ab) = -j|\mathcal{E}_0| \qquad (3.57)$$

Note when V is large and negative, y is also large and negative. Therefore, the complex potential corresponds to the side where $y < y[t]$. Replacing b with $1/(a|\mathcal{E}_0|)$ in (3.55) gives

$$z = aj\left((\Phi/(a\,|\mathcal{E}_0|)) + 1 - e^{\Phi/(a|\mathcal{E}_0|)}\right) \qquad (3.58)$$

which can be solved numerically to plot a field map or to obtain the complex potential at a given location. To obtain the surface charge density, we use

$$\sigma_f = \epsilon \operatorname{Im}[\mathcal{E}^* dz/|dz|] = \pm\epsilon\left|\frac{d\Psi}{dz}\right| \qquad (3.59)$$

where the sign of the (3.59) is determined by the physics of the problem. For this case, $\mathcal{E}$ is in the negative y direction requiring a positive surface charge. On the surface,

$$\Phi = j\Psi, \quad x = a\sin\left[\frac{\Psi}{a|\mathcal{E}_0|}\right] - \frac{\Psi}{|\mathcal{E}_0|}, \quad y = a\left(1 - \cos\left[\frac{\Psi}{a|\mathcal{E}_0|}\right]\right) \qquad (3.60)$$

and (3.58) becomes

$$z = aj \left((j\Psi/(a|\mathcal{E}_0|)) + 1 - e^{j\Psi/(a|\mathcal{E}_0|)} \right) \tag{3.61}$$

Differentiating (3.61) allows evaluating $|d\Psi|/|dz|$ which leads to a surface charge density of

$$\sigma_f = +\epsilon \left| \frac{d\Psi}{dz} \right| = \epsilon |\mathcal{E}_0| \left| \left(1 - e^{j\Psi/(a|\mathcal{E}_0|)} \right)^{-1} \right| = \epsilon |\mathcal{E}_0| \left(\frac{a}{2y} \right)^{1/2} \tag{3.62}$$

The field map for the corrugated sheet of metal is shown in Fig. 3.9 (b). Care should be exercised when using this method as the solution may not always be a physical one. For example, the complex potential for top side $(y > y[t])$ of the cycloid obtained by this method is not a physical solution.

3.5 GREEN'S FUNCTION

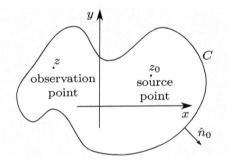

FIGURE 3.10 The contour C encloses an area where Green's function is used to determine the electrostatic potential. The unit vector $\hat{n}_0$ is normal to C and points in the direction away from the enclosed region as shown.

We define Green's function in 2D, G (unit: m/F), as the sum of two functions

$$G[x, y, x_0, y_0] = G_0[x, y, x_0, y_0] + G_L[x, y, x_0, y_0] \tag{3.63}$$

where G_L is required to be twice continuously differentiable in the bounded domain and satisfies Laplace's equation

$$\nabla_0^2 G_L = 0 \tag{3.64}$$

The subscript 0 in (3.64) refers to the (x_0, y_0) variables. As shown in Fig. 3.10, the z_0 location is sometimes labeled the source point and the z location the observation point [2]. The function G_0 satisfies the equation

$$\nabla_0^2 G = \nabla_0^2 G_0 = -\delta[z_0 - z]/\epsilon \qquad (3.65)$$

where we have assumed ϵ is a constant. As discussed in Appendix B, $\delta[z_0 - z]$ is the 2D δ-function defined in Cartesian coordinates as

$$\delta[z_0 - z] = \delta[x_0 - x]\,\delta[y_0 - y] \qquad (3.66)$$

with

$$\delta[z_0 - z] = 0, \quad z_0 \neq z$$

$$\int_{area} \delta[z_0 - z]\, da_0 = \begin{cases} 0, & z \text{ outside of area} \\ 1, & z \text{ inside of area} \end{cases} \qquad (3.67)$$

where $da_0 = dx_0 dy_0$. To determine G_0 we note by the divergence theorem and Cauchy's second integral theorem that

$$\int_{area} \nabla_0^2 \left(\ln[|z - z_0|] \right) da_0 = \text{Im} \left[\oint_C \frac{dz_0}{z_0 - z} \right] = \begin{cases} 0 , & z \text{ outside area} \\ 2\pi, & z \text{ inside area} \end{cases} \qquad (3.68)$$

The properties of (3.68) are the same as the 2D δ-function. Thus, we can formally write

$$\nabla_0^2 \left(\ln[|z - z_0|] \right) = 2\pi\delta[z - z_0] \qquad (3.69)$$

and

$$G_0[x, y, x_0, y_0] = -\frac{1}{2\pi\epsilon} \ln[|z - z_0|] \qquad (3.70)$$

in 2D. This means G_0 is the electrostatic potential of a unit line charge located at z_0, independent of the boundary conditions and G_L tied to boundary conditions. Therefore, G_L is the function that changes in G when boundary conditions are changed.

To determine how boundary conditions define G_L, we start with Green's second identity (1.151) when ϵ is a constant. The power of Green's identities is that the G and V in them can represent any function. Our goal is to determine the electrostatic potential $V[x, y]$ by letting (3.63) represent the G and the electrostatic potential represent the V in (1.151). Integrating over the source point variable in (1.151) gives

$$\int_{area} \left(-G\frac{\rho_f}{\epsilon} + V\frac{\delta[z_0 - z]}{\epsilon} \right) da_0 = \oint_C \left(G\frac{\partial V}{\partial n_0} - V\frac{\partial G}{\partial n_0} \right) ds_0 \qquad (3.71)$$

where $ds_0 = |dz_0|$. Note that (3.65) and

$$\nabla_0^2 V = -\rho_f[x_0, y_0] / \epsilon \tag{3.72}$$

were used on the left side of (3.71). The second term in the first integral of (3.71) can be simplified using the δ-function property

$$V[x, y] / \epsilon = \int_{area} V \left(\frac{\delta[z_0 - z]}{\epsilon} \right) da_0 \tag{3.73}$$

when the area of integration contains z. Substituting (3.73) into (3.71) and rearranging gives

$$V[x, y] = \int_{area} G\rho_f da_0 + \oint_C \epsilon \left(G \frac{\partial V}{\partial n_0} - V \frac{\partial G}{\partial n_0} \right) ds_0 \tag{3.74}$$

Next we use G_L in (3.63) to eliminate one of the terms in the boundary integral of (3.74). The boundary term eliminated depends on the type of boundary condition given.

First consider Green's function G_D for Dirichlet boundary conditions. We force G_D to have the property

$$G_D = 0, \quad \text{on boundary} \tag{3.75}$$

Then the first term in the boundary integral of (3.74) is zero and $V[x, y]$ becomes

$$V[x, y] = \int_{area} G_D \rho_f da_0 - \oint_C \epsilon \left(V \frac{\partial G_D}{\partial n_0} \right) ds_0 \tag{3.76}$$

Now consider Green's function G_N for Neumann boundary conditions. We cannot choose $\partial G_N / \partial n_0 = 0$ for finite C since the divergence theorem combined with (3.65) give

$$\int_{area} \nabla_0 \cdot (\nabla_0 G_N) \, da_0 = \oint_C \frac{\partial G_N}{\partial n_0} ds_0 = -\frac{1}{\epsilon} \tag{3.77}$$

Instead we choose

$$\frac{\partial G_N}{\partial n_0} = -\frac{1}{\epsilon C} \tag{3.78}$$

where C is the contour length. Plugging (3.78) into (3.74) gives

$$V[x, y] = \int_{area} G_N \rho_f \, da_0 + \langle V \rangle_C + \oint_C G_N \sigma_f \, ds_0 \tag{3.79}$$

where $\langle V \rangle_C$ is the average potential on contour C and $\partial V/\partial n_0 = \sigma_f/\epsilon$ (no minus sign since the normal vector that determines σ_f points into the region of interest, but $\hat{n}_0$ points in the opposite direction) was used with σ_f as the free charge density on the contour.

A useful property of Green's functions that are solutions to Poisson's equation with Dirichlet or Neumann (with a boundary infinitely long) boundary conditions is

$$G[x, y, x_0, y_0] = G[x_0, y_0, x, y] \tag{3.80}$$

In other words, interchange of (x, y) and (x_0, y_0) does not change G. To prove this, we consider two Green's functions G_1 and G_2 that are subject to the same boundary conditions but the location of the sources are z_1 and z_2. For G_1, (3.65) gives

$$\nabla_0^2 G_1[x, y, x_1, y_1] = \nabla_0^2 G_1 = -\delta[z_1 - z]/\epsilon \tag{3.81}$$

and for G_2, (3.65) gives

$$\nabla_0^2 G_2[x, y, x_2, y_2] = \nabla_0^2 G_2 = -\delta[z_2 - z]/\epsilon \tag{3.82}$$

Multiplying (3.81) by G_2 and (3.82) by G_1 then subtracting the two resulting equations gives

$$G_2 \nabla_0^2 G_1 - G_1 \nabla_0^2 G_2 = (-G_2 \delta[z_1 - z] + G_1 \delta[z_2 - z])/\epsilon \tag{3.83}$$

Letting $G_1 = V$ and $G_2 = G$ in Green's second identity with constant ϵ (1.151) gives

$$\int_{area} \left(G_2 \nabla_0^2 G_1 - G_1 \nabla_0^2 G_2 \right) da_0 = \oint_C \left(G_2 \frac{\partial G_1}{\partial n_0} - G_1 \frac{\partial G_2}{\partial n_0} \right) ds_0 \tag{3.84}$$

Since $G_1 = G_2 = 0$ on C for Dirichlet boundary conditions and $\partial G_1/\partial n_0 = \partial G_2/\partial n_0 = 0$ for Neumann boundary conditions with an infinite boundary, the right hand side of (3.84) is zero for Dirichlet and Neumann boundary conditions. The left hand side of (3.84) can be evaluated using (3.83) to obtain

$$-G[x_1, y_1, x_2, y_2] + G[x_2, y_2, x_1, y_1] = 0 \tag{3.85}$$

Since z_1 and z_2 are arbitrary points, we can let $z_1 = z$ and $z_2 = z_0$ resulting in (3.80).

3.6 METHOD OF IMAGES AND GREEN'S FUNCTIONS

The main technique we use to determine Green's function is the method of images. The general approach for the method of images is to introduce image charges outside the bounded region of interest in order to satisfy the boundary conditions. The complex potential for the region is then the superposition of the complex potential due to the true charges in the region and the image charges outside of the region used to satisfy the boundary conditions. All the complex potentials developed so far in this chapter can be viewed as using the method images. For example, charges of opposite sign were used to determine the complex potential of two cylinders in Section 3.3. The charges were located inside the cylinders and we were interested in the electrostatic potential and electric field between the cylinders. Thus, the charges are image charges located outside the region of interest used to satisfy the boundary conditions. When finding Green's function using the method of images, we first determine the complex potential due to a unit charge in the region of interest and the image charges outside the region used to satisfy the boundary conditions. Green's function is then the real part of the complex potential.

3.7 GREEN'S FUNCTION FOR A CONDUCTIVE CYLINDER

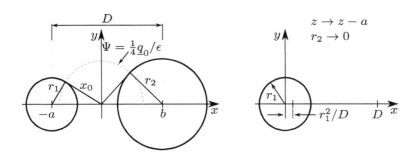

FIGURE 3.11 The complex potential of two conductive cylinders external to each other transformed into a conductive cylinder and line charge by letting $r_2 \to 0$.

In Section 3.3, it was shown that two line charges of opposite sign and equal magnitude form circular equipotential contours. We can view one of the line charges as the true charge in the region of interest and the other line charge as an image charge used to satisfy the boundary conditions for a grounded cylinder. The equations of Section 3.3 are still valid if we let the image and true charges have their locations as $z = -x_0$

and $z = x_0$, respectively. First we consider the case where the region of interest is outside of cylinder 1 as shown in Fig. 3.11. We translate the origin so that is coincides with the center of cylinder 1 in Fig. 3.7 (a) by replacing z with $z - a$ in all equations. Letting $r_2 \to 0$ gives the following relationships

$$a = \frac{D^2 + r_1^2}{2D}, \quad b = x_0, \quad x_0 = \frac{D^2 - r_1^2}{2D} \tag{3.86}$$

where D has become the location of the unit charge $(D = z_0)$ described by G_0. The image charge is now located at

$$z_1 = a - x_0 = r_1^2/D \tag{3.87}$$

With the location of the true charge of value 1 and image charge of value -1 known, the complex potential is

$$\Phi[z] = -\frac{1}{2\pi\epsilon} \ln\left[\frac{z - D}{z - (r_1^2/D)}\right] + C_0 \tag{3.88}$$

The $\mathrm{Re}[C_0]$ is determined by requiring the electrostatic potential to equal zero on the cylinder

$$V\left[r_1 e^{j\theta}\right] = \frac{1}{2\pi\epsilon} \ln\left[\frac{r_1}{D}\right] + \mathrm{Re}[C_0] = 0, \quad \mathrm{Re}[C_0] = -\frac{1}{2\pi\epsilon} \ln\left[\frac{r_1}{D}\right] \tag{3.89}$$

Before we can obtain Green's function for a conductive cylinder, we need to generalize the location of the line charge from being on the real axis to an arbitrary point z_0. Clearly, a rotation of the coordinate axes by an angle $-\theta_0$ leaves the center of the cylinder at the origin, but now the true charge and image charge are located at

$$z_0 = De^{j\theta_0}, \quad z_1 = r_1^2 e^{j\theta_0}/D = r_1^2/z_0^* \tag{3.90}$$

respectively. The complex potential for the generalized location of the true charge is

$$\Phi[z] = \frac{1}{2\pi\epsilon} \ln\left[\frac{|z_0|}{r_1} \frac{z - (r_1^2/z_0^*)}{z - z_0}\right] + j\mathrm{Im}[C_0] \tag{3.91}$$

Thus, Green's function for the region outside a grounded cylinder of radius r_1 centered at the origin is the real part of (3.91)

$$G_D[x, y, x_0, y_0] = \frac{1}{2\pi\epsilon} \ln\left[\frac{|z_0|}{r_1} \left|\frac{z - (r_1^2/z_0^*)}{z - z_0}\right|\right] \tag{3.92}$$

Now consider the case where the region of interest is inside cylinder 1. The charge inside the cylinder is now the true charge with value 1 and the charge outside the cylinder is now the image charge with value -1. Let the location of the true charge be $z = z_0$. The image charge is then located at $z = r_1^2/z_0^*$. Relabeling the location and changing the signs of the charges results in the same complex potential and Green's function given by (3.91) and (3.92) (see Fig. 3.12).

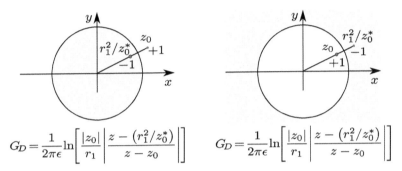

$$G_D = \frac{1}{2\pi\epsilon} \ln\left[\frac{|z_0|}{r_1} \left|\frac{z - (r_1^2/z_0^*)}{z - z_0}\right|\right] \qquad G_D = \frac{1}{2\pi\epsilon} \ln\left[\frac{|z_0|}{r_1} \left|\frac{z - (r_1^2/z_0^*)}{z - z_0}\right|\right]$$

FIGURE 3.12 Green's function for the region internal to a conductive cylinder or external to the conductive cylinder are the same if in each case, the location of the true charge is labeled z_0.

3.8   GREEN'S FUNCTION FOR A CONDUCTIVE PLANE

In Section 3.3, it was shown that the plane equidistant and between two line charges of equal magnitude but opposite sign was an equipotential plane. We can view one of the line charges as the true charge in the region of interest and the other charge as an image charge used to achieve the boundary conditions for the grounded plane. If we let the conducting plane correspond to the imaginary axis of the z-plane and the location of the true charge of unit value as $z = z_0$ (see Fig. 3.13), the complex potential is

$$\Phi[z] = -\frac{1}{2\pi\epsilon} \ln\left[\frac{z - z_0}{z + z_0^*}\right] \tag{3.93}$$

Green's function is the real part of (3.93)

$$G_D[x, y, x_0, y_0] = -\frac{1}{2\pi\epsilon} \ln\left[\frac{|z - z_0|}{|z + z_0^*|}\right] \tag{3.94}$$

Since G_0 is given by (3.70), G_L is

$$G_L[x, y, x_0, y_0] = \frac{1}{2\pi\epsilon} \ln[|z + z_0^*|] \tag{3.95}$$

Note that the region of interest contains z_0 and does not contain z_0^*. Thus, G_L is twice continuously differentiable in the region of interest as required.

If the grounded plane is the real axis, we rotate the axis by $\pi/2$ by replacing z with $-jz$ and z_0 with $-jz_0$ in (3.93) and the complex potential becomes

$$\Phi[z] = -\frac{1}{2\pi\epsilon}\ln\left[\frac{z-z_0}{z-z_0^*}\right] \tag{3.96}$$

Green's function is

$$G_D[x, y, x_0, y_0] = \frac{1}{2\pi\epsilon}\ln\left[\left|\frac{z-z_0^*}{z-z_0}\right|\right] \tag{3.97}$$

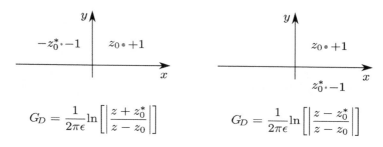

FIGURE 3.13   Green's function for a conductive plane.

3.9   GREEN'S FUNCTION FOR TWO CONDUCTING PLANES

Let two grounded conducting planes intersect at an angle π/m where $m > 1$ is an integer. We orient the axes so that one of the conducting planes lies along the x-axis. The other plane then makes an angle π/m with the x-axis. Let $z_0 = r_0 e^{j\theta_0}$ be the location of the true line charge and we are interested in the region between $0 \le \theta \le \pi/m$. Based on the analysis of Section 3.8, each plane requires an image charge due to the true charge to satisfy the boundary condition as shown in step 1 of Fig. 3.14 for the $m = 3$ case. The image charges introduced to satisfy the boundary conditions for each plane do not satisfy the boundary condition on the other plane. Therefore, additional image charges need to be introduced to satisfy the boundary conditions at the other plane as shown in step 2 of Fig. 3.14. If the image charges introduced in step 2 do not satisfy the boundary conditions on both planes for all charges, then additional image charges need to be added until the boundary conditions are satisfied for all charges. In general, $2m-1$ image charges (or m steps as outlined in Fig. 3.14) are required to satisfy the boundary conditions at both planes when m is an integer. There are $m-1$ image charges with the same charge value as the true charge and are located at

$$z = z_0 e^{2\pi jn/m}, \quad n = 1, 2, \cdots, m-1 \tag{3.98}$$

The other m image charges have the opposite sign as the true charge and are located at

$$z = z_0^* e^{2\pi jn/m}, \quad n = 1, 2, \cdots, m \tag{3.99}$$

Note that the true and all image charges lie on a circle centered at the origin of radius r_0. The complex potential is

$$\Phi[z] = \frac{1}{2\pi\epsilon} \sum_{n=1}^{m} \ln\left[\frac{z - z_0^* e^{2\pi jn/m}}{z - z_0 e^{2\pi j(n-1)/m}}\right] \tag{3.100}$$

Green's function is the real part of (3.100)

$$G_D[x, y, x_0, y_0] = \frac{1}{2\pi\epsilon} \sum_{n=1}^{m} \ln\left[\left|\frac{z - z_0^* e^{2\pi jn/m}}{z - z_0 e^{2\pi j(n-1)/m}}\right|\right] \tag{3.101}$$

A field map for $m = 3$ is shown in Fig. 5.18.

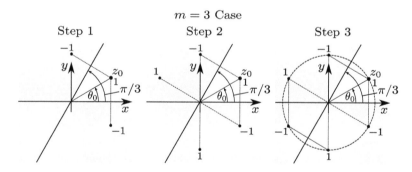

FIGURE 3.14 Green's function for two conducting planes that intersect at an angle π/m where $m > 1$ is an integer.

3.10 RAY TRACING FOR PLANAR DIELECTRIC BOUNDARIES

The largest obstacle when applying the method of images is determining the location and value of each image charge. For planar boundaries, a geometric approach has been developed to overcome this obstacle [3]. Although this approach is only useful for planar boundaries, the development of conformal mapping in subsequent chapters allows extending the results to other geometries.

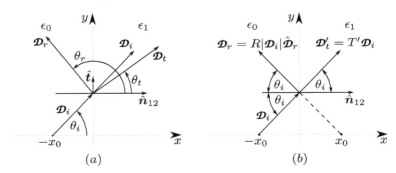

FIGURE 3.15 (a) Definition of an incident electric flux density ray $\mathcal{D}_i$, a reflected electric flux density ray $\mathcal{D}_r$, a transmitted electric flux density ray $\mathcal{D}_t$, the incident angle θ_i, the reflected angle θ_r, and the transmitted angle θ_t. (b) Resulting angles required for the boundary dielectric boundary condition to be satisfied.

Let the imaginary axis in the z-plane represent the boundary between a dielectric material and free space with no free charge at the interface. We want to determine if the dielectric boundary conditions for a line charge located at $z = -x_0$ can be satisfied through the combination of the potential due to the true charge at $z = -x_0$ and image charges. First we postulate that in addition to the true line charge (located in free space), the potential for each region ($x < 0$ and $x > 0$) can be described by one additional image line charge. Let the image line charge located in the region $x > 0$ have a value q_r and let the image line charge located in the region $x < 0$ have a value q_t. Due to symmetry, the image charges are located somewhere on the real axis which we label $x_t < 0$ and $x_r > 0$. To determine the location of the image charges, we imagine that line charges emit equal rays of flux density vector $\mathcal{D}$ radially. Let an emitted ray $\mathcal{D}_i$ that is directed toward the dielectric boundary with angle θ_i from a line charge located at $z = -x_0$ meet the boundary at $z = y$. The postulated image charge located in the region $x > 0$ emits a ray $\mathcal{D}_r$ that intersects the boundary at $z = y$. The total flux density vector in region 1 at $z = y$ is the vector sum of these two rays

$$\mathcal{D}_1 = \mathcal{D}_i + \mathcal{D}_r \qquad (3.102)$$

For region 2, we have an image charge located in the region $x < 0$ that emits a ray $\mathcal{D}_t$ that intersects the boundary at $z = y$ resulting in the total flux density vector in region 2 at $z = y$ of

$$\boldsymbol{D}_2 = \boldsymbol{D}_i + \boldsymbol{D}_t \tag{3.103}$$

Let

$$|\boldsymbol{D}_r| = R\,|\boldsymbol{D}_i|, \quad |\boldsymbol{D}_t| = T\,|\boldsymbol{D}_i| \tag{3.104}$$

The boundary conditions of (2.71) and (2.73) give

$$(\boldsymbol{D}_2 - \boldsymbol{D}_1) \cdot \hat{n}_{12} = |\boldsymbol{D}_i|\,(T\cos[\theta_t] - R\cos[\theta_r]) = 0 \tag{3.105}$$

and

$$(\boldsymbol{E}_2 - \boldsymbol{E}_1) \cdot \hat{t} = |\boldsymbol{D}_i|\left(\frac{\sin[\theta_i] + T\sin[\theta_t]}{\epsilon_1} - \frac{\sin[\theta_i] + R\sin[\theta_r]}{\epsilon_0}\right) = 0 \tag{3.106}$$

Since $|\boldsymbol{D}_i|$, $|\boldsymbol{D}_r|$, and $|\boldsymbol{D}_t|$ are independent of angle, R and T should also be independent of angle. From (3.105), T is independent of angle if $\theta_r = \theta_t$ or $\theta_r = \pi \pm \theta_t$. In order for $\boldsymbol{D}_r$ to transmit into region 1 and $\boldsymbol{D}_t$ to transmit into region 2, $\theta_r = \pi \pm \theta_t$ are the only possible solutions. Solving for T in (3.105), inserting the result into (3.106), and using both possible solutions $\theta_r = \pi \pm \theta_t$ gives

$$R = \left(\frac{\sin[\theta_i]}{\sin[\theta_t]}\right)\left(\frac{-\epsilon_0 + \epsilon_1}{-\epsilon_0 \pm \epsilon_1}\right) \tag{3.107}$$

In order for R to be angle independent and $\boldsymbol{D}_t$ to transmit into region 2, $\theta_t = \theta_i$. In order for R not be a constant, $\theta_r = \pi - \theta_t$. Thus, the only physical solutions are

$$R = \frac{\epsilon_0 - \epsilon_1}{\epsilon_0 + \epsilon_1}, \quad T = -R = \frac{\epsilon_1 - \epsilon_0}{\epsilon_0 + \epsilon_1}, \quad \theta_t = \theta_i, \quad \theta_r = \pi - \theta_i \tag{3.108}$$

With θ_r and θ_t determined, the location of the image charges can be obtained by tracing the flux density rays due to the image charges back to the real axis. The locations for the image charges are

$$x_r = x_0, \quad x_t = -x_0 \tag{3.109}$$

Since q_0 provided a flux density magnitude of $|\boldsymbol{D}_i|$, the value of the image charges are

$$q_r = Rq_0, \quad q_t = Tq_0 \tag{3.110}$$

With the image charge q_t occupying the same location as the true charge, the potential of region 2 can be determined by a single image charge $\underline{q}_t$ of value

$$\underline{q}'_t = \underline{q}_0 + T\underline{q}_0 = T'\underline{q}_0 = \frac{2\epsilon_1}{\epsilon_0 + \epsilon_1}\underline{q}_0 \tag{3.111}$$

Let us now summarize and generalize the findings. When a flux density vector ray $\mathbf{D}_i$ encounters a planar dielectric boundary, it can be viewed as being partly reflected and partly transmitted. The transmitted ray has a transmission coefficient of T' given by (3.111) and is in the same direction as the incident ray. The reflected ray has a reflection coefficient R given by (3.108) and is in the direction of $\pi - \theta_i$. The transmitted ray can be viewed as being generated by an image charge with value $\underline{q}'_t$ given by (3.111) located at the same location as the true charge. The reflected ray can be viewed as being generated by an image charge with value $\underline{q}_r$ given by (3.110) located on the perpendicular line to the boundary that goes through the true charge an equal distance from the boundary as the true charge but on the opposite side.

In general, the method of images allows writing the complex potential of region k in terms of true and image charges as

$$\Phi_k = -\frac{1}{2\pi\epsilon_k}\sum_i \underline{q}_i \ln[z - z_i] + C_k \tag{3.112}$$

where the sum is over all true and image charges that contribute flux to region k, ϵ_k is the permittivity of region k, and C_k is a complex constant that is not arbitrary. Note that the permittivity to use for the factor in front of the sum is the permittivity for the region of interest and does not depend on the permittivity of the region the image charge is located. To determine the value of C_k, we apply the dielectric boundary conditions for the complex potential given by (2.136) and (2.139). The complex potential for each region in Fig. 3.15 is

$$\Phi[z] = \begin{cases} -\dfrac{\underline{q}_0}{2\pi\epsilon_0}\left(\ln[z + x_0] + \dfrac{\epsilon_0 - \epsilon_1}{\epsilon_0 + \epsilon_1}\ln[x_0 - z]\right), & x < 0 \\ -\dfrac{\underline{q}_0}{\pi(\epsilon_0 + \epsilon_1)}\ln[z + x_0] & , \quad x > 0 \end{cases} \tag{3.113}$$

The total charge at the dielectric boundary and at infinity can be calculated using ME(1a). First we recall that

$$\int_{s_1}^{s_2} \mathbf{E}\cdot\hat{n}\,ds = \int_{z_1}^{z_2} \mathrm{Im}\left[-\left(\frac{\partial\Phi}{\partial z}\right)\right]dz = \Psi[z_1] - \Psi[z_2] \tag{3.114}$$

Next, we enclose the boundary with a loop that has two sides parallel to the boundary and two sides perpendicular to the boundary. If we let the length of the perpendicular sides go to zero, then

$$\oint_C \boldsymbol{E} \cdot \hat{\boldsymbol{n}} \, ds = \Psi_{x>0}[-\infty] - \Psi_{x>0}[\infty] + \Psi_{x<0}[\infty] - \Psi_{x<0}[-\infty] \quad (3.115)$$

where $\Psi_{x>0}$ and $\Psi_{x<0}$ are obtained from the imaginary part of (3.113). From ME(1a), the net charge (bound and free charge) at the dielectric boundary is

$$\underline{q}_{net} = \epsilon_0 \oint_C \boldsymbol{E} \cdot \hat{\boldsymbol{n}} \, ds = \underline{q}_0 \left(\frac{\epsilon_0 - \epsilon_1}{\epsilon_0 + \epsilon_1} \right) \quad (3.116)$$

As discussed at the end of Section 2.5, the entire system must be charge neutral. Thus, the total charge located at infinity ($\underline{q}_\infty$) must have the magnitude of the line charge plus the charge at the dielectric boundary but of opposite polarity

$$\underline{q}_\infty = -\underline{q}_0 - \underline{q}_0 \left(\frac{\epsilon_0 - \epsilon_1}{\epsilon_0 + \epsilon_1} \right) \quad (3.117)$$

Although the field maps for each region can be plotted, the discontinuities in the lines of force at the boundary no longer allow a simple interpretation of the resulting plot. Therefore, equipotential contours and constant flux contours ($\epsilon\Psi = $ constant) are drawn instead. Separate plots of each are given to prevent associating the diagrams with field maps. The constant potential contours and constant flux contours for the complex potential (3.113) with $\epsilon_1 = 3\epsilon_0$ are shown in Fig. 3.16. The constant potential contours are compressed near the boundary compared to the case of no boundary. This compression leads to the curvature observed in the constant flux contours compared to the straight lines emanating from $-x_0$ in the case of no boundary.

3.11 RAY TRACING FOR PLANAR CONDUCTOR BOUNDARIES

If we let $\epsilon_1 \to \infty$ in Fig. 3.15, then $R \to -1$ as shown in Fig. 3.17. From (3.102) $\boldsymbol{E}_1$ at the boundary is

$$\boldsymbol{E}_1 = (|\boldsymbol{D}_i|/\epsilon_0) \left(\hat{\boldsymbol{D}}_i - \hat{\boldsymbol{D}}_r \right) = (2|\boldsymbol{D}_i|/\epsilon_0) \cos[\theta_i] \quad (3.118)$$

which is perpendicular to the boundary ($\text{Im}[\boldsymbol{E}_1] = 0$). From Section 2.20, a perpendicular $\boldsymbol{E}$ is the boundary condition for a conductor. Thus, the

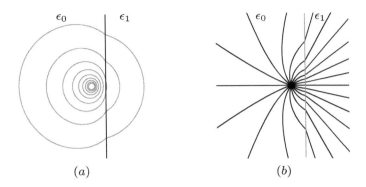

FIGURE 3.16 (a) Constant potential contours and (b) constant flux contours for a line charge in vacuum in front of a dielectric half plane. The permittivity of the dielectric half plane is $\epsilon_1 = 3\epsilon_0$.

boundary can be replaced by a conductor and the complex potential for $x < 0$ reduces to the potential due to a positive and negative charge of same magnitude equidistant from the boundary but on opposite sides. This is consistent with the method of images for a grounded conducting plane derived in Section 3.8.

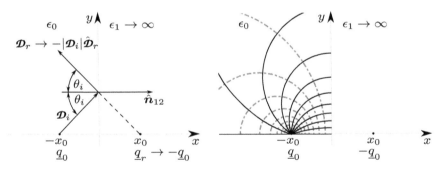

FIGURE 3.17 Conductor boundary condition obtained by letting $\epsilon_1 \to \infty$ for a dielectric half-plane.

3.12 RAY TRACING FOR PLANAR LINE OF FORCE BOUNDARIES

From Section 3.11, it was shown that a positive and negative charge of same magnitude equidistant from the boundary but on opposite sides produces an electric field perpendicular to the interface at the boundary. It is reasonable to postulate that two charges of same magnitude equidistant from the boundary but on opposite sides produce an electric field parallel to the interface at the boundary. An image charge of same sign is equivalent to $R = 1$ in (3.105) and from (3.102) $\boldsymbol{\mathcal{E}}_1$ at the boundary is

$$\boldsymbol{\mathcal{E}}_1 = (|\boldsymbol{D}_i|/\epsilon_0)\left(\hat{\boldsymbol{D}}_i + \hat{\boldsymbol{D}}_r\right) = (2j|\boldsymbol{D}_i|/\epsilon_0)\sin[\theta_i] \qquad (3.119)$$

which is parallel to the boundary. Thus, the boundary is a line of force contour as shown in Fig. 3.18.

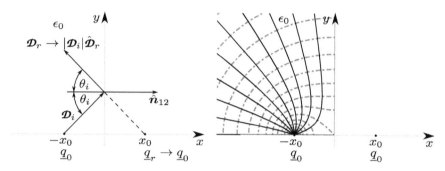

FIGURE 3.18   Planar line of force boundary condition by equating $\underline{q}_r = \underline{q}_0$.

3.13   RAY TRACING FOR MULTIPLE PLANAR BOUNDARIES

The value of the ray tracing method becomes clear when extended to multiple dielectric regions with planar boundaries. For example, consider a dielectric of thickness a separating two semi-infinite dielectric regions as shown in Fig. 3.19. Let region 1 have a dielectric constant ϵ_0 and occupy the space $x > 0$. Let region 2 have a dielectric constant ϵ_1 and occupy the space $-a < x < 0$. Let region 3 have a dielectric constant ϵ_2 and occupy the space $x < -a$. Let the location of a line charge q_0 be $z = b$. Due to the two boundaries, a $\boldsymbol{D}$ ray from the line charge that enters region 2 is reflected an infinite number of times at each region 2 boundary. At each of these reflections, a $\boldsymbol{D}$ ray is transmitted into either region 1 or region 3. The infinite transmissions and reflections translate into an infinite number of image charges for all three regions. The locations of the image charges are found by tracing the $\boldsymbol{D}$ rays back to the real axis as shown in Fig. 3.19. To determine the values of the image charges, we first define the transmission coefficient T'_{ij} and reflection coefficient R_{ij} as

$$T'_{ij} = \frac{2\epsilon_j}{\epsilon_i + \epsilon_j}, \quad R_{ij} = \frac{\epsilon_i - \epsilon_j}{\epsilon_i + \epsilon_j} \qquad (3.120)$$

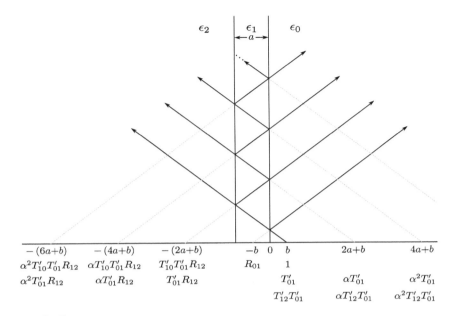

FIGURE 3.19 Location and ratio of image charge to true charge for a planar dielectric slab of permittivity ϵ_1 located between vacuum and an infinite dielectric region of permittivity ϵ_2.

with i being the region the ray is in before reaching the boundary and j being the region on the other side of the boundary. The values for the image charges are found by multiplying the original $\mathcal{D}$ ray by the appropriate transmission coefficient T'_{ij} or reflection coefficient R_{ij} each time the ray encounters a boundary. The three rows of values below the location of the image charges in Fig. 3.19 tabulate the values of the first few image charges divided by the original line charge for each region. The highlighted region indicate the region for which the image charges on that line are used to determine complex potential. Note that none of the image charges are located in the region for which they describe the potential as required. Taking into account all of the image charges based on Fig. 3.19 and applying the dielectric boundary conditions, the complex potential for region 1 $(x > 0)$ is

$$\Phi_1[z] = -\frac{q_0}{2\pi\epsilon_0}\left(\ln[b-z] + R_{01}\ln[z+b] \right.$$

$$\left. + \sum_{n=1}^{\infty} T'_{01}T'_{10}R_{12}\alpha^{n-1}\ln[z+2na+b] \right)$$

(3.121)

for region 2 $(-a < x < 0)$ is

$$\Phi_2[z] = -\frac{q_0}{2\pi\epsilon_1}\sum_{n=0}^{\infty}T'_{01}a^n\left(\ln[2na+b-z]\right. \tag{3.122}$$
$$\left.+ R_{12}\ln[z+2(n+1)a+b]\right)$$

and for region 3 $(x < -a)$ is

$$\Phi_3[z] = -\frac{q_0}{2\pi\epsilon_2}\sum_{n=0}^{\infty}T'_{12}T'_{01}a^n\ln[2na+b-z] \tag{3.123}$$

A plot of the equipotential contours and constant flux contours are shown in Fig. 3.20 for $\epsilon_2 = 2\epsilon_1$ and $\epsilon_1 = 2\epsilon_0$.

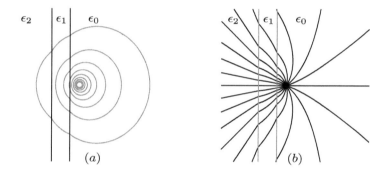

(a) (b)

FIGURE 3.20 (a) Constant potential contours and (b) constant flux contours for a line charge in vacuum in front of a dielectric slab of permittivity $\epsilon_1 = 2\epsilon_0$ and an infinite dielectric region of permittivity $\epsilon_2 = 2\epsilon_1$.

3.14 1D ARRAY OF LINE CHARGES

The complex potential due to n charges is given by (3.1). If all the charges have the same magnitude, then q_i can be brought outside the summation and (3.1) becomes

$$\Phi[z] = -\frac{q}{2\pi\epsilon}\ln\left[\prod_{i=1}^{n}(z-z_i)\right] + C_0 = -\frac{q}{2\pi\epsilon}\ln[f[z]] + C_0 \tag{3.124}$$

where $f[z]$ is an analytic function (except at the locations of negative line charge) with simple zeros and poles at the location of positive and negative line charges, respectively. Now consider an infinite 1D array of line charges that all have the same value with equal spacing a between

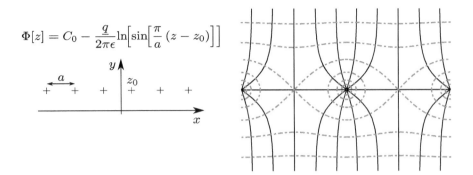

$$\Phi[z] = C_0 - \frac{q}{2\pi\epsilon}\ln\left[\sin\left[\frac{\pi}{a}(z - z_0)\right]\right]$$

FIGURE 3.21 Complex potential for an infinite 1D array of positive line charges with spacing between charges of a. The field map for the region enclosed by the gray rectangle is shown on the right. $\Delta V = \Delta \Psi = 0.1 \underline{q}_0/\epsilon$.

them. The complex potential still has the form given by (3.124) with n replaced by infinity. If we can find an analytic function $f[z]$ that has zeros at the location of charges, then the infinite product of (3.124) can be replaced by $f[z]$. The function $\sin[z + \delta]$ meets these requirements. Let the array of line charges be oriented parallel to the real axis with one of line charges located at z_0 as shown in Fig. 3.21. Then the complex potential is

$$\Phi[z] = -\frac{q}{2\pi\epsilon}\ln[\sin[\pi(z - z_0)/a]] + C_0 \qquad (3.125)$$

Note that the vertical line midway between line charges corresponds to

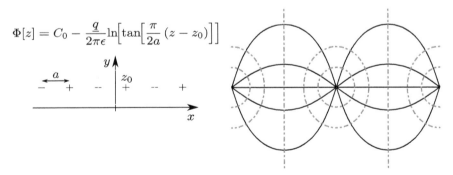

$$\Phi[z] = C_0 - \frac{q}{2\pi\epsilon}\ln\left[\tan\left[\frac{\pi}{2a}(z - z_0)\right]\right]$$

FIGURE 3.22 Complex potential for an infinite 1D array of alternating sign line charges with spacing between charges of a. The field map for the region enclosed by the gray rectangle is shown on the right. $\Delta V = \Delta \Psi = 0.1 \underline{q}_0/\epsilon$.

a constant line of force. Thus, (3.125) is the complex potential for a line charge located midway between two infinitely long vertical boundaries with constant lines of force boundary conditions.

If the infinite 1D array of line charges is composed of equal magnitude charges but of alternating sign, then the function has poles and zeros. The function $\tan[z + \delta]$ meets these requirements. Let the 1D array of equal magnitude but alternating sign line charges be oriented parallel to the real axis with one of the positive line charges located at z_0 (see Fig. 3.22). Then the complex potential is

$$\Phi[z] = -\frac{q}{2\pi\epsilon} \ln[\tan[\pi (z - z_0) / (2a)]] + C_0 \qquad (3.126)$$

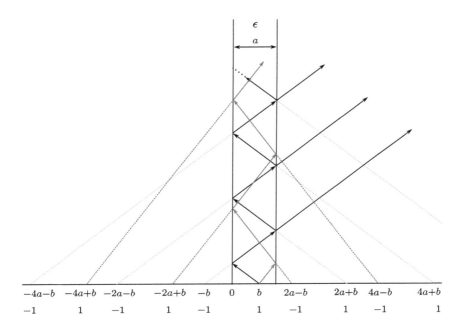

FIGURE 3.23 Location and ratio of image charges to the true line charge located at $z = b$ between two planar grounded conductors.

As we have seen in Section 3.13, an infinite array of 1D image line charges occurs when multiple planar boundaries are present. If the boundaries are conductors, then the images have equal magnitudes and the above analysis can be applied. Consider the case where a line charge is located between two infinite vertical conductors a distance a apart. Let the left boundary cross the real axis at $x = 0$, the right boundary then crosses the real axis at $x = a$. Let the line charge have a value of q_0 and location $z = b < a$ on the real axis. Since there are two boundaries, we have to trace two rays of $\mathcal{D}$ originating from the true line charge, one toward each boundary. The location and values of the first few image charges are shown in Fig. 3.23. The image charges can be viewed as two sets of infinite 1D array of line charges. Each line charge in the first set has a value of q_0, a distance between charges of $2a$, and are located on the real axis at $x = 2na + b$ where n is an integer. Each line charge in the second set has a value of $-q_0$, a distance between charges of $2a$, and are located on the real axis at $x = -(2na+b)$ where n is an integer. The complex potential is

$$\Phi[z] = -\frac{q_0}{2\pi\epsilon} \ln\left[\frac{\sin[\pi\,(z-b)\,/\,(2a)]}{\sin[\pi\,(z+b)\,/\,(2a)]}\right] + C_0 \qquad (3.127)$$

The field map for (3.127) with $b = a/5$ is shown in Fig. 3.24.

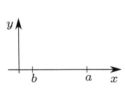

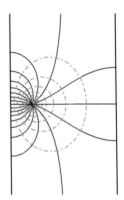

$$\Phi[z] = C_0 - \frac{q_0}{2\pi\epsilon}\ln\left[\frac{\sin[\pi\,(z-b)\,/\,(2a)]}{\sin[\pi\,(z+b)\,/\,(2a)]}\right]$$

FIGURE 3.24   Field map of a line charge located between two ground planes. The distance between ground planes is a and the line charge is located $b = a/5$ away from the closest ground plane. $\Delta V = \Delta \Psi = 0.05q_0/\epsilon$.

3.15   2D ARRAY OF LINE CHARGES

Similar to an infinite 1D array of line charges, the complex potential due to an infinite 2D array of line charges with the same magnitude can be expressed by the last form of (3.124) if an analytic function $f[z]$ can be found that has simple zeros and poles at the location of positive and negative charges, respectively. A class of functions that meet these requirements are called elliptic functions. For example, the Jacobian elliptic function $\text{sn}[u, m]$ (see Appendix D) has simple zeros at

$$u = 2nK[m] + 2jpK'[m], \quad n, p = 0, \pm 1, \pm 2, \ldots \qquad (3.128)$$

poles • and zeros □ of sn[z, m]   poles • and zeros □ of cn[z, m]

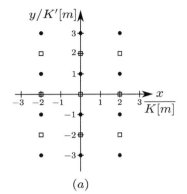
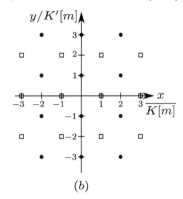

(a)   (b)

FIGURE 3.25 (a) Poles and zeros of the sn[z, m] function. (b) Poles and zeros of the cn[z, m] function.

and simple poles at

$$u = 2nK[m] + j(2p+1)K'[m], \quad n, p = 0, \pm 1, \pm 2, \dots \qquad (3.129)$$

as shown in Fig. 3.25 (a). The Jacobian elliptic function cn[u, m] has simple zeros at

$$u = (2(n+p)+1)K[m] + 2jpK'[m], \quad n, p = 0, \pm 1, \pm 2, \dots \qquad (3.130)$$

and simple poles at

$$u = 2nK[m] + j(2(p+n)+1)K'[m], \quad n, p = 0, \pm 1, \pm 2, \dots \qquad (3.131)$$

as shown in Fig. 3.25 (b). If the poles and zeros are made to align with the location of positive and negative charges, then $f[z]$ can be written in terms of these functions.

The infinite 2D array composed of alternating rows of positive line charges and negative line charges as shown in Fig. 3.26 can be built up by combining the pole/zero pattern of sn$[K[m]z/a, m]$ function with sn$[K[m](z+a)/a, m]$ (the same function just shifted by a in the horizontal direction). The value of m is determined by the ratio of the vertical to horizontal spacing through the relationship

$$K'[m]/K[m] = b/a \qquad (3.132)$$

Let z_0 correspond to the position of a positive charge, then the complex

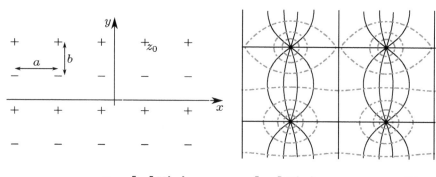

$$\Phi[z] = C_0 - \frac{q_0}{2\pi\epsilon}\ln\left[\operatorname{sn}\left[\frac{K[m]}{a}(z - z_0), m\right]\operatorname{sn}\left[\frac{K[m]}{a}(z - z_0 + a), m\right]\right]$$

FIGURE 3.26 Complex potential for an infinite 2D array of line charges where charges on the same row have the same value, but charges on adjacent rows have the opposite sign. Field map on the right corresponds to the region enclosed in the gray rectangle in the diagram on the left. $\Delta V = \Delta \Psi = 0.1 q_0/\epsilon$.

potential for Fig. 3.26 is

$$\Phi[z] = -\frac{q_0}{2\pi\epsilon}\ln\left[\operatorname{sn}\left[\frac{K[m]}{a}(z - z_0), m\right]\right]$$
$$- \frac{q_0}{2\pi\epsilon}\ln\left[\operatorname{sn}\left[\frac{K[m]}{a}(z - z_0 + a), m\right]\right] + C_0 \tag{3.133}$$

To obtain the alternating pattern of zeros and poles shown in Fig. 3.27, we combine the pole/zero pattern of $\operatorname{sn}[K[m] z/a, m]$ function with $\operatorname{sn}[K[m](z + a + jb)/a, m]$ (the same function just shifted by $a + jb$). If a positive charge is located at z_0, the complex potential for Fig. 3.27 is

$$\Phi[z] = -\frac{q_0}{2\pi\epsilon}\ln\left[\operatorname{sn}\left[\frac{K[m]}{a}(z - z_0), m\right]\right]$$
$$- \frac{q_0}{2\pi\epsilon}\ln\left[\operatorname{sn}\left[\frac{K[m]}{a}(z - z_0 + a + jb), m\right]\right] + C_0 \tag{3.134}$$

where m is still obtained by (3.132). A pattern of zeros and poles as shown in Fig. 3.28 is the same as the pole/zero pattern for the elliptic function $\operatorname{cn}[z, m]$ shown in Fig. 3.25 (b). The complex potential is

$$\Phi[z] = -\frac{q_0}{2\pi\epsilon}\ln\left[\operatorname{cn}\left[\frac{K[m]}{a}(z - z_0 + a), m\right]\right] + C_0 \tag{3.135}$$

where z_0 is the location of a positive charge and the value of m is obtained from (3.132).

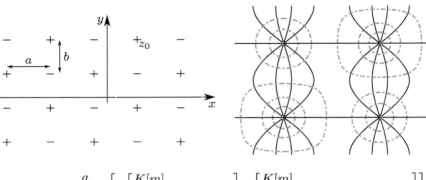

$$\Phi[z] = C_0 - \frac{q_0}{2\pi\epsilon}\ln\left[\mathrm{sn}\left[\frac{K[m]}{a}(z - z_0), m\right]\mathrm{sn}\left[\frac{K[m]}{a}(z - z_0 + a + jb), m\right]\right]$$

FIGURE 3.27 Complex potential for an infinite 2D array of line charges where charges on the same row have the same magnitude but alternate in polarity. Field map on the right corresponds to region enclosed in the gray rectangle in the diagram on the left. $\Delta V = \Delta \Psi = 0.1 q_0/\epsilon$.

Engineering the location of the zeros and poles with elliptic functions requires a deeper understanding of elliptic functions even in simple cases. For example, consider a line charge located inside a grounded conducting cylinder of infinite length with rectangular cross-section as shown in Fig. 3.29. The infinite array of 2D image charges shown in Fig. 3.29 is required to satisfy the boundary conditions [1]. Positive charges are located at

$$z = \begin{cases} 4an + j2bp + z_0 \\ 2a(2n+1) + j2bp - z_0, \end{cases} \quad n, p = 0, \pm 1, \pm 2, \ldots \qquad (3.136)$$

The negative charges are located at

$$z = \begin{cases} 4an + j2bp + z_0^* \\ 2a(2n+1) + j2bp - z_0^*, \end{cases} \quad n, p = 0, \pm 1, \pm 2, \ldots \qquad (3.137)$$

In the horizontal or vertical directions, the distance between consecutive same sign charges is a constant, but the distance between positive and negative charges has two values. The functions $\mathrm{sn}[u, m]$ and $\mathrm{cn}[u, m]$ have constant distances between poles and zeros. How to use elliptic functions to meet the image charge locations is not obvious. It turns out that the complex potential is

$$\Phi[z] = -\frac{q_0}{2\pi\epsilon}\ln\left[\frac{\mathrm{sn}[K[m]z/a] - \mathrm{sn}[K[m]z_0/a]}{\mathrm{sn}[K[m]z/a] - \mathrm{sn}[K[m]z_0^*/a]}\right] + C_0 \qquad (3.138)$$

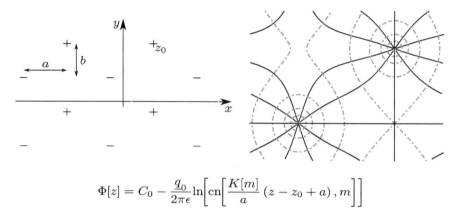

$$\Phi[z] = C_0 - \frac{q_0}{2\pi\epsilon}\ln\left[\mathrm{cn}\left[\frac{K[m]}{a}\left(z - z_0 + a\right), m\right]\right]$$

FIGURE 3.28 Complex potential for an infinite 2D array of line charges where charges on the same row have the same value, but charges on adjacent rows have the opposite sign. Field map on the right corresponds to region enclosed in the gray rectangle in the diagram on the left. $\Delta V = \Delta \Psi = 0.1q_0/\epsilon$.

where m is obtained with (3.132). To understand how to arrive at (3.138), we need to realize that subtracting a constant term from $\mathrm{sn}[u, m]$ changes the location of the zeros but the poles stay at their original location. Subtraction of a real number $\mathrm{sn}[K[m]x_0/a, m]$ from $\mathrm{sn}[K[m]z/a, m]$ results in zeros located at $4an + j2bp$ to be shifted by x_0

$$4an + j2bp \rightarrow 4an + x_0 + j2bp \tag{3.139}$$

and zeros located at $2a(2n + 1) + j2bp$ to be shifted by $-x_0$

$$2a(2n + 1) + j2bp \rightarrow 2a(2n + 1) - x_0 + j2bp \tag{3.140}$$

The subtraction of an imaginary number $\mathrm{sn}[K[m]jy_0/a, m]$ from $\mathrm{sn}[K[m]z/a, m]$ results in zeros located at $4an + j2bp$ to be shifted by jy_0

$$4an + j2bp \rightarrow 4an + j(2bp + y_0) \tag{3.141}$$

and zeros located at $2a(2n + 1) + j2bp$ to be shifted by $-jy_0$

$$2a(2n + 1) + j2bp \rightarrow 2a(2n + 1) + j(2bp - y_0) \tag{3.142}$$

Thus, we can see that the zeros and poles of (3.138) do align with the positive and negative charges as required. In Chapter 5, conformal mapping is used to solve this problem which does not require engineering the poles and zeros of elliptic functions. The field map for (3.138) with $b = a$ and $z_0 = \frac{1}{2}a + j\frac{3}{4}b$ is shown in Fig. 3.30.

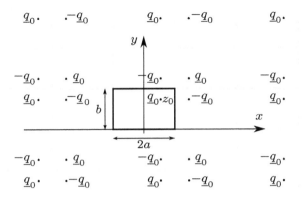

FIGURE 3.29 2D array of image charges for a line charge located inside a grounded $2a \times b$ rectangular cylinder.

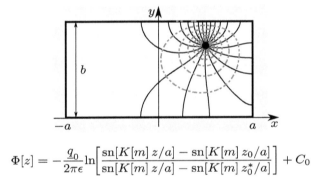

$$\Phi[z] = -\frac{q_0}{2\pi\epsilon}\ln\left[\frac{sn[K[m]\,z/a] - sn[K[m]\,z_0/a]}{sn[K[m]\,z/a] - sn[K[m]\,z_0^*/a]}\right] + C_0$$

FIGURE 3.30 Field map of a line charge inside of a grounded $2a \times b$ rectangular cylinder. Line charge is located at $z_0 = \frac{1}{2}a + j\frac{3}{4}b$, $b = a$, and $\Delta V = \Delta \Psi = 0.05 q_0/\epsilon$.

3.16 LINE CHARGE BETWEEN A GROUNDED AND FLOATING CYLINDER

When using the method of images, the resulting boundary conditions due to the image charges should be verified to be the same as the desired boundary conditions. To show this is not always the case, consider a line charge q_0 located at $w = u_0$ between two conductive concentric cylinders. Let cylinder 1 have a radius r_1 and cylinder 2 have a radius $r_2 > r_1$ as shown in Fig. 3.31. Let the center of the cylinders coincide with the origin of the w-plane. Based on the results from Section 3.7, an image charge $-q_0$ is required at $w = r_1^2/u_0$ to make cylinder 1 a grounded cylinder and

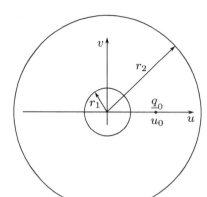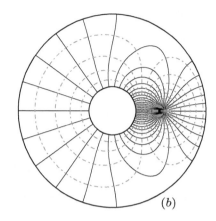

FIGURE 3.31 (a) Line charge located between two conductive coaxial cylinders. (b) Field map for a line charge located between two conductive coaxial cylinders. The smaller cylinder is grounded and the larger cylinder is floating. $r_2 = 4r_1$, $u_0 = r_2^{0.525}$, and $\Delta V = \Delta \Psi = 0.025 q_0/\epsilon$.

an image charge at $w = r_2^2/u_0$ to make cylinder 2 a grounded conductor. These images then require additional image charges leading to an infinite array of image charges. The positive charges (including the original true charge) are located at

$$w = u_0 \left(r_1/r_2\right)^{2n}, \quad n = 0, \pm 1, \pm 2, \ldots \quad (3.143)$$

The negative charges are located at

$$w = \left(r_1^2/u_0\right)\left(r_1/r_2\right)^{2n}, \quad n = 0, \pm 1, \pm 2, \ldots \quad (3.144)$$

Up to this point, it would seem that the method of images is leading to the solution for a line charge located between two grounded conductive cylinders, but this is incorrect. The method of images produces equipotential boundaries for the two conductive cylinders, but the charge on cylinder 1 is $-q_0$ and the charge on cylinder 2 is zero. Therefore, the resulting solution is actually the solution to a line charge located between a grounded cylinder 1 and an uncharged floating cylinder 2. Determining the net charge on each conductor is performed in the following manner. After the first image charge is placed at $w = r_1^2/u_0$, the system is charge neutral and cylinder 1 has a net charge of $-q_0$ and cylinder 2 has a net charge of zero. The next image charge is placed at $w = r_2^2/u_0$. Now system has a net charge $-q_0$ in finite space. Therefore, the charge at infinity must be q_0. The charge at infinity results in a net zero charge on cylinder 2 and a net charge of $-q_0$ on cylinder 1. As more image charges are added, the net charge remains zero on cylinder 2 and $-q_0$ and cylinder 1 when the charge at infinity is taken into account. Although the first image charge was placed at $w = r_1^2/u_0$, cylinder 1 still has a net charge of $-q_0$ and cylinder 2 has a net charge of zero if the first image charge is placed at $w = r_2^2/u_0$ when the same number of image charges are considered for each cylinder. The complex potential for a line charge q_0 located at $w = u_0$ between concentric conductive cylinders with cylinder 1 grounded and cylinder 2 floating is

$$\Phi[w] = -\frac{q_0}{2\pi\epsilon} \ln\left[\prod_{n=-\infty}^{\infty} \left(\frac{w - u_0(r_1/r_2)^{2n}}{w - (r_1^2/u_0)\,(r_1/r_2)^{2n}} \right) \right]$$
$$- \frac{q_0}{2\pi\epsilon} \ln\left[\prod_{n=-\infty}^{\infty} \left(\frac{1 - (r_1/u_0)\,(r_1/r_2)^{2n}}{1 - u_0(r_1/r_2)^{2n}} \right) \right]$$

$$(3.145)$$

The complex potential reference point is $z = r_1$.

So why does the method of images produce two grounded planes when used to analyze a line charge between conductive planes but a grounded cylinder and a floating cylinder when used to analyze a line charge between conductive cylinders? The reason lies with the charge at infinity. For two conductive planes, the charge at infinity can contribute charge to either conductor and does so in the appropriate amount to make them both grounded. For the concentric conductive cylinders, the charge at infinity only contributes to the net charge on the larger cylinder.

3.17 LINE CHARGE BETWEEN TWO GROUNDED CONCEN-TRIC CYLINDER

The complex potential for a line charge between two grounded concentric cylinders can be obtained from (3.145) by superimposing a potential difference between the outer and inner cylinders. Let V_f be the potential cylinder 2 obtains if it is floating. Then adding the complex potential

$$\Phi[w] = -V_f \ln[w/r_1] / \ln[r_2/r_1] \qquad (3.146)$$

to (3.145) forces the potential on cylinder 2 to zero while cylinder 1 remains at zero potential. Green's reciprocation theorem (2.177) can be used to obtain the value of V_f. First, we surround the point $z = x_0$ with a small conducting cylinder (conductor 3) so that it takes the value of the potential and charge at x_0. Let the unprimed case have cylinder 1 at

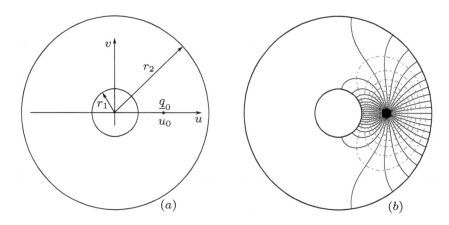

FIGURE 3.32 (a) Line charge located between two conductive coaxial cylinders. (b) Field map for a line charge located between two grounded conductive coaxial cylinders. $r_2 = 4r_1$, $u_0 = r_2^{0.525}$, and $\Delta V = \Delta \Psi = 0.025q_0/\epsilon$.

potential $V_1 = 0$ with charge $Q_1 = -q_0$, cylinder 2 floating at potential $V_2 = V_f$ with charge $Q_2 = 0$, and conductor 3 at potential $V_3 = V_3$ with charge $Q_3 = q_0$. Let the primed case have cylinder 1 at potential $V_1' = 0$ with charge $Q_1' = -q_0$, cylinder 2 at potential $V_2' = V_2'$ with charge $Q_2' = q_0$, and conductor 3 floating at potential $V_3' = V_3'$ with charge $Q_3' = 0$. Using (2.177) we obtain

$$V_f = V_3' \tag{3.147}$$

The complex potential for the primed case is given by (3.5) with $V_1 = 0$

$$\Phi[w] = \frac{q_0}{2\pi\epsilon} \ln[w/r_1] \tag{3.148}$$

Thus,

$$V_f = \text{Re}[\Phi[u_0]] = \frac{q_0}{2\pi\epsilon} \ln[u_0/r_1] \tag{3.149}$$

and the complex potential for a line charge q_0 located at $w = u_0$ between two grounded cylinders is

$$\begin{aligned}
\Phi = &-\frac{q_0}{2\pi\epsilon} \frac{\ln[u_0/r_1]}{\ln[r_2/r_1]} \ln\left[\frac{w}{r_1}\right] \\
&-\frac{q_0}{2\pi\epsilon} \ln\left[\prod_{n=-\infty}^{\infty} \left(\frac{w - u_0(r_1/r_2)^{2n}}{w - (r_1^2/u_0)(r_1/r_2)^{2n}}\right)\right] \\
&-\frac{q_0}{2\pi\epsilon} \ln\left[\prod_{n=-\infty}^{\infty} \left(\frac{1 - (r_1/u_0)(r_1/r_2)^{2n}}{1 - u_0(r_1/r_2)^{2n}}\right)\right]
\end{aligned} \tag{3.150}$$

The field map for a line charge between two grounded cylinders is shown in Fig. 3.32.

EXERCISES

3.1 Determine the complex potential for the upper half plane when the real axis consists of a conductor with the potential distribution given by

$$V[x, 0] = \begin{cases} 0, & |x| > a \text{ or } |x| < b \\ V_0, & \text{otherwise} \end{cases}$$

where $a > b$

3.2 Determine the complex potential for the region $x > 0$ and $y > 0$ when the real axis and imaginary axis are grounded conductors and a line charge q_0 is located at $z = a + jb$ where $a > 0$ and $b > 0$.

3.3 A semi-infinite grounded conductor lie along the negative real axis and a line charge $q_0 > 0$ is located at $z = a$ where $a > 0$. Assume the real axis corresponds to the boundary. Map the complex potential along the boundary to the complex potential plane. Label the points that correspond to the start and end of the conductor, both sides of the line charge and the distance between any resulting parallel lines. Let the zero of the complex potential be $z = 0$.

3.4 An n-cusped hypocycloid has the parametric equation

$$x[t] = (a/n) ((n - 1) \cos[t] - \cos[(n - 1) t])$$
$$y[t] = (a/n) ((n - 1) \sin[t] + \sin[(n - 1) t])$$

with $0 \leq t \leq 2\pi$. If a conductor has the shape of an n-cusped hypocycloid and a total charge per unit length of q_0, what is the complex potential for any point outside the conductor?

3.5 The real axis is a grounded conductor and a charge distribution

$$\rho_f[r_0, \theta_0] = \begin{cases} \sigma_0 (\delta[\theta_0 - \theta_1] / r_0), & r_1 \leq r_0 \leq r_2 \\ 0 & , \text{otherwise} \end{cases}$$

exist in the upper half of the z-plane. The delta function in the theta direction $\delta[\theta_0 - \theta_1]/r_0$ signifies charge exist only along the $\theta_0 = \theta_1$ line. Assume the permittivity ϵ is a constant and calculate the electrostatic potential for the upper half plane.

3.6 Show the complex potential for an infinite planar grounded conductor located along the $x = -a$ line, coated with a planar dielectric of thickness a and permittivity ϵ_1, and a line charge q_0 located in vacuum at $z = b > a$ is

$$\Phi_0 = -\frac{q_0}{2\pi\epsilon_0}\left(\ln[b-z] + R_{01}\ln[z+b]\right)$$

$$+ \frac{q_0}{2\pi\epsilon_0}\sum_{n=1}^{\infty} T'_{01}T'_{10}(-R_{10})^{n-1}\ln[z+2na+b]$$

in the dielectric and

$$\Phi_1 = -\frac{q_0}{2\pi\epsilon_1}\sum_{n=0}^{\infty} T'_{01}(-R_{10})^n \frac{\ln[2na+b-z]}{\ln[z+2(n+1)a+b]}$$

in vacuum with

$$T'_{ij} = \frac{2\epsilon_j}{\epsilon_i+\epsilon_j}, \quad R_{ij} = \frac{\epsilon_i-\epsilon_j}{\epsilon_i+\epsilon_j}$$

3.7 A hollow grounded conductive cylinder of radius r_1 has its center at the origin of the z-plane. A uniform sheet charge density σ_z lie along the real axis between $-b \geq x \geq b$ where $0 < b < r_1$. Determine the complex potential inside the cylinder.

3.8 The region $y \geq 0$ bounded by the imaginary axis, the real axis, and the line $x = a$ has a line charge q_0 located at $z = z_0$. If the walls of the bounded region are grounded, show that the complex potential is

$$\Phi = \frac{q_0}{2\pi\epsilon}\ln\left[\frac{\sin[\pi(z-z_0^*)/(2a)]\sin[\pi(z+z_0^*)/(2a)]}{\sin[\pi(z-z_0)/(2a)]\sin[\pi(z+z_0)/(2a)]}\right]$$

REFERENCES

[1] Jakob Kunz and P. L. Bayley. Some applications of the method of images—i. *Physical Review*, 17(2):147–156, February 1921.

[2] Philip M. Morse and Herman Feshbach. *Methods of Theoretical Physics, Part I*. McGraw-HillBook Company, Inc., 1953.

[3] P. Silvester. TEM wave properties of microstrip transmission lines. *Proceedings of the Institution of Electrical Engineers*, 115(1):43, January 1968.

[4] William R. Smythe. *Static and Dynamic Electricity*. McGraw-Hill Book Company, 3rd edition, 1968.

CHAPTER 4

Conformal Mapping I

In this chapter, we begin our study of conformal mapping. With the exception of the logarithm transformation, all of the conformal mappings developed in this chapter are full plane mapping. In other words, all the points in the complex z-plane are mapped into another full complex plane we designate as the w-plane. The full plane conformal mappings to be developed consist of translation, expansion (or contraction), rotation, complex inversion, and the Möbius or bilinear transformations.

4.1 DEFINING CONFORMAL TRANSFORMATIONS

To define conformal mapping, first consider a curve in the z-plane and an analytic mapping $w[z]$ such that each point in the z-plane corresponds to at least one point in the w-plane. As a curve in the z-plane is traced out, a curve in the w-plane is also traced out. An infinitesimal section dz of the curve in the z-plane has a corresponding infinitesimal section dw in the w-plane of

$$dw = he^{j\alpha}dz \tag{4.1}$$

where

$$h = \left|\frac{dw}{dz}\right|, \quad \alpha = \arg\left[\frac{dw}{dz}\right] \tag{4.2}$$

Thus, h measures the local magnification and α measures the local rotation from the z-plane to the w-plane which are typically functions of position.

Next we form an infinitesimal triangle created by the intersection of three curves in the z-plane. Let the length of the sides be

$$|dz|_1, \quad |dz|_2, \quad |dz|_3 \tag{4.3}$$

A corresponding infinitesimal triangle in the w-plane also exists with sides

$$|dw|_1 = h|dz|_1, \quad |dw|_2 = h|dz|_2, \quad |dw|_3 = h|dz|_3 \qquad (4.4)$$

The two triangles are similar since

$$|dz|_1 : |dz|_2 : |dz|_3 = |dw|_1 : |dw|_2 : |dw|_3 \qquad (4.5)$$

requiring the angles of the intersections of the corresponding curves in the two planes to be equal. An alternative argument for equal angles is that each side of the infinitesimal triangle is rotated by the same amount α when going from the z-plane to the w-plane. Therefore, the angles of the intersections of the corresponding curves in the two planes must be equal. Transformations that conserve both the magnitude and the sense of the angle are called conformal transformations or conformal mappings.

4.2 TRANSFORMING COMPLEX POTENTIALS

Conformal mapping is a powerful technique for solving 2D electrostatic problems. The concept is to transform the geometry where the complex potential is unknown into a geometry where the complex potential is either known or can be more easily solved. For example, let the complex potential $\Phi[w]$ in the w-plane be known and the goal is to determine the complex potential $\Phi[z]$ of a different geometry in the z-plane. To obtain $\Phi[z]$, we must first determine a conformal mapping function $w[z]$ that maps the boundaries in the z-plane onto the boundaries in the w-plane. When the boundaries are mapped, the boundary conditions are also mapped. If $\Phi[w]$ is a solution to the mapped boundary conditions, then

$$\Phi[z] = \Phi[w[z]] \qquad (4.6)$$

and we are done.

The concept is straightforward although some steps in the procedure may pose some difficulty in practice. One potential difficulty is determining the appropriate conformal mapping function and is the focus of the current chapter and Chapter 5. Another potential difficulty is determining $\Phi[w]$. Although the transformed geometry in the w-plane may be an easier geometry to solve than in the z-plane, the boundary conditions must also be met. A clear understanding of how boundary conditions are mapped is therefore needed.

As discussed previously (see Section 2.19), boundary conditions can be grouped into three categories: Dirichlet boundary conditions where the potential on the boundary is specified, Neumann boundary conditions where the charge on the boundary is specified, and mixed boundary conditions where portions of the boundary are specified with Dirichlet boundary conditions and other portions are specified with Neumann boundary conditions. We investigate conformal mapping effects on Dirichlet and Neumann boundary conditions.

When a geometry is conformally mapped from the z-plane into the w-plane, points on the boundary in one plane have corresponding points on the boundary in the other plane. From (4.6), the corresponding points in the two planes are at the same potential. Therefore, Dirichlet boundary conditions in the z-plane are transformed into Dirichlet boundary conditions in the w-plane where corresponding sections of boundaries between the two planes are at the same potential.

When Neumann boundary conditions are conformally mapped, we have to consider how the charge density is modified by the mapping. In general, the distance between two points on the boundary is modified by the conformal mapping function and therefore we would expect the charge density to change. Conservation of charge requires the total charge between two points in the z-plane to be the same as the charge between the two corresponding points in the w-plane. Charge density along the boundary is related to flux in the z-plane and w-plane as

$$\epsilon d\Psi = \sigma_z \left| dz \right|, \quad \epsilon d\Psi = \sigma_w \left| dw \right| \tag{4.7}$$

Requiring charge to be conserved between the two planes gives

$$\sigma_w = \sigma_z \frac{\left| dz \right|}{\left| dw \right|} = \sigma_z / h \tag{4.8}$$

As previously discussed in this section, the scaling factor h determines the magnification of the mapping. If $h > 1$, the distance between two points in the w-plane is greater than the distance between the corresponding points in the z-plane and charge per unit distance should decrease as established by (4.8). Thus, Neumann boundary conditions map to Neumann boundary conditions, but the charge density is different between the two planes.

Since the electric field is related to charge density through (2.144), the electric field is different in the two planes. The complex potential is analytic in both planes allowing the electric field in the z-plane to be written in terms of the electric field in the w-plane as

$$\mathcal{E}_z = -\left(\frac{\partial \Phi}{\partial z} \right)^* = -\left(\frac{\partial \Phi}{\partial w} \right)^* \left(\frac{dw}{dz} \right)^* = \mathcal{E}_w \left(\frac{dw}{dz} \right)^* \tag{4.9}$$

As a consequence of conserving the potential difference between points in both planes and the flux between points in both planes, the capacitance of a region or structure can be computed in either plane. Therefore, conformal mapping may be used to simplify capacitance calculations as well.

4.3 TRANSLATION

One of the simplest mappings is translation without rotation. Translation is equivalent to selecting a new origin for the plane. Since translation does not change any angles it is conformal. Additionally, any symmetry with respect to a line in the plane is also preserved. Although translation is a simple mapping, it is often very helpful in simplifying mathematical notation. For example, the complex potential for a line charge q_0 at z_1 in the z-plane is

$$\Phi[z] = -\frac{q_0}{2\pi\epsilon} \ln[z - z_1] \tag{4.10}$$

To translate the origin in the z-plane to z_1, we use the transformation

$$w[z] = z - z_1, \quad z[w] = w + z_1 \tag{4.11}$$

The complex potential in the w-plane is obtained by replacing z with $z[w]$

$$\Phi[w] = -\frac{q_0}{2\pi\epsilon} \ln[w] \tag{4.12}$$

Fig. 4.1 is an example of translation.

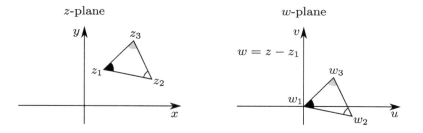

FIGURE 4.1 An example of translation.

4.4 MAGNIFICATION AND ROTATION

Two additional simple mappings are magnification and rotation. Both can be accomplished by multiplying z by a complex number. In section

4.1, we established that magnification and rotation between the z-plane and w-plane were given by $|dw/dz|$ and $\arg[dw/dz]$, respectively. If we have the mapping

$$w = mz \tag{4.13}$$

then the magnification is given by

$$\left|\frac{dw}{dz}\right| = |m| \tag{4.14}$$

and the rotation by

$$\arg\left[\frac{dw}{dz}\right] = \arg[m] \tag{4.15}$$

Thus all points in the w-plane are rotated by an $\arg[m]$ compared to their corresponding points in the z-plane. Since both magnification and rotation do not change the angle between curves, the transformation is conformal. Additionally, any symmetry with respect to a line in the plane is also preserved as this line undergoes the same magnification and rotation as shown in Fig. 4.2.

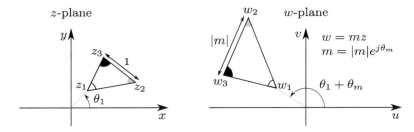

FIGURE 4.2 An example of magnification and rotation.

4.5 COMPLEX INVERSION AND INVERSION

Next we study inversion and complex inversion. Although complex inversion is just the complex conjugate of inversion, this difference has some impact when transforming a complex potential as discussed at the end of this section. When defining complex inversion, the center and radius of the circle of inversion must be defined. For example, if the unit circle centered at the origin is used as the circle of inversion, complex inversion is defined as

$$w[z] = 1/z \tag{4.16}$$

Using (4.16), a point $z_0 = r_0 e^{j\theta_0}$ is transformed to the point $w_0 = (1/r_0)e^{-j\theta_0}$. The location of this point under goes two transformations (see Fig. 4.3). The distance from the origin in the w-plane becomes the reciprocal of the distance from the origin in the z-plane. This is the process of inversion and the point $1/z_0^*$ is referred to as the inverse point of z_0. Note that the inverse point for a point on the circle of inversion is itself. Additionally, the angle in the w-plane becomes the negative of the angle in the z-plane. This second transformation is the difference between complex inversion and inversion. Thus every point outside the unit circle is transformed to a point inside the unit circle and vice versa. Note that the circle of inversion (the unit circle in this case) defines the boundary between points increasing or decreasing in magnitude after inversion. Performing the complex conjugate operation is a reflection about the real axis. Thus the interesting properties of complex inversion come from inversion. We develop the properties of inversion and then expand those properties to complex inversion by taking the complex conjugate.

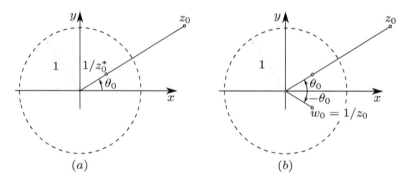

FIGURE 4.3 The process of (a) inversion and (b) complex inversion with the circle of inversion as the unit circle centered at the origin.

Inversion can be generalized to any circle of inversion C_I by requiring that the properties obtained from (4.16) for the unit circle are preserved. These properties are

1. Points outside the circle of inversion transform into points inside the circle of inversion and vice versa.

2. Points on the circle of inversion occupy the same locations after inversion.

3. If the distance of a point from the center of inversion is r_0, then

after inversion the distance from the center of inversion is proportional to $1/r_0$. We will show that the proportionality constant is r_I^2 where r_I is the radius of the circle of inversion.

If we take an arbitrary circle C_I with center z_c and radius r_I as the circle of inversion, we can determine the general inversion formula by first translating the center of C_I to the origin of the w_1-plane with the transformation $w_1 = z - z_c$. Next we hypothesize that the inversion formula for a circle of radius r_I centered at the origin differs from the conjugate of (4.16) by only a factor α and perform inversion of the w_1-plane about C_I into the w_2 plane as

$$w_2[w_1] = \alpha/w_1^* \tag{4.17}$$

To determine α, we use Property 2 of inversion that points on the circle of inversion occupy the same location after inversion. The points on C_I in the w_1-plane are given by $w_1 = r_I e^{j\theta}$ and the transformation (4.17) then requires $\alpha = r_I^2$. The final transformation is to translate the center of C_I in the w_2-plane back to its original location in the z-plane so that Property 2 of inversion is satisfied for the original z-plane,

$$w_3[w_2] = w_2 + z_c \tag{4.18}$$

Combining all the transformations leads to the generalized inversion equation for an arbitrary circle of inversion centered at z_c with radius r_I of

$$w = \frac{r_I^2}{z^* - z_c^*} + z_c \quad \text{(Inversion)} \tag{4.19}$$

The generalized complex inversion equation is the complex conjugate of (4.19)

$$w = \frac{r_I^2}{z - z_c} + z_c^* \quad \text{(Complex Inversion)} \tag{4.20}$$

Now that the general complex inversion transformation has been obtain, it is easily seen that it is an analytic mapping (except at the center of inversion) and therefore a conformal transformation. Since inversion is the complex conjugate of complex inversion, the sign of the angle is opposite to complex inversion which makes inversion an anti-conformal transformation. The complex potential has information in both the real and imaginary parts of the function. Therefore, we want to only use conformal transformations when transforming the complex potential from one plane to another. Thus, the complex potential derived from inversion gives the correct electrostatic potential (as the real part of a function is not changed by taking the complex conjugate), but the imaginary part has the opposite sign.

4.6 INVERSION OF A POINT

Visualizing how different geometries transform under inversion helps bring further insight into the inversion process. Let the circle of inversion have its center at z_c and radius r_I. Since the circle of inversion stays in the same location after inversion, the center of the circle of inversion in the z-plane and in the w-plane have the same value ($z_c = w_c$). This does not mean that z_c transforms to w_c, just that the center of the circle of inversion is the same in both planes. We can now write (4.19) in a slightly different form of

$$w - w_c = \frac{r_I^2}{z^* - z_c^*} \tag{4.21}$$

which highlights the simple relationship between the distances from the center of inversion between the two planes.

The first property of inversion we investigate is how a point transforms given a general circle of inversion. Since r_I is a positive real number, (4.21) requires $w - w_c$ and $z - z_c$ to have the same angle. As the circle of inversion does not change after inversion, we can represent the z-plane and the w-plane all in one diagram with point labels signifying whether they are in the z-plane or the w-plane. When this is done, we have

Any point z_0, its inverse w_0 and the center of inversion all line on the same line. (I1)

4.7 INVERSION OF A TRIANGLE WITH VERTEX AT Z_C

Let a triangle be formed with two arbitrary points in the z-plane (z_1 and z_2) and the center of inversion z_c (see Fig. 4.4 (a)). Another triangle can be formed in the w-plane by w_c and the inverse points w_1 and w_2. Due to (I1), the angle between z_1 and z_2 must equal the angle between w_1 and w_2. From (4.21), the lengths of two of the sides have the relationship

$$\frac{|z_1 - z_c|}{|w_2 - w_c|} = \frac{|z_2 - z_c|}{|w_1 - w_c|} \tag{4.22}$$

Since two sides of the triangles have the same ratio and the angle between these two sides is the same for both triangles, we have our next property of inversion

A triangle formed by two points and the center of inversion in the z-plane is similar to the triangle formed by the inverse points and the center of inversion in the w-plane. (I2)

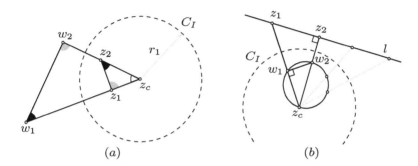

(a) (b)

FIGURE 4.4 (a) A triangle formed with one vertex as the center of inversion is similar to the resulting triangle formed by the center of inversion and the image points. (b) Lines not passing through the center of inversion transform into circles passing through the center of inversion.

4.8 INVERSION OF A LINE

Consider a line that passes through the center of inversion. (I1) requires all the inverse points to lie on this same line, but points internal to the circle of inversion are exchanged with points outside the circle of inversion. In general,

Lines passing through center of inversion transform into themselves. (I3)

Now consider a line l that does not pass through the center of inversion (see Fig. 4.4 (b)). Let z_2 represent the closest point on l to the center of inversion. A line drawn through z_2 and z_c must be perpendicular to l by the definition of point closest to the center of inversion. Since the denominator of (4.21) is minimized at z_2, the inverse point w_2 is the inverse point with the largest distance from w_c. The denominator of (4.21) is maximized by the point at infinity on l and transforms to the center of the circle of inversion. Now take an arbitrary point z_1 on l. The triangle $z_1 z_2 z_c$ is a right triangle. (I2) requires $z_1 z_2 z_c$ and $w_2 w_1 w_c$ to be similar and therefore triangle $w_2 w_1 w_c$ is a right triangle. The hypotenuse of triangle $w_2 w_1 w_c$ is located between w_c and w_2. Since z_1 can represent any point on l (excluding z_2 and ∞), every inverse point of line l forms a right triangle with hypotenuse located between w_c and w_2. This is the property of a circle [3] and leads to

> *Lines not passing through the center of inversion transform into circles passing through the center of inversion. The diameter of the circle is $r_I^2/|z_2 - z_c|$, where z_2 is the point on the line closest to the center of inversion.* (I4)

4.9 INVERSION OF A CIRCLE

All the steps used to show lines not passing through the center of inversion transform to circles passing through the center of inversion can be reversed to show

> *Circles passing through the center of inversion transform into lines not passing through the center of inversion. The line has a minimum distance of r_I^2/D from the center of inversion, where D is the diameter of the circle. The line is perpendicular to the line formed by the center of inversion and the point on the circle farthest from the center of inversion.* (I5)

Next we consider a circle C_z not passing through the center of inversion. Let C_z have a center z_0 and radius r_0 in the z-plane. Any point z_s on C_z satisfies

$$|z_s - z_0|^2 = |z_s|^2 - 2\mathrm{Re}[z_s z_0^*] + |z_0|^2 = r_0^2 \tag{4.23}$$

From (4.21), a circle of inversion centered at z_c with radius r_I transforms z_s to

$$z_s - z_c = \frac{r_I^2}{(w_s - w_c)^*} \tag{4.24}$$

where $w_c = z_c$. Since $|z_s - z_0| = |z_s - z_c - (z_0 - z_c)|$, we can re-write (4.23) as

$$\left| \frac{r_I^2}{(w_s - w_c)^*} \right|^2 - 2\mathrm{Re}\left[\left(\frac{r_I^2}{(w_s - w_c)^*} \right)(z_0 - z_c)^* \right] + |z_0 - z_c|^2 = r_0^2 \tag{4.25}$$

Multiply both sides of (4.25) by $|w_s - w_c|^2$ and simplifying leads to

$$|w_s - w_c|^2 - 2\mathrm{Re}\left[(w_s - w_c)\left(\frac{r_I^2(z_0 - z_c)^*}{|z_0 - z_c|^2 - r_0^2}\right)\right] = \frac{-r_I^4}{|z_0 - z_c|^2 - r_0^2} \quad (4.26)$$

Adding $\left(\frac{r_I^2|z_0 - z_c|}{|z_0 - z_c|^2 - r_0^2}\right)^2$ to both sides of (4.26) and simplifying gives

$$|w_s - w_1|^2 = r_1^2 \quad (4.27)$$

where

$$w_1 = w_c + \frac{r_I^2(z_0 - z_c)}{|z_0 - z_c|^2 - r_0^2}, \quad r_1 = \frac{r_0 r_I^2}{\left||z_0 - z_c|^2 - r_0^2\right|} \quad (4.28)$$

Thus, we have

Circles not passing through the center of inversion transform into circles not passing through the center of inversion. (I6)

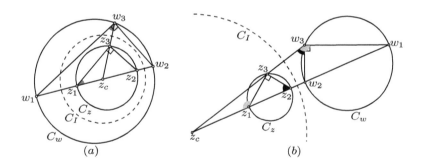

FIGURE 4.5 (a) Transformation of a circle contain the center of inversion. (b) Transformation of a circle with the center of inversion outside the circle.

An alternative visual derivation of (I6) is also possible. Two figures are needed where one figure has the center of the circle of inversion (z_c) inside C_z and the other figure has z_c outside C_z as shown in Fig. 4.5. Draw a line through the center of inversion z_c, the point on the circle C_z closest to z_c which is labeled z_1 and the point on C_z farthest from z_c which is labeled z_2. The diameter of C_z is then $|z_2 - z_1|$. Choose a third point on C_z and label it z_3. The angle at the vertex z_3 in the triangle $z_1 z_2 z_3$ must be a right angle [3]. In Fig. 4.5 (a), the dark shaded angle at vertex z_1 and the lighter shaded angle at vertex z_2 then must sum to $\pi/2$. From the similar triangles $z_1 z_3 z_c$ and $w_3 w_1 w_c$ and the similar triangles $z_2 z_3 z_c$ and $w_3 w_2 w_c$, the angle at the vertex of w_3 for triangle $w_1 w_2 w_3$ in Fig. 4.5 (a) is the sum of the dark and lighter shaded angles and therefore $\pi/2$. In Fig. 4.5 (b) we still use the same similar triangles as before, but the angle at vertex w_3 in the triangle $w_1 w_2 w_3$ is the difference between the lightly shaded angle at vertex z_1 and the dark shaded angle at vertex z_2. Since the lightly shaded angle at vertex z_1 is an external angle to triangle $z_1 z_2 z_3$, it must equal the sum of the internal angles at vertices z_2 and z_3. Therefore, the angle at the vertex of w_3 for triangle $w_1 w_2 w_3$ must be $\pi/2$. Since z_3 can represent any point (excluding z_1 and z_2) on C_z, every inverse point forms a right triangle with hypotenuse located between w_1 and w_2. The contour in the w-plane is therefore a circle.

4.10 INVERSION OF ORTHOGONAL CIRCLES

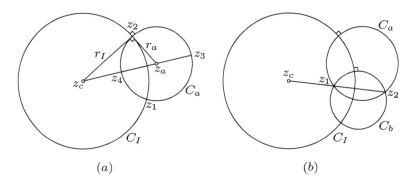

(a)   (b)

FIGURE 4.6 (a) C_a is an orthogonal circle to the circle of inversion C_I. The points z_3 and z_4 are inverse points under inversion. Thus, under inversion, orthogonal circles map to themselves. (b) The intersection of two circles orthogonal to the circle of inversion (z_1 and z_2) are inverse points.

Orthogonal circles are circles that intersect each other at right angles. Figure 4.6 (a) shows an orthogonal circle C_a to the circle of inversion C_I that intersect at z_1 and z_2. Draw a line that goes through the center of C_I and the center of C_a. This line intersect C_a at points z_3 and z_4 in Fig. 4.6 (a). The triangle formed by $z_c z_2 z_a$ is a right triangle giving

$$r_I^2 = |z_a - z_c|^2 - r_a^2 = (|z_a - z_c| + r_a)(|z_a - z_c| - r_a) \qquad (4.29)$$

From Fig. 4.6 (a), it is easily observed that

$$|z_4 - z_c| = |z_a - z_c| - r_a \quad \text{and} \quad |z_3 - z_c| = |z_a - z_c| + r_a \qquad (4.30)$$

Combining (4.29) and (4.30) gives

$$|z_4 - z_c| = \frac{r_I^2}{|z_3 - z_c|} \tag{4.31}$$

Thus, z_3 and z_4 are inverse points under inversion in C_I. Additionally, z_1 and z_2 are mapped to the same location in the w-plane under inversion in C_I. We have four points on C_a that under inversion in C_I are still on C_a which means

> *Under inversion in C_I, every circle orthogonal to C_I is mapped to itself* (I7)

Since C_a maps to itself, the shaded region of C_a in Fig. 4.6 (a) maps to the unshaded region of C_a and vice versa. Also, the intersection points in C_a of any line drawn through the center of C_I that passes through C_a are inverse points. An even more general statement is

> *The intersections of two circles orthogonal to the circle of inversion are inverse points* (I8)

To prove (I8), draw a circle C_a that is orthogonal to the circle of inversion C_I. Then draw a line from the center of inversion through C_a as shown in Fig. 4.6 (b). Due to (I7), the intersection of this line and C_a are inverse points. Now any circle C_b drawn through these two inverse points also intersects C_I. Under inversion, the inverse points exchange locations on C_b and the locations on C_b that cross C_I stay fixed. Therefore, C_b is mapped to itself which means it is an orthogonal circle with respect to C_I.

4.11 SYMMETRY PRESERVATION WITH INVERSION

A discussion on reflection symmetry typically involves defining symmetric points with respect to a line of symmetry. For simplicity, let the line of symmetry coincide with the real axis. Then z^* is the reflection image of z with respect to the line of symmetry and the two points are declared symmetric with respect to the real axis. A broader definition of reflection symmetry that includes a circle of symmetry is [4]

> *Two points z and z_s are symmetric with respect to a circle or straight line L if all circles and straight lines passing through z and z_s are orthogonal to L*

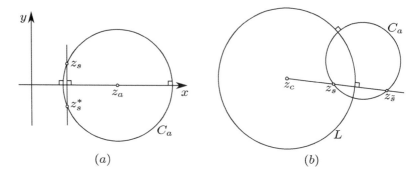

FIGURE 4.7 (a) Reflection symmetry about a line that coincides with the real axis. The line and any circle that goes through any two symmetric points z_s and z_s^* are orthogonal to the real axis. (b) Inverse points to any circle are considered symmetric with respect to the circle due to the boarder definition of reflection symmetry.

We focus first on a line of symmetry L and again let the real axis coincide with L. Then two symmetric points about L are conjugates of each other which we label z_s and z_s^*. Since $\text{Re}[z_s] = \text{Re}[z_s^*]$, the line going through both points is a vertical line and perpendicular to L (see Fig. 4.7 (a)). Now let a circle C_a with center z_a and radius r_a go through z_s and z_s^*. Based on the equation of the circle, we have

$$|z_s - z_a|^2 = |z_s^* - z_a|^2 = r_a^2 \tag{4.32}$$

which requires $\text{Im}[z_a] = 0$. In other words, the center of any circle that goes through z_s and z_s^* lies on the line of symmetry. Therefore, the circle and the line of symmetry have orthogonal intersections as shown in Fig. 4.7 (a) satisfying the broader definition of reflection symmetry.

From Section 4.10, inverse points with respect to the circle of inversion C_I occur at the intersections of an orthogonal circle C_a to C_I and a line that goes through the center of inversion and C_a. These points are labeled z_s and $z_{\bar{s}}$ in Fig. 4.7 (b). Clearly any line that passes through the center of inversion is orthogonal to C_I. By the broader definition of reflection symmetry, inverse points are considered symmetric points with respect to the circle of inversion. Thus, inverse points to any circle L are considered symmetric points with respect to L and L is considered the circle of symmetry.

Since inversion maintains the magnitude of the angle between curves from the original plane to the transformed plane, circles orthogonal to a circle of symmetry (which does not need to be the circle of inversion) map to circles that are still orthogonal to the new circle of symmetry (which may be a line). The inverse points defined by the intersections of the orthogonal circles are mapped to the intersection of the mapped orthogonal circles and therefore are symmetric with respect to the new circle of symmetry. Based on this, inversion (and complex inversion) preserve reflection symmetry with respect to circles and lines.

4.12 MÖBIUS TRANSFORM

The Möbius transform (or bilinear transform) is a mapping of the form

$$w = \frac{az + b}{cz + d}, \quad ad \neq bc \tag{4.33}$$

where a, b, c, and d are complex constants. Solving for $z[w]$ in (4.33) gives

$$z = \frac{wd - b}{-cw + a} \tag{4.34}$$

which is also a Möbius transformation. Thus the inverse of a Möbius transformation is another Möbius transformation. Although (4.33) may seem like a completely new transformation, it can be decomposed into four steps [2]

(i) $z_1 = z + (d/c)$, translation

(ii) $z_2 = 1/z_1 = \dfrac{c}{cz + d}$, complex inversion

(iii) $z_3 = \dfrac{(bc - ad)}{c^2} z_2 = \dfrac{(bc - ad)}{c\,(cz + d)}$, expansion and rotation

(iv) $w = z_3 + (a/c) = \dfrac{az + b}{cz + d}$, translation

Note if $ad = bc$, then $z_3 = 0$ and w is equal to a constant. Therefore, this case is excluded as a mapping. Based on previous analysis in this chapter, we can immediately list several important properties of Möbius transforms

Möbius transformations map circles to circles (MT1)

Möbius transformations are conformal (MT2)

Möbius transformations preserve reflection symmetry with respect to a circle (MT3)

If all of the constants in (4.33) are multiplied by an arbitrary complex number $k \neq 0$,

$$w = \frac{kaz + kb}{kcz + kd} = \frac{az + b}{cz + d} \tag{4.35}$$

and the mapping is unchanged. In other words, the ratios of coefficients are important, not their values alone. The fixed points z_f of a general Möbius transformation are the solutions of

$$z_f = \frac{az_f + b}{cz_f + d} \tag{4.36}$$

which results in a quadratic equation. Thus,

With the exception of the identity mapping, a Möbius transformation has at most two fixed points (MT4)

Solving the quadratic equation gives

$$z_f = \frac{a - d \pm \sqrt{(d - a)^2 + 4bc}}{2c} \tag{4.37}$$

The final Möbius transformation property we discuss is called the cross-ratio. The vector between two image points w_i and w_j is

$$w_i - w_j = \frac{(ad - bc)(z_i - z_j)}{(cz_i + d)(cz_j + d)} \tag{4.38}$$

Using (4.38), the product of two vectors defined by four finite distinct points z_1, z_2, z_3, z_4 in the z-plane can be computed as

$$(w_1 - w_4)(w_3 - w_2) = \frac{(ad - bc)^2}{\prod\limits_{i=1}^{4}(cz_i + d)}(z_1 - z_4)(z_3 - z_2) \tag{4.39}$$

and

$$(w_1 - w_2)(w_3 - w_4) = \frac{(ad - bc)^2}{\prod\limits_{i=1}^{4}(cz_i + d)}(z_1 - z_2)(z_3 - z_4) \tag{4.40}$$

Taking the ratio of (4.39) and (4.40) gives

$$\frac{(w_1 - w_4)(w_3 - w_2)}{(w_1 - w_2)(w_3 - w_4)} = \frac{(z_1 - z_4)(z_3 - z_2)}{(z_1 - z_2)(z_3 - z_4)} \tag{4.41}$$

The cross-ratio is defined as

$$\frac{(z_1 - z_4)(z_3 - z_2)}{(z_1 - z_2)(z_3 - z_4)} \tag{4.42}$$

and we have

The cross-ratio is invariant under a Möbius transformation (MT5)

If one of the points, say z_1, is the point at infinity, then the cross-ratio reduces to

$$\frac{z_3 - z_2}{z_3 - z_4} \tag{4.43}$$

The invariance of the cross-ratio leads to

There exist a unique Möbius transformation sending any three points to any other three points (MT6)

Letting $z = z_4$ and $w = w_4$ in (4.41) gives the Möbius transformation that maps z_1, z_2, and z_3 onto w_1, w_2, and w_3 as

$$\frac{(w_1 - w)(w_3 - w_2)}{(w_1 - w_2)(w_3 - w)} = \frac{(z_1 - z)(z_3 - z_2)}{(z_1 - z_2)(z_3 - z)} \tag{4.44}$$

Since translation, rotation, expansion, and complex inversion are all Möbius transformations, the additional properties (MT4) through (MT6) also apply to them.

4.13   LOGARITHM TRANSFORMATION

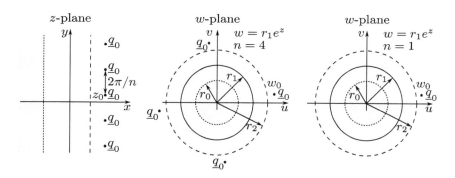

FIGURE 4.8 Logarithm transformation $z = \ln[w/r_1]$ between the z-plane and w-plane.

The logarithmic transformation can be viewed as a full plane transform and is useful when solving problems involving single cylinders or concentric cylinders. Let the w-plane consists of circular concentric cylinders as shown in Fig. 4.8. The logarithmic transformation

$$z = \ln[w/r_1] \qquad (4.45)$$

transforms the circles in the w-plane into vertical lines in the z-plane. A feature of this transformation that often leads to confusion is its multi-valued nature. If we restrict the value of θ in the w-plane to $0 \le \theta < 2\pi$, then the full w-plane is transformed into the strip located between $y = 0$ and $y = 2\pi$ in the z-plane. If we restrict θ to $2\pi \le \theta < 4\pi$, then the full w-plane is transformed into the strip located between $y = 2\pi$ and $y = 4\pi$. Thus, the vertical lines in the z-plane correspond to an infinite number of traces of the corresponding circles in the w-plane. Each trace increases θ by 2π. The imaginary axis in the z-plane corresponds to the cylinder of radius r_1 in the w-plane. Any vertical line with $x < 0$ in the z-plane is transformed into a cylinder with radius less than r_1 in the w-plane. Thus, $z = -\infty$ transforms into the origin of the w-plane. Additionally, a line charge in the w-plane transforms into an infinite 1D array of line charges in the vertical direction with a distance 2π between line charges in the z-plane. A vertical 1D array of line charges with distance $\Delta y = 2\pi/n$ (where n is a positive integer) in the z-plane transforms into n line charges equally spaced on a circle in the w-plane as shown in Fig. 4.8.

Two important items should be noted when using the logarithm transformation to transform complex potentials between the z-plane and w-plane. First, the potential of the infinite strip in the z-plane that corresponds to the full w-plane depends on all the charges in the z-plane, not just the charges in the strip. Therefore, the complex potential due to an infinite 1D line charge in the z-plane is required to obtain the complex potential for even a single line charge in the w-plane. Second, only charges located in the strip of the z-plane that corresponds to the full w-plane are transferred into the w-plane. Since complex potentials based on the method of images can be obtained by determining the location of where the image charges are transformed (including any charges at infinity), we only need to consider the w-plane locations of the charges in the corresponding strip of the z-plane.

4.14   RIEMANN SPHERE

In terms of conformal mapping, it is important to understand that curves that go to infinity can meet at infinity despite arriving at infinity from

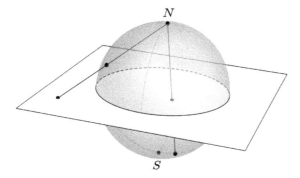

FIGURE 4.9 Stereographic projection of a sphere of unit radius to the points in a plane. Points on the plane inside the equator of the sphere are mapped onto the southern hemisphere of the sphere. The equator is mapped to itself. Points exterior to the equator are mapped onto the northern hemisphere.

different directions. To help visualize this, we introduce a one-to-one mapping of points on a sphere of unit radius to the points in a plane called stereographic projection [2]. A 3D drawing of stereographic projection is shown in Fig. 4.9. To determine the location of a corresponding point on the sphere to a point in the plane, a line is drawn through the north pole (N) of the sphere and the point in the plane. The intersection of this line with the sphere is the corresponding point on the sphere (see Fig. 4.9). Three key items result from this procedure. First, points inside the equator of the sphere are mapped onto the southern hemisphere of the sphere with zero being mapped to the south pole (S). Second, the equator is mapped to itself. Third, points exterior to the equator are mapped onto the northern hemisphere. As the point in the plane moves further away from the sphere (in any direction), the corresponding point on the sphere gets closer to the north pole. Thus, the point at infinity in the plane approached from any direction maps to a single point (the north pole) on the sphere. When this model is used to produce a one-to-one mapping of the unit sphere with the complex plane, the sphere is called the Riemann sphere.

The Riemann sphere has many uses [2], but we focus on visualizing lines meeting at infinity. We start by mapping a line L to the Riemann sphere Σ. The line connecting the north pole of Σ and a point p on L produces a plane as p moves along L. The intersection of a plane and a sphere is a circle on the surface of the sphere. Therefore, L is mapped to a circle that goes through N on Σ as shown in Fig. 4.10. This can be viewed as the two ends of the line in the plane that go to infinity in opposite direction meet at infinity. The mapping of two parallel lines results in two circles that meet at the north pole on the Riemann sphere. This can be viewed as parallel lines in the plane meeting at infinity. Thus, curves that go to infinity from any direction in the complex plane can be viewed as meeting at infinity. This concept is exploited in Chapter 5 when treating lines that go off to infinity in any direction as forming a vertex at infinity to form a closed polygon. This concept also helps visualize why mapping a vertex at infinity of a polygon to a finite point results in a continuous boundary at the new finite vertex.

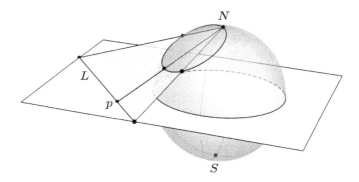

FIGURE 4.10 Mapping of a line in the complex plane to the Riemann sphere. The result is a circle that goes through N.

4.15   CHARGES AT INFINITY

As discussed in Section 2.5, when the sum of all charges over finite space is not zero,

$$\sum_i q_i \neq 0, \tag{4.46}$$

the net charge per unit length at infinity (q_∞) is

$$q_\infty = -\sum_i q_i \tag{4.47}$$

If the net charge in finite space is zero, then $q_\infty = 0$. Use of conformal mappings that transform the point at infinity to finite space requires knowing how to properly transform charge at infinity. For example, to make a one to one correspondence between the original plane (the z-plane) and the transformed plane (the w-plane) with complex inversion, the center of inversion z_c and infinity in the z-plane correspond to infinity and z_c in the w-plane, respectively. Thus, a common assumption is that complex inversion would send any net charge at z_c in the z-plane to infinity in the w-plane and any net charge at infinity in the z-plane is sent to z_c in the w-plane. This is only true for cases consisting of charge distributions only and are boundary free. When no boundaries are present, the complex potential is given by (2.89). From (4.20), the general relationship between z and w for complex inversion is

$$z[w] = \frac{r_I^2}{w - z_c} + z_c^* \tag{4.48}$$

To obtain the potential in the w-plane, we replace z with $z[w]$ and z_i with $z[w_i]$ in (2.89) to obtain

$$\Phi[w] = -\frac{1}{2\pi\epsilon}\left(-\sum_{i=1}^{n} \underline{q}_i\right)\ln[w - z_c]$$
$$-\sum_{i=1}^{n} \frac{\underline{q}_i}{2\pi\epsilon}\left(\ln[w - w_i] + \ln\left[\frac{r_I^2}{z_c - w_i}\right]\right) + C_0 \tag{4.49}$$

The first term of (4.49) is the complex potential due to a line charge $\underline{q}_\infty$ at $w = z_c$ (the image location of $z = \infty$ in the w-plane). The second term is the complex potential due to line charges $\underline{q}_i$ at w_i. The third term is just a constant that can be combined into C_0 to give

$$\Phi[w] = -\frac{\underline{q}_\infty}{2\pi\epsilon}\ln[w - z_c] - \frac{1}{2\pi\epsilon}\sum_{i=1}^{n}\underline{q}_i\ln[w - w_i] + C_0' \tag{4.50}$$

Thus, when there are no boundaries, any charge at infinity in the z-plane is sent to the center of inversion in the w-plane by complex inversion. When boundaries are present, boundary conditions must be satisfied and the details of the problem determine how to handle the charge at infinity.

When transforming charge from infinity in the z-plane, a line charge is introduced at the image point for infinity (w_∞) in the w-plane. The first question to answer is if the boundary conditions in the w-plane can be met with a line charge at w_∞. If the answer is no, then the conformal mapping being used is not valid for the problem. If the answer is yes, then the value of the line charge and whether it is true charge or image charge then needs to be determined. When boundaries are present, charge transformed from infinity often becomes image charge. Using the method of images implies that some regions in the w-plane can be excluded and image charge can exist in these excluded regions. Examples of excluded regions include the inside of conductors or the region of dielectric 1 when determining the complex potential in a region of dielectric 2. Section 4.16 demonstrates this procedure by using conformal mapping to determine the complex potential of a line charge next to a dielectric cylinder.

4.16 DIELECTRIC CYLINDER AND LINE CHARGE

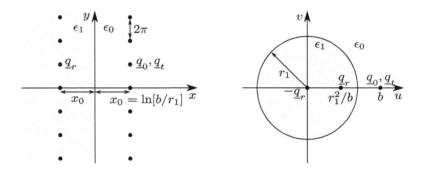

FIGURE 4.11 Logarithm transformation of a line charge in front of a dielectric cylinder in the w-plane into an infinite array of line charges with planar boundaries in the z-plane.

When boundaries are present, details of the problem determine how to transform charge at infinity. For example, consider an uncharged dielectric cylinder of radius r_1 with permittivity ϵ_1 surrounded by a material of permittivity ϵ_0 centered at the origin of the w-plane. Let a line charge $\underline{q}_0$ be located at $w = b > r_1$. If the w-plane can be transformed into the planar boundary problem of the z-plane shown in Fig. 4.11, the image charges could be found using the ray tracing method of Section 3.10 and the problem would be solved except for the charge at infinity. Two candidates to use as conformal transformations are complex inversion and the logarithm transformation. Complex inversion can be ruled out since the circle of inversion needed to transform the boundary of the cylinder into a planar boundary transforms the point at infinity in the z-plane onto the boundary of the dielectric cylinder in the w-plane. The boundary conditions cannot be satisfied with a line charge on the boundary. With the logarithm transformation, the point at infinity in the z-plane is transformed to the center of the cylinder and the imaginary axis of the z-plane becomes the boundary of the cylinder. Therefore, any charge transformed from infinity can be used as image charge to determine the complex potential external to the dielectric cylinder.

To determine how the charge at infinity transforms to the center of the dielectric cylinder, we must use the physics of the problem. The dielectric cylinder is charge free and total net polarization charge is always zero. To make the net charge of the dielectric cylinder zero when $r > r_1$, an amount of

$$-\underline{q}_r = -\left(\frac{\epsilon_0 - \epsilon_1}{\epsilon_0 + \epsilon_1}\right) q_0 \qquad (4.51)$$

is transferred from infinity in the z-plane to the center of the cylinder in the w-plane as image charge. Note $-q_0$ is located at infinity in the w-plane. Thus, not all of the charge at infinity in the z-plane was transferred from infinity to the center of the dielectric cylinder.

With the ray tracing approach, we can determine the image charges in the z-plane. Then the locations of the image charges in the w-plane are obtained with conformal mapping. The complex potential in the w-plane is then determined based on locations of the image charges in the w-plane and charge transferred from infinity. The true charge in the w-plane results in the infinite 1D array of line charges in the z-plane shown in Fig. 4.11. To determine the complex potential when $x > 0$, each line charge has an image charge q_r associated with it also shown in Fig. 4.11. To determine the complex potential when $x < 0$, the 1D array of true charges located at $z_0 + j2n\pi$ are replaced by a 1D array of image charges at the same location with value

$$\underline{q}_t = \left(\frac{2\epsilon_1}{\epsilon_0 + \epsilon_1}\right) q_0 \qquad (4.52)$$

We associate the infinite strip between $-\pi < y \le \pi$ with the entire w-plane. In each region ($r < r_1$ and $r > r_1$), there is only one image charge located between $-\pi < y \le \pi$ to the transform back to the w-plane. In the w-plane, the image charge located at $-x_0$ is located at

$$w_r = r_1 e^{-\ln[b/r_1]} = r_1^2/b \qquad (4.53)$$

Taking into account the image charges, true charge, charge from infinity and requiring the flux line to be continuous at $w = 0$, the complex potential in the w-plane for $|w| \ge r_1$ is

$$\Phi = \frac{q_0}{2\pi\epsilon_0}\left(\frac{\epsilon_1 - \epsilon_0}{\epsilon_0 + \epsilon_1}\ln\left[\frac{w - (r_1^2/b)}{r_1 - (r_1^2/b)}\frac{r_1(b - r_1)}{w}\right] - \ln[b - w]\right) \qquad (4.54)$$

and for $|w| \le r_1$ the complex potential is

$$\Phi = -\frac{q_0}{\pi \left(\epsilon_0 + \epsilon_1\right)} \ln[b - w] \qquad (4.55)$$

It is interesting to note that the location of the image charge q_r would be the same if complex inversion was used as the transformation. The difference between complex inversion and the logarithm transform is the location of the image of infinity in the w-plane. The equipotential contours and constant flux contours for a cylinder with $b = 1.25r_1$ and $\epsilon_1 = 3\epsilon_0$ are shown in Fig. 4.12.

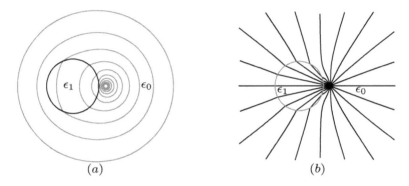

FIGURE 4.12 (a) Constant potential contours and (b) constant flux contours for a line charge in vacuum in front of a dielectric cylinder of permittivity $\epsilon_1 = 3\epsilon_0$.

4.17 FLOATING CONDUCTIVE CYLINDER AND LINE CHARGE

Letting the permittivity of the dielectric cylinder from Section 4.16 go to infinity results in an uncharged floating conductive cylinder. The complex potential for a floating conductive cylinder located at the origin of the w-plane and a line charge located at $w = b$ is

$$\Phi = \frac{q_0}{2\pi\epsilon_0} \left(\ln\left[\frac{w - (r_1^2/b)}{r_1 - (r_1^2/b)} \frac{r_1 (b - r_1)}{w} \right] - \ln[b - w] \right) \qquad (4.56)$$

The portion of the complex potential due to the floating cylinder is equivalent to the complex potential due to a line charge q_0 located at $w = 0$ and a line charge $-q_0$ located at $w = r_1^2/b$. The field map for a floating cylinder and line charge is shown in Fig. 4.13. For comparison, the field map for a grounded cylinder and line charge is also shown. Note that all the field lines originating from the line charge terminate completely on the grounded cylinder, but some field lines terminate on the right side of the floating cylinder only to re-emerge from the left side of the floating cylinder.

floating cylinder   grounded cylinder

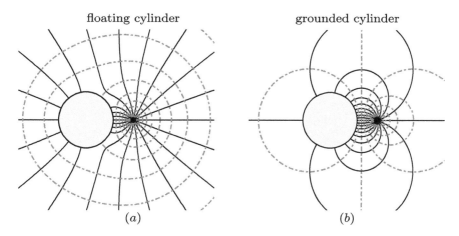

(a)   (b)

FIGURE 4.13   Field map of a line charge in front of a (a) floating conductive cylinder and (b) grounded conductive cylinder. For both cases, the location of the line charge is $b = 1.75r_1$ where r_1 is the radius of the cylinder and $\Delta V = \Delta \Psi = 0.05q_0/\epsilon$.

4.18   LINE CHARGE BETWEEN TWO CONDUCTIVE CONCEN-TRIC CYLINDERS REVISITED

In Sections 3.16 and 3.17, the complex potentials for a line charge between conductive cylinders with one cylinder grounded and the other cylinder either floating or grounded were obtained. With the logarithm transformation

$$z = \ln[w/r_1] \tag{4.57}$$

we can now better understand how the charge at infinity impacts which complex potential is the final solution. The image charges transformed from the z-plane with planar boundaries to the w-plane with concentric cylinders are almost independent of the potentials on the cylinders. The only difference is the amount of charge at infinity in the z-plane that is transferred to the origin of the w-plane. If no charge at infinity is transferred to the origin of the w-plane, then the complex potential corresponds to a line charge between a smaller grounded cylinder and a larger floating cylinder. If all the charge at infinity is transferred to the origin of the w-plane, then the complex potential corresponds to a line charge between a larger grounded cylinder and a smaller floating cylinder. If $\ln[u_0/r_1]/\ln[r_2/r_1]$ of the charge at infinity is transferred to the origin of the w-plane (where u_0 is the location of the line charge in the w-plane), then the complex potential corresponds to a line charge between two grounded cylinders. Thus, how the charge at infinity transforms can lead to very different physical situations.

4.19 NONCONCENTRIC CYLINDERS TO CONCENTRIC CYLIN- DERS

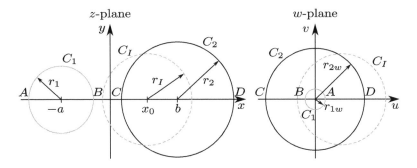

FIGURE 4.14 Nonconcentric cylinders in the z-plane transformed by inversion into concentric cylinders in the w-plane.

Complex inversion can be used to transform two nonconcentric cylinders that do not intersect into concentric cylinders. We assume the complex potential of two concentric cylinders is known and the complex potential for the nonconcentric cylinders is the solution we seek. From Section 3.2, the complex potential for concentric cylinders was obtained by considering a line charge located at a finite z value and an equal but opposite polarity line charge at infinity. From Section 3.3, the complex potential for nonconcentric cylinders was obtained by considering two equal but opposite polarity line charges not located at infinity. Based on this analysis, we would expect using complex inversion to send one of the line charges for the nonconcentric cylinders case to infinity would result in the conformal transformation that transforms nonconcentric cylinders into concentric cylinders. This indeed is the case. To show this, let Fig. 3.7 (a) be our starting point where the relevant portions along with a circle of inversion has been redrawn in Fig. 4.14 (a). The line charges are located at $z = \pm x_0$. To send one of the line charges to infinity, the center of inversion can be located at $z_c = x_0$ or $z_c = -x_0$. Using (4.20) with $z_c = x_0$ to send the negative charge to infinity and adding a translation of $(r_I^2 - 2x_0^2)/(2x_0)$ to send the positive charge and the centers of the concentric cylinders to the origin of the w-plane gives

$$w = \frac{r_I^2}{2x_0}\left(\frac{z + x_0}{z - x_0}\right) \tag{4.58}$$

To transform the complex potential in the w-plane given by (3.5) into the z-plane, we replace w with (4.58) to obtain

$$\Phi[z] = \frac{(V_2 - V_1)}{\ln[r_{2,w}/r_{1,w}]}\ln\left[\frac{r_I^2}{2x_0 r_{1w}}\left(\frac{z + x_0}{z - x_0}\right)\right] + V_1 \tag{4.59}$$

where r_{1w} and r_{2w} are the radii of the concentric cylinders in the w-plane. To write $\Phi[z]$ in terms of r_1 and r_2 instead of r_{1w} and r_{2w}, we use the correspondence of points between the two planes. By choosing $z_c = x_0$, cylinder 2 is always closer to the center of inversion and is transformed into the larger of two the cylinders in the w-plane. Using the points labeled A and D gives

$$r_{1w} = \frac{r_I^2}{2x_0}\left(\frac{-r_1 - a + x_0}{-r_1 - a - x_0}\right) = \frac{r_I^2}{2x_0}\left(\frac{r_1}{x_0 + a}\right) \tag{4.60}$$

and

$$r_{2w} = \frac{r_I^2}{2x_0}\left(\frac{r_2 + b + x_0}{r_2 + b - x_0}\right) = \frac{r_I^2}{2x_0}\left(\frac{x_0 + b}{r_2}\right) \tag{4.61}$$

where a and b are given by (3.23). The complex potential in the z-plane can be written as

$$\Phi[z] = \frac{(V_2 - V_1)}{\ln[(x_0 + b)(x_0 + a)/(r_2 r_1)]}\ln\left[\frac{a + x_0}{r_1}\left(\frac{z + x_0}{z - x_0}\right)\right] + V_1 \tag{4.62}$$

Other transformations based on complex inversion can be found in [1].

EXERCISES

4.1 Let the negative real axis in the z-plane have a constant charge density σ_z and the positive real axis be a grounded conductor. This problem can be transformed into a sheet charge σ_w located along the line $\text{Im}[w] = a$ and an infinite grounded conductor that lies along the real axis with the transformation

$$w = \frac{a}{\pi}\ln[z]$$

Determine the charge density in the w-plane.

4.2 A grounded conductor lies along the entire real axis of the w-plane. A sheet charge

$$\sigma_w = \frac{\sigma_0}{1 + \sin[\theta_0]}$$

lies along the circle of radius r_1 centered at the origin between the angles $\theta_1 \leq \theta \leq \pi - \theta_1$.

a. What is the Möbius transformation that transforms the upper half plane onto the inside of a circular disk of radius r_1 centered at the origin of the z-plane. Let $w_1 = 0$, $w_2 = r_1$, and $w_3 = \infty$ map to $z_1 = -jr_1$, $z_2 = r_1$, and $z_3 = jr_1$, respectively.

b. What is the charge density in the z-plane?

c. Determine the complex potential and electric field in the w-plane.

4.3 Let a line charge q_0 be located at $z = -a$ where $a > 0$. Let a semi-infinite grounded conductor lie along the positive real axis. Show the Möbius transformation that transforms this problem into a line charge located $w = -b$ and a finite conductor located between $0 < w < c$ where b and c are positive real values is

$$w = \frac{bcz}{(ca + b(a + z))}$$

4.4 A line charge q_0 is located at $z = ja$ where $a > 0$. A grounded conductor lies along the positive real axis occupying $-b \leq x \leq b$. Show the Möbius transformation that transforms this problem into a line charge located at $w = jc$ and a grounded circular conducting arc of radius c symmetric about the imaginary axis is

$$w = \frac{j2cz}{z + ja}$$

What is the central angle subtended by the grounded arc?

4.5 A circular dielectric cylinder of radius r_0 and permittivity ϵ_1 is conformally coated with another dielectric of thickness $r_1 - r_0$ and permittivity ϵ_2. Let the center of the dielectric cylinder coincide with the origin of the w-plane and the material surrounding the coated dielectric cylinder be vacuum. If a line charge q_0 is located on the positive real axis at $w = u_0 > r_1$, determine the complex potential for $|w| < r_0$, $r_0 < |w| < r_1$, and $|w| > r_1$.

4.6 The impedance Z_p of any electrical circuit composed of linear resistors, capacitors and inductors can be described by a complex function of frequency with $\mathrm{Re}[Z_p] \geq 0$ and $-\infty \geq \mathrm{Im}[Z_p] \geq \infty$. Show the Möbius transformation that maps the portion of the complex plane containing Z_p onto the unit disk is

$$w = \frac{z_1 - 1}{z_1 + 1}$$

where $z_1 = Z_p/Z_0$. Assume $Z_p = 0$, $Z_p = Z_0$, and $Z_p = \infty$ are mapped to $w = -1$, $w = 0$, and $w = 1$, respectively. What are the shapes contours of constant $\mathrm{Re}[Z_p]$ and constant $\mathrm{Im}[Z_p]$ make in the w-plane?

REFERENCES

[1] A. E. H. Love. Some electrostatic distributions in two dimensions. *Proceedings of the London Mathematical Society*, s2-22(1):337–369, 1924.

[2] Tristan Needham. *Visual Complex Analysis*. Oxford University Press, 1998.

[3] Werner Oeschslin. *Oliver Byrne: The First Six Books of The Elements of Euclid*. Taschen, 2013.

[4] Eric W. Weisstein. Symmetric points. From MathWorld–A Wolfram Web Resource. https://mathworld.wolfram.com/SymmetricPoints. html.

CHAPTER 5

Conformal Mapping II

In this chapter, we develop a conformal transformation that transforms the region inside an arbitrary closed polygon in the z-plane onto the upper half of the w-plane. The transformation is called the Schwarz-Christoffel transformation and is a powerful method for solving electrostatic problems or for determining Green's function for the region.

5.1   RIEMANN MAPPING THEOREM

Before developing the Schwarz-Christoffel transformation, we derive the Riemann mapping theorem (RMT) which states [8]

A simply connected domain S in the z-plane bounded by a simple closed curve C has a unique analytic function $w[z]$ that is single valued, conformally maps S onto the unit disk, transforms the point $z = z_0$ within C into the origin, and transforms a given direction at $z = z_0$ into the positive direction of the real axis

(RMT)

The physicist's proof of the theorem given in [8, 2] provides a connection between the conformal mapping function and the 2D Dirichlet Green's function for electrostatic problems. This connection proves useful when determining properties of the mapping function without knowledge of the exact mapping.

Let the transformed plane containing the unit disk be the w-plane as shown in Fig. 5.1 and assume the conformal map $w[z]$ exist as a unique analytic function, regular in S. Mapping z_0 to the origin allows writing

$$w[z] = (z - z_0)\, e^{\psi[z]} \tag{5.1}$$

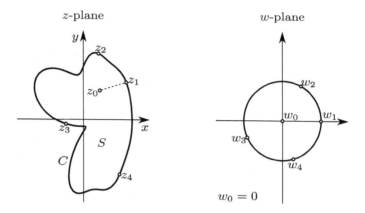

FIGURE 5.1 Mapping of region enclosed by C in the z-plane onto the unit disk in the w-plane.

where $\psi[z]$ is regular in S. On the boundary C, $|w[z]| = 1$ and the real part of $\psi[z]$ must satisfy

$$\ln[|z - z_0|] + \text{Re}[\psi[z]] = 0 \quad \text{on } C \tag{5.2}$$

Since $\psi[z]$ is an analytic function, $\text{Re}[\psi]$ and $\text{Im}[\psi]$ satisfy Laplace's equation. Multiplying both sides of (5.2) by $-1/(2\pi\epsilon)$ puts it in a form that based on the analysis in Section 3.5 is recognized as the 2D Dirichlet Green's function for the region S bounded by the contour C. Thus,

$$G_L[x, y, x_0, y_0] = -\frac{1}{2\pi\epsilon}\text{Re}[\psi[z]] \tag{5.3}$$

is the solution of Laplace's equation within S caused by the sources external to S that make Green's function G_D vanish on the boundary C.

The existence and uniqueness of Green's function would then lead to the existence and uniqueness of $\text{Re}[\psi]$. The uniqueness of solutions to Poisson's and Laplace's equations was established in Section 2.19. The existence of Green's function can be argued on physical grounds, but has also been established mathematically [11]. In principle, $\text{Re}[\psi]$ has been determined. From Section 1.4, $\text{Im}[\psi]$ can be determined from $\text{Re}[\psi]$ to within a constant. The value of this constant can be chosen to make a given direction in the z-plane map into the positive direction of the real axis in the w-plane. With $\psi[z]$ determined, the conformal mapping between the two planes is then obtained from (5.1).

Based on the above derivation, finding the conformal transformation for a region enclosed by a simple curve to the unit disk is equivalent to finding the 2D Dirichlet Green's function for the region or the complex potential due to a unit charge located at z_0 inside the region with the boundary grounded. We can state this as

$$w[z] = \exp[-2\pi\epsilon\,(\Phi[z] - jC_0)], \quad \Phi[z] = -\frac{1}{2\pi\epsilon}\ln[w[z]] + jC_0 \quad (5.4)$$

where $\Phi[z]$ is the complex potential of a unit charge located at z_0, $w[z]$ is the conformal mapping function, and C_0 is a real constant that maps a given direction in the z-plane into the positive direction of the real axis in the w-plane. This is consistent with replacing w in the complex potential of the w-plane with $w[z]$ to obtain the complex potential in the z-plane as discussed in Section 4.2.

Although region S was only mapped to the unit disk, a mapping to another region S' is also guaranteed to exist. By the Riemann mapping theorem, a conformal mapping exist that maps region S' to the unit disk. The inverse mapping that maps the unit disk to S' can be used to map S from the unit disk to S'. Therefore, region S can be mapped to region S'. A very common mapping that is heavily used in this book is to map a region S in the z-plane onto the upper half of the w-plane. Based on these arguments, the Riemann mapping theorem states a conformal transformation exist that maps a simply enclosed region in the z-plane to the upper half of the w-plane.

5.2 SYMMETRY OF CONFORMAL MAPS

If a geometry in the z-plane has reflection symmetry, does its image in the w-plane maintain this symmetry when mapped to the unit disk or upper half plane? To answer this question, we first define reflection symmetry with respect to a line in terms of conjugate coordinates. Let the line of symmetry go through point z_0 at an angle α with the real axis (see Fig. 5.2). The corresponding symmetric point z_s to a point z with respect to the line of symmetry is

$$z_s = e^{2\alpha j}\,(z - z_0)^* + z_0 \quad (5.5)$$

For example, let the line of symmetry be the real axis. Then $\alpha = 0$, $\mathrm{Im}[z_0] = 0$, and (5.5) reduces to

$$z_s = z^* \quad (5.6)$$

If the line of symmetry is the imaginary axis, then $\alpha = \pi/2$, $\mathrm{Re}[z_0] = 0$ and (5.5) reduces to

$$z_s = -z^* \quad (5.7)$$

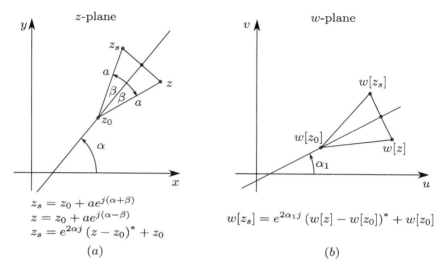

$$z_s = z_0 + ae^{j(\alpha+\beta)}$$
$$z = z_0 + ae^{j(\alpha-\beta)}$$
$$z_s = e^{2\alpha j}(z - z_0)^* + z_0$$
(a)

$$w[z_s] = e^{2\alpha_1 j}(w[z] - w[z_0])^* + w[z_0]$$
(b)

FIGURE 5.2 (a) Definition of symmetric points in the z-plane for a line of symmetry that goes through z_0 and makes an angle α with the real axis. (b) Conformal mapping $w[z]$ that preserves reflection symmetric with respect to a line.

A conformal mapping $w[z]$ produces a symmetric mapping if

$$w[z_s] = e^{2\alpha_1 j}(w[z] - w[z_0])^* + w[z_0] \tag{5.8}$$

where the line of symmetry in the w-plane goes through point $w[z_0]$ at an angle α_1 with the real axis. For example, if the line of symmetry in the z-plane is the imaginary axis and this line of symmetry is mapped to the imaginary axis of the w-plane, then combining (5.7) and (5.8) gives

$$w[-z^*] = -w[z]^* \tag{5.9}$$

Having established the equivalence of complex potentials and conformal transformations to the unit disk allows discussing reflection symmetry of conformal transformations based on physical considerations for the complex potential. In electrostatics, the environmental variables that determine the electrostatic potential (and electric field) for a region are boundary conditions and the charge distribution within the region. As was shown in Section 5.1, the conformal transformation problem is equivalent finding the complex potential for a region with a grounded boundary and a unit charge located at z_0. If the boundary has reflection symmetry with respect to a line, then the complex potential for a line charge on the line of symmetry can be forced to have certain properties. These properties are then passed on to the conformal mapping function.

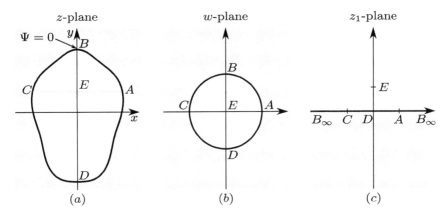

FIGURE 5.3 Conserving reflection symmetry in the z-plane by mapping points where the line of symmetry intersects the boundary (points B and D) onto opposite ends of a diameter for the unit disk in the w-plane or to the origin and $j\infty$ for the upper half plane of the z_1-plane.

For example, consider the geometry shown in Fig. 5.3 (a). This geometry has reflection symmetry with respect to the imaginary axis. If we let z_0 (location of the line charge) be a point on the line of symmetry, then the electrostatic environment from a point on the imaginary axis in the positive x direction is the same as the electrostatic environment in the negative x direction. The electrostatic potential must be the same in both directions (symmetric about the imaginary axis)

$$\text{Re}[\Phi[z]] = \text{Re}[\Phi[-z^*]] \qquad (5.10)$$

With the value of C_0 in (5.4), we can set the location for the zero of flux. If we set the zero of flux at one of the points on the boundary that crosses the line of symmetry, say B in Fig. 5.3 (a), then the flux from B to a point z on the boundary is the same as the flux from $-z^*$ to B resulting in

$$\text{Im}[\Phi[z]] = -\text{Im}[\Phi[-z^*]] \qquad (5.11)$$

Multiplying (5.11) by j and then adding it to (5.10) gives

$$\Phi[z] = (\Phi[-z^*])^* \qquad (5.12)$$

The relationship between $w[z]$ and $w[-z^*]$ is now easily established by replacing z with $-z^*$ in (5.4) and then taking the complex conjugate to obtain

$$(w[-z^*])^* = \left(e^{-2\pi\epsilon(\Phi[-z^*]-jC_0)}\right)^* = e^{-2\pi\epsilon(\Phi[z]+jC_0)} = e^{-4\pi\epsilon jC_0}w[z] \quad (5.13)$$

Taking the complex conjugate of (5.13) and comparing to (5.8) shows that $w[z]$ is symmetric with a line of symmetry that goes through the origin of the w-plane and forms an angle $\alpha_1 = 2\pi C_0$ with the real axis. The value of C_0 is determined by

$$\Psi[z_B] = 0 = -\frac{1}{2\pi\epsilon}\ln[w[z_B]] + jC_0 \tag{5.14}$$

as

$$C_0 = \frac{1}{2\pi\epsilon}\arg[w[z_B]] \tag{5.15}$$

Specifying $w[z_B]$ (the location of point B in the w-plane) is equivalent to specifying the mapping of a given direction in the z-plane into a given direction in the w-plane. Thus, we are free to choose the value of $w[z_B]$. Choosing $w[z_B] = j$ gives $C_0 = 1/(4\epsilon)$ and (5.13) becomes

$$(w[-z^*])^* = -w[z] \tag{5.16}$$

This choice of C_0 results in $w[z]$ having reflection symmetry about the imaginary axis.

We can now determine how to preserve symmetry when mapping to the upper half plane. To map the unit disk in the w-plane to the upper half of the z_1-plane, we translate the unit disk by j, and perform complex inversion using a circle of inversion with $w_c = 2j$ and $r_I = 2$. The resulting conformal transformation is

$$z_1[w] = 2j\left(\frac{1 - jw}{1 + jw}\right) \tag{5.17}$$

To obtain the conformal mapping between z_1 and z, we replace w with $w[z]$ in (5.17) to obtain

$$z_1[z] = 2j\left(\frac{1 - jw[z]}{1 + jw[z]}\right) \tag{5.18}$$

Replacing z with $-z^*$ in (5.18) and taking the complex conjugate gives

$$(z_1[-z^*])^* = -2j\left(\frac{1 + j(w[-z^*])^*}{1 - j(w[-z^*])^*}\right)$$
$$= -2j\left(\frac{1 - jw[z](-e^{-4\pi\epsilon jC_0})}{1 + jw[z](-e^{-4\pi\epsilon jC_0})}\right) \tag{5.19}$$

In order for the right hand side of (5.19) to be proportional to $z_1[z]$, we need

$$e^{-4\pi\epsilon jC_0} = -1 \quad\text{or}\quad C_0 = \pm 1/(4\epsilon) \tag{5.20}$$

Using (5.20) in (5.19) gives

$$z_1[z] = -(z_1[-z^*])^* \tag{5.21}$$

The line of symmetry in the z-plane is mapped onto the imaginary axis of the z_1-plane and the mapping is symmetric with respect to the imaginary axis. In general, mapping the two points where the line of symmetry crosses the boundary (B and D in Fig. 5.3) to the ends of the line of symmetry for the mapped region results in a symmetric mapping. The unit disk centered on the origin has many possible lines of symmetry (any line that forms a diameter), but the upper half plane only has one (the imaginary axis).

5.3 VAN DER PAUW THEOREM

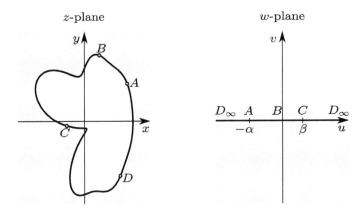

FIGURE 5.4 Conformal mapping of the region enclosed in the z-plane onto the upper half of the w-plane to establish van der Pauw's theorem for electrostatics.

Even if the mapping between planes is unknown, just knowing a conformal mapping exist can lead to useful results. For example, consider the general enclosed geometry shown in Fig. 5.4. By the Riemann mapping theorem, there is a conformal mapping that maps the enclosed region onto the upper half of the w-plane. We can specify the location of up to three different points between the planes. Let point A map to $w = -\alpha$, point B map to $w = 0$, and point D to $w = \infty$. Point C is mapped to $w = \beta$ where β is a positive real number. In order to determine the value of β, further details of the geometry being mapped need to be known. Let the boundary between points A and B in z-plane of Fig. 5.4 have a potential V_0 and the remaining portion of the boundary

grounded. The w-plane has an infinite flat plate that lies along the real
axis that is grounded except for the portion between $-\alpha$ and 0 which has
a potential V_0. From Section 3.3, the complex potential in the w-plane
is j times the complex potential of a line charge $q = 2\epsilon V_0$ located at
$w = -\alpha$ and a line charge $q = -2\epsilon V_0$ located at $w = 0$

$$\Phi[w] = -j\frac{V_0}{\pi}\left(\ln[w + \alpha] - \ln[w]\right) \tag{5.22}$$

The boundary in the z-plane can be viewed as having four segments
located between the four points $ABCD$. We define the cross-capacitance
per unit length as minus the charge per unit length on the grounded
segment not adjacent to the segment at potential V_0 divided by V_0. The
cross-capacitance per unit length for the segment AB at potential V_0 is

$$\underline{C}_1 = -\frac{\epsilon\left(\Psi[\infty] - \Psi[\beta]\right)}{V_0} = \frac{\epsilon}{\pi}\left(\ln[\beta] - \ln[\beta + \alpha]\right) \tag{5.23}$$

We can define a second cross-capacitance per unit length by setting the
potential of segment BC to V_0 and grounded the rest of the boundary.
For this case, the complex potential is

$$\Phi[w] = -j\frac{V_0}{\pi}\left(\ln[w] - \ln[w - \beta]\right) \tag{5.24}$$

and the cross-capacitance per unit length is

$$\underline{C}_2 = -\frac{\epsilon\left(\Psi[-\alpha] - \Psi[-\infty]\right)}{V_0} = -\frac{\epsilon}{\pi}\left(\ln[\alpha] - \ln[\alpha + \beta]\right) \tag{5.25}$$

Subtracting (5.23) from (5.25) and rearranging gives

$$e^{-\pi\underline{C}_1/\epsilon} + e^{-\pi\underline{C}_2/\epsilon} = 1 \tag{5.26}$$

In other words,

> *When the boundary of a region free of charge and filled with a
> dielectric material of uniform permittivity ϵ is divided into four
> segments, it has two cross-capacitances $\underline{C}_1$ and $\underline{C}_2$. Independent
> of which points are chosen for dividing the boundary into four
> parts, the cross-capacitances have the relationship $e^{-\pi\underline{C}_1/\epsilon} +$
> $e^{-\pi\underline{C}_2/\epsilon} = 1$*

(VDPT)

We refer to this result as van der Pauw's theorem since the analogous
form of (5.26) in terms of resistivity and resistance was first published
by van der Pauw [14]. The original form is derived in Chapter 7 where
conformal mapping is extended to areas of physics outside of electrostatics.

5.4   THOMPSON-LAMPARD THEOREM

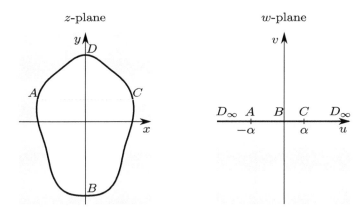

FIGURE 5.5   Conformal mapping of a geometry with reflection symmetry onto the upper half of the w-plane in order to establish the Thompson-Lampard theorem.

When reflection symmetry about a line is present in the region to be mapped, van der Pauw's relationship (5.26) reduces to a very simple form. Consider the symmetric structure in the z-plane shown in Fig. 5.5. Mapping the line of symmetry in the z-plane to the imaginary axis of the w-plane produces a symmetric mapping with respect to the imaginary axis in the w-plane. For the symmetric case, $\beta = \alpha$ in Fig. 5.4 resulting in

$$\underline{C}_1 = \underline{C}_2 = \underline{C}_s \tag{5.27}$$

Combining (5.26) and (5.27) gives

$$\underline{C}_s = \epsilon \ln[2]\,/\pi \tag{5.28}$$

In other words,

> Let the line of symmetry and a line perpendicular to the line of symmetry divide the boundary of a region with reflection symmetry into four segments. The cross-capacitance per unit length has a value of $\epsilon \ln[2]\,/\pi$ where ϵ is the permittivity of the region

which is known as the Thompson-Lampard theorem (TLT) [13].

5.5   SCHWARZ-CHRISTOFFEL TRANSFORMATION

Although the Riemann mapping theorem ensures a conformal mapping $w[z]$ exist, it does not provide a method for finding $w[z]$. The Schwarz-Christoffel transformations fills part of this gap by providing a method

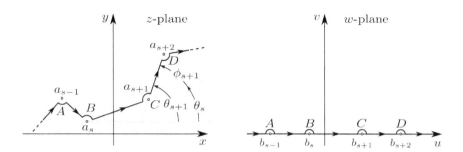

FIGURE 5.6 Definition of a_s, b_s, θ_s, and ϕ_s used to derive the Schwarz-Christoffel transformation.

for finding $w[z]$ that maps the region inside an arbitrary polygon in the z-plane onto the w-plane bounded by the real axis and an arc of infinite radius (the upper half plane). We start in the w-plane and deform the real axis and infinite arc (if needed) so that they are transformed into the polygon of the z-plane. To do this, we investigate the complex derivative

$$\frac{dz}{dw} = A_1 \prod_{i=1}^{n} (w - b_i)^{\beta_i} = A_1 (w - b_1)^{\beta_1} (w - b_2)^{\beta_2} \cdots (w - b_n)^{\beta_n} \quad (5.29)$$

where the b_i and β_i are real numbers, A_1 is a complex constant, and $b_n > b_{n-1} > \cdots > b_1$. When $w = b_i$, (5.29) is either zero or infinity depending if β_i is greater than or less than zero. Therefore, we avoid b_i using a small semicircle in the upper half plane centered at each b_i as shown in Fig. 5.6. Rewriting (5.29) in polar representation gives

$$\frac{dz}{dw} = |A_1| e^{j \arg[A_1]} \prod_{i=1}^{n} |w - b_i| e^{j\beta_i \arg[w - b_i]} \quad (5.30)$$

The argument (or phase) of (5.30) is

$$\arg\left[\frac{dz}{dw}\right] = \arg[A_1] + \sum_{i=1}^{n} \beta_i \arg[w - b_i] \quad (5.31)$$

When $dw = du$, it is an element of the real axis in the w-plane and we can write (5.31) as

$$\arg\left[\frac{dz}{dw}\right] = \arg\left[\frac{dx + jdy}{du}\right] = \tan^{-1}\left[\frac{dy}{dx}\right] \quad (5.32)$$

Thus (5.32) is the angle that the corresponding element dz makes with the positive real axis in the z-plane.

Now choose a point that lies on the real axis in the w-plane between b_s and b_{s+1}. All the factors in (5.31) where $b_i \leq b_s$ are real positive numbers with arguments of zero and all the factors where $b_i \geq b_{s+1}$ are negative numbers with arguments of π. The angle in the z-plane from (5.31) and (5.32) is then

$$\theta_s = \tan^{-1}\left[\frac{dy}{dx}\right] = \arg[A_1] + (\beta_{s+1} + \beta_{s+2} + \cdots + \beta_n)\pi \qquad (5.33)$$

Therefore, all elements on the real axis which lie between b_s and b_{s+1} have the same direction forming a straight line with slope given by (5.33). We now see that (5.29) has transformed the portion of the real axis between b_s and b_{s+1} into a straight line in the z-plane with angle θ_s measured from the positive real axis. Since b_s can represent any b_i, the segment between b_{s+1} and b_{s+2} is also a straight line with angle

$$\theta_{s+1} = \tan^{-1}\left[\frac{dy}{dx}\right] = \arg[A_1] + (\beta_{s+2} + \beta_{s+3} + \cdots + \beta_n)\pi \qquad (5.34)$$

The angle between two lines is

$$\phi_{s+1} = \theta_{s+1} - \theta_s = -\pi\beta_{s+1} \qquad (5.35)$$

and is called the exterior angle at vertex $s + 1$ of the polygon in the z-plane. The range of possible ϕ_i and β_i values can be determined with the help of Fig. 5.7. From Fig. 5.7 (a), the maximum positive ϕ for a vertex at a finite z value must be less than π as no interior of the polygon exist between segments AB and BC when $\phi = \pi$. From Fig. 5.7 (b), the minimum negative ϕ for a vertex at a finite z value is $-\pi$ as the interior of the polygon is bounded by the segments AB and BC when $\phi = -\pi$. This gives the limits of ϕ_i and β_i at finite vertices as

$$-\pi \leq \phi_i < \pi, \quad -1 < \beta_i \leq 1 \qquad (5.36)$$

Additionally, exterior angles for any closed polygon must sum to 2π requiring

$$\sum_{i=1}^{n}\phi_i = 2\pi, \quad \sum_{i=1}^{n}\beta_i = -2 \qquad (5.37)$$

Next we investigate the behavior of the contour near b_s. On the semicircle near b_s, we have

$$w - b_s = re^{j\theta}, \quad dw = jre^{j\theta}d\theta \qquad (5.38)$$

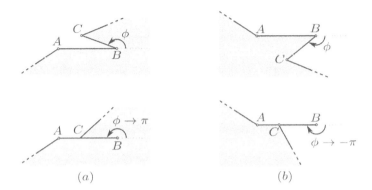

FIGURE 5.7 Defining the limits of an exterior angle with a finite vertex for (a) $\phi \to \pi$ and (b) $\phi \to -\pi$. These limits do not apply to exterior angles for vertices at infinity.

For r infinitesimal, we can make the following approximations

$$w - b_i \approx \begin{cases} b_s - b_i, & i \neq s \\ re^{j\theta} & i = s \end{cases} \tag{5.39}$$

Substituting the approximations of (5.39) into (5.29) gives

$$\frac{dz}{jre^{j\theta}d\theta} = A_1\left(re^{j\theta}\right)^{\beta_s} \prod_{i \neq s}^{n} (b_s - b_i)^{\beta_i} \tag{5.40}$$

or

$$\int dz = jC_1 \int e^{j\theta(1+\beta_s)}d\theta, \quad C_1 = A_1 r^{(1+\beta_s)} \prod_{i \neq s}^{n} (b_s - b_i)^{\beta_i} \tag{5.41}$$

Performing the integration gives

$$z = z_1 + C_1\frac{e^{j\theta(1+\beta_s)}}{(1+\beta_s)} \tag{5.42}$$

where z_1 is a complex constant. We rewrite (5.42) as

$$z - z_1 = C_2 e^{j\theta(1+\beta_s)}, \quad C_2 = C_1/(1+\beta_s) \tag{5.43}$$

which is the equation of a circle centered at z_1 with radius $|C_2|$ that transverses an angle $-(\beta_s + 1)\pi$ when θ goes from π to 0. Since $\beta_s > -1$, we can take the limit of (5.42) as $r \to 0$ resulting in

$$z = z_1 \tag{5.44}$$

Thus the two lines formed in the z-plane meet at the center of the circle defined by (5.43) and we have now formed two sides of the polygon. The procedure to form the remaining sides is the same as outlined, but we still need to understand how (5.29) handles the infinite arc in the w-plane.

To investigate the behavior near the infinite arc, let $w = re^{j\theta}$ and $dw = jre^{j\theta}d\theta$ where r is now very large so that the following approximation holds for all b_i

$$w - b_i \approx re^{j\theta} \tag{5.45}$$

Substituting (5.45) into (5.29) gives

$$\frac{dz}{jre^{j\theta}d\theta} = A_1 \prod_{i=1}^{n} w^{\beta_i} = A_1 \left(re^{j\theta}\right)^{-2} \tag{5.46}$$

or

$$\int dz = (jA_1/r) \int e^{-j\theta}d\theta \tag{5.47}$$

Note that (5.37) was used to eliminate β_i in (5.46). Performing the integration gives

$$z = z_1 + C_1 e^{-j\theta}, \quad C_1 = -A_1/r \tag{5.48}$$

where z_1 is a complex constant. Clearly (5.48) is the equation of a circle with center z_1 and radius $|C_1|$ that transverses an angle $-\pi$ when θ goes from 0 to π. Taking the limit as $r \to \infty$ gives $z = z_1$. Therefore, the arc of infinite radius in the w-plane converges to the center of the circle defined by (5.48) in the z-plane if z_1 is finite. The polygon in the z-plane can also be a polygon with an arc of infinite radius and z_1 is not always finite.

We can now write (5.29) in final form as

$$\frac{dz}{dw} = A_1 \prod_{i=1}^{n} (w - b_i)^{-\phi_i/\pi} \tag{5.49}$$

Integrating (5.49) gives

$$z = A_1 \int \prod_{i=1}^{n} (w - b_i)^{-\phi_i/\pi} dw + B_1 \tag{5.50}$$

which is known as the Schwarz-Christoffel (S-C) transformation.

In (5.50), the magnitude of A_1 allows scaling the polygon to the desired size and the argument of A_1 allows producing the desired orientation of the polygon. B_1 is an arbitrary complex constant that can be used to translate (without rotation) the polygon to the desired location in the z-plane.

Although we derived (5.50) by transforming the real axis of the w-plane into a polygon in the z-plane, typically we are presented with the opposite problem of transforming the inside of a polygon in the z-plane onto the upper half of the w-plane. With (5.50), the transformation is performed with the following steps. First, all the vertices in the z-plane are labeled with capital letters. This helps to identify common points between the w-plane and z-plane. Then the exterior angle (ϕ_i) for each vertex in the z-plane is determine. A useful check once this step is completed is to ensure that the sum of the exterior angles is 2π. Next, the ordering of the b_i along the real axis in the w-plane is determined and we are free to choose the location of three b_i. The final steps are to perform the integral (if possible), then to determine A_1, B_1, and the remaining b_i from the constraints of the polygon being transformed.

5.6 S-C TRANSFORMATION WITH $B_N = \infty$

For our analysis, b_1 is always finite, but b_n may be finite or infinite. Often it is useful to place b_n at ∞. To understand how an infinite b_n changes (5.49), we first rewrite (5.49) in terms of a new constant $\left(A_1(-b_n)^{\phi_n/\pi}\right)$ as

$$\frac{dz}{dw} = A_1(-b_n)^{\phi_n/\pi}(w - b_n)^{-\phi_n/\pi} \prod_{i=1}^{n-1}(w - b_i)^{-\phi_i/\pi} \qquad (5.51)$$

In the limit

$$\lim_{b_n \to \infty}\left[(-b_n)^{\phi_n/\pi}(w - b_n)^{-\phi_n/\pi}\right] = 1 \qquad (5.52)$$

and (5.51) becomes

$$\frac{dz}{dw} = A_1 \prod_{i=1}^{n-1}(w - b_i)^{-\phi_i/\pi} \qquad (5.53)$$

Therefore, the factor in (5.49) associated with a vertex at ∞ does not contribute to the integral of (5.50). The price paid for this simplification is that now we are free to specify the location of only two b_i in (5.50).

5.7   S-C TRANSFORMATION ONTO A UNIT DISK

In addition to the upper half-plane, the S-C transformation can be re-formulated to map a polygon onto other domains such as a disk, a strip, and a rectangle [3]. We only derive the S-C transformation for mapping the interior of a polygon onto the unit disk centered at the origin of the z_1-plane. The Möbius transformation that maps $z_a = 0$, $z_b = 1$, and $z_c = \infty$ to $w_a = -j$, $w_b = 1$, and $w_c = j$, respectively, gives

$$w[z_1] = j \left(\frac{z_1 - j}{z_1 + j} \right) \tag{5.54}$$

Replacing w with $w[z_1]$ and b_i with $w[c_i]$ in (5.50) gives

$$z = -2A_1 \int \prod_{i=1}^{n} \left(j \frac{z_1 - j}{z_1 + j} - j \frac{c_i - j}{c_i + j} \right)^{-\phi_i/\pi} \frac{dz_1}{(z_1 + j)^2} + B_1 \tag{5.55}$$

Since the sum of the external angles ϕ_i is 2π, we have

$$\prod_{i=1}^{n} \left(\frac{j}{z_1 + j} \right)^{-\phi/\pi} = -(z_1 + j)^2 \tag{5.56}$$

With the vertices mapped to the unit circle,

$$c_i = e^{j\alpha_i} \tag{5.57}$$

Using (5.56) and (5.57) to simplify (5.55) and then absorbing all the factors not dependent on z_1 into a new constant A_2 allows writing (5.55) as

$$z = A_2 \int \prod_{i=1}^{n} \left(1 - e^{-j\alpha_i} z_1 \right)^{-\phi_i/\pi} dz_1 + B_1 \tag{5.58}$$

This is the S-C transformation for mapping the interior of a polygon onto the unit disk.

5.8   PHASE OF A_1

The phase of A_1 is determined by the orientation of the polygon in the z-plane and whether b_n is finite or infinite. Thus, the phase of A_1 can be determined without having to perform the integration of (5.50). With the phase of A_1 known, we can determine if A_1 has a real, imaginary, or complex value which can often simplify the process of determining the remaining b_i, B_1 and even the value of A_1. When b_n is finite, (5.33) states

$$\arg[A_1] = \theta_n, \quad b_n \text{ finite} \tag{5.59}$$

where θ_n is the angle in the z-plane that the portion of the polygon between vertex n and vertex 1 makes with the real axis. For b_n finite, A_1 has a real value if the portion of the polygon located between vertex n and vertex 1 is parallel to the real axis.

When b_n is infinite, the term associated with ϕ_n is not part of (5.33) resulting in

$$\arg[A_1] = \theta_{n-1}, \quad b_n \text{ infinite} \tag{5.60}$$

For b_n infinite, A_1 has a real value if the portion of the polygon located between vertex $n-1$ and vertex n is parallel to the real axis.

5.9 EXTERIOR ANGLE FOR A VERTEX AT INFINITY

The limits on the exterior angle ϕ_i at a finite vertex given in (5.36) do not apply to exterior angles for lines that meet at infinity. To determine the range for exterior angles at infinity, let line 1 be aligned with the real axis and go to infinity in the $+x$ direction. Line 2 that emerges from infinity must form an angle with the real axis that is greater than or equal to π otherwise line 1 and line 2 intersect before infinity. Let the angle γ be defined as the angle formed by connecting the finite vertex of line 1 to the finite vertex of line 2 as shown in Fig. 5.8. The exterior angle at infinity is then defined as

$$\phi_{i,\infty} = \gamma + \pi \tag{5.61}$$

Since γ has a range of $0 \leq \gamma \leq 2\pi$, exterior angles at infinity have a range of

$$\pi \leq \phi_{i,\infty} \leq 3\pi \tag{5.62}$$

5.10 BOUNDARY CONDITION FOR PARALLEL LINES THAT MEET AT INFINITY

5.10.1 $\phi_{s,\infty} = \pi$

When applying the S-C transformation to polygons with vertices at infinity, knowledge of the boundary conditions for parallel lines that meet at infinity is often useful. There are four cases to consider. First is the

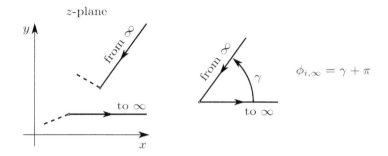

FIGURE 5.8 Definition of an exterior angle at infinity.

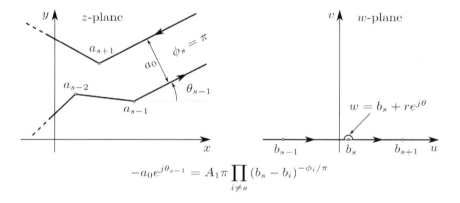

$$-a_0 e^{j\theta_{s-1}} = A_1 \pi \prod_{i \neq s} (b_s - b_i)^{-\phi_i/\pi}$$

FIGURE 5.9 Arbitrary polygon with parallel lines that meet at infinity, form an exterior angle of π, and the vertex is mapped to a finite b_s.

case where two parallel lines have their vertex mapped to b_s which is finite and have an external angle $\phi_s = \pi$. We can write (5.49) as

$$\int_{z[w_1]}^{z[w_2]} dz = A_1 \int_{w_1}^{w_2} (w - b_s)^{-1} \prod_{i \neq s} (w - b_i)^{-\phi_i/\pi} dw \qquad (5.63)$$

Let line 1 be the line between vertices a_{s-1} and a_s and line 2 be the line between vertices a_s and a_{s+1} in the z-plane. Let line 1 go to infinity in the $+x$ direction at an angle θ_{s-1} (see Fig. 5.9). Let the distance between lines 1 and 2 be a_0. Integrating along a small semicircle centered at b_s in the w-plane allows writing $w = b_s + re^{j\theta}$ and (5.63) becomes

$$\int_{z[w_1]}^{z[w_2]} dz = A_1 \int_0^\pi \prod_{i \neq s} \left(b_s + re^{j\theta} - b_i \right)^{-\phi_i/\pi} j\, d\theta \qquad (5.64)$$

Let $r \ll |b_s - b_i|$ for all $b_i \neq b_s$ allowing the approximation

$$b_s + re^{j\theta} - b_i \approx b_s - b_i, \quad i \neq s \tag{5.65}$$

Since the image of w_2 is on line 1 and the image of w_1 is on line 2, we can write

$$z[w_2] = a_{s-1} + r_1 e^{j\theta_{s-1}}, \quad z[w_1] = a_{s-1} + ja_0 e^{j\theta_{s-1}} + r_2 e^{j\theta_{s-1}} \tag{5.66}$$

Substituting (5.65) and (5.66) into (5.64) and performing the integration gives

$$(r_1 - r_2) e^{j\theta_{s-1}} - ja_0 e^{j\theta_{s-1}} \approx A_1 \prod_{i \neq s} (b_s - b_i)^{-\phi_i/\pi} j\pi \tag{5.67}$$

In the limit that $r \to 0$, $r_2 \to r_1$ resulting in the boundary condition

$$-a_0 e^{j\theta_{s-1}} = A_1 \pi \prod_{i \neq s} (b_s - b_i)^{-\phi_i/\pi} \tag{5.68}$$

5.10.2 $\phi_{n,\infty} = \pi$

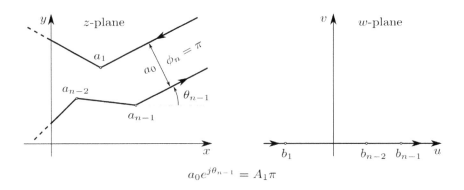

$$a_0 e^{j\theta_{n-1}} = A_1 \pi$$

FIGURE 5.10 Arbitrary polygon with parallel lines that meet at infinity, form an exterior angle of π, and the vertex is mapped to infinity.

The second case is where two parallel lines have their vertex mapped to infinity ($b_n = \infty$) and have an external angle $\phi_n = \pi$. For the integration in the w-plane, we use a large semicircle centered at the origin. We can write $w = re^{j\theta}$ where $r \gg |b_i|$ for $b_i \neq b_n$ allowing the approximation

$$w - b_i \approx re^{j\theta} \tag{5.69}$$

From (5.37), the sum of all the external angles is 2π. Since we are excluding $\phi_n = \pi$, we have

$$\sum_{i\neq n} \phi_i = 2\pi - \phi_n = \pi \tag{5.70}$$

Using (5.69) and (5.70), (5.53) becomes

$$\int_{z[w_1]}^{z[w_2]} dz \approx A_1 \int_0^\pi \prod_{i\neq n} \left(re^{j\theta}\right)^{-\phi_i/\pi} jre^{j\theta} d\theta = A_1 \int_0^\pi jd\theta \tag{5.71}$$

Let line 1 be the line between vertices a_{n-1} and a_n and line 2 be the line between vertices a_n and a_1 in the z-plane. Let line 1 go to infinity in the $+x$ direction at an angle θ_{n-1} (see Fig. 5.10). Let the distance between lines 1 and 2 be a_0. Since the image of w_1 is on line 1 and the image of w_2 is on line 2, we can write

$$z[w_1] = a_{n-1} + r_1 e^{j\theta_{n-1}}, \quad z[w_2] = a_{n-1} + ja_0 e^{j\theta_{n-1}} + r_2 e^{j\theta_{n-1}} \tag{5.72}$$

Substituting (5.72) into (5.71) and performing the integration gives

$$(r_2 - r_1) e^{j\theta_{n-1}} + ja_0 e^{j\theta_{n-1}} \approx A_1 j\pi \tag{5.73}$$

In the limit that $r \to \infty$, $r_2 \to r_1$ resulting in the boundary condition

$$a_0 e^{j\theta_{n-1}} = A_1 \pi \tag{5.74}$$

5.10.3 $\phi_{s,\infty} = 2\pi$

The third case is where two parallel lines have their vertex mapped to b_s which is finite and have an external angle $\phi_s = 2\pi$. We can write (5.49) as

$$\int_{z[w_1]}^{z[w_2]} dz = A_1 \int_{w_1}^{w_2} (w - b_s)^{-2} \prod_{i\neq s} (w - b_i)^{-\phi_i/\pi} dw \tag{5.75}$$

Let line 1 be the line between vertices a_{s-1} and a_s and line 2 be the line between vertices a_s and a_{s+1} in the z-plane. Let line 1 go to infinity in the $+x$ direction at an angle θ_{s-1} (see Fig. 5.11). Let the distance between lines 1 and 2 be a_0. Integrating along a small semicircle centered at b_s in the w-plane allows writing $w = b_s + re^{j\theta}$ and (5.75) becomes

$$\int_{z[w_1]}^{z[w_2]} dz = A_1 \int_0^\pi \left(re^{j\theta}\right)^{-1} \prod_{i\neq s} \left(b_s + re^{j\theta} - b_i\right)^{-\phi_i/\pi} jd\theta \tag{5.76}$$

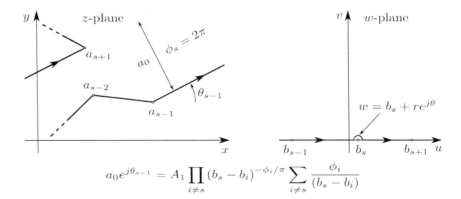

$$a_0 e^{j\theta_{s-1}} = A_1 \prod_{i \neq s} (b_s - b_i)^{-\phi_i/\pi} \sum_{i \neq s} \frac{\phi_i}{(b_s - b_i)}$$

FIGURE 5.11 Arbitrary polygon with parallel lines that meet at infinity, form an exterior angle of 2π, and the vertex is mapped to a finite b_s.

Let $r \ll |b_s - b_i|$ for all $b_i \neq b_s$ allowing the approximation

$$\left(b_s + r e^{j\theta} - b_i\right)^{-\phi_i/\pi} \approx (b_s - b_i)^{-\phi_i/\pi} \left(1 - \frac{\phi_i r e^{j\theta}}{\pi (b_s - b_i)} + O\left[r^2\right]\right) \quad (5.77)$$

where $O[r^2]$ signifies that all higher order terms of the series expansion decrease at least as fast as r^2 as r is reduced. Since the image of w_2 is on line 1 and the image of w_1 is on line 2, we can write

$$z[w_2] = a_{s-1} + r_1 e^{j\theta_{s-1}}, \quad z[w_1] = a_{s-1} + j a_0 e^{j\theta_{s-1}} - r_2 e^{j\theta_{s-1}} \quad (5.78)$$

Substituting (5.77) and (5.78) into (5.76) and performing the integration gives

$$(r_1 + r_2) e^{j\theta_{s-1}} - j a_0 e^{j\theta_{s-1}}$$
$$\approx A_1 \left(\prod_{i \neq s} (b_s - b_i)^{-\phi_i/\pi}\right) \left(\frac{2}{r} - \sum_{i \neq s} \frac{j\phi_i}{(b_s - b_i)} + \int_0^\pi O[r] j d\theta\right) \quad (5.79)$$

The phase of A_1 is given by (5.33) and since $\theta_s = \theta_{s-1}$,

$$\arg\left[A_1 \prod_{i \neq s} (b_s - b_i)^{-\phi_i/\pi}\right] = \arg[A_1] - \sum_{i > s} \phi_i = \theta_{s-1} \quad (5.80)$$

Therefore,

$$A_1 \prod_{i \neq s} (b_s - b_i)^{-\phi_i/\pi} = |A_1| \left|\prod_{i \neq s} (b_s - b_i)^{-\phi_i/\pi}\right| e^{j\theta_{s-1}} \quad (5.81)$$

and (5.79) can be written as

$$(r_1 + r_2) - ja_0$$

$$\approx |A_1| \left| \prod_{i \neq s} (b_s - b_i)^{-\phi_i/\pi} \right| \left(\frac{2}{r} - \sum_{i \neq s} \frac{j\phi_i}{(b_s - b_i)} + \int_0^\pi O[r] jd\theta \right) \quad (5.82)$$

In the limit that $r \to 0$, the imaginary part of (5.82) remains finite giving

$$a_0 = |A_1| \left| \prod_{i \neq s} (b_s - b_i)^{-\phi_i/\pi} \right| \left| \sum_{i \neq s} \frac{\phi_i}{(b_s - b_i)} \right| \quad (5.83)$$

or

$$a_0 e^{j\theta_{s-1}} = A_1 \prod_{i \neq s} (b_s - b_i)^{-\phi_i/\pi} \sum_{i \neq s} \frac{\phi_i}{(b_s - b_i)} \quad (5.84)$$

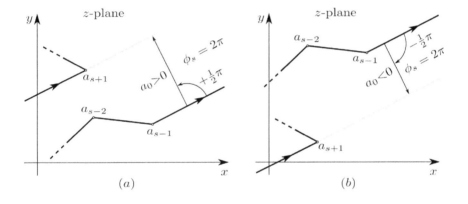

(a)

(b)

FIGURE 5.12 Method for determining the sign of a_0 when parallel lines that meet at infinity have an external angle of 2π. (a) The vector pointing from a_{s-1} to a_s needs to rotate by $\pi/2$ to form a perpendicular intersection with the line emerging from infinity resulting in $a_0 > 0$. (b) The vector pointing from a_{s-1} to a_s needs to rotate by $-\pi/2$ to form a perpendicular intersection with the line emerging from infinity resulting in $a_0 < 0$.

Although we defined a_0 as the distance between the parallel lines, it can have a negative or positive value. To determine the sign of a_0, we define a vector that points in the direction of the line going to infinity and draw the two parallel lines long enough so they overlap as shown in Fig. 5.12. This vector needs to rotate $\frac{1}{2}\pi$ or $-\frac{1}{2}\pi$ to form a perpendicular intersection with the line emerging from infinity. The sign of this rotation determines the sign of a_0. A negative a_0 is only possible for parallel lines that meet at infinity when $\phi = 2\pi$. When $\phi = \pi$, a_0 is always positive.

5.10.4 $\phi_{n,\infty} = 2\pi$

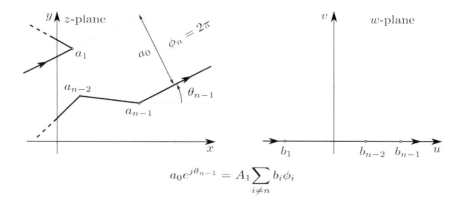

$$a_0 e^{j\theta_{n-1}} = A_1 \sum_{i \neq n} b_i \phi_i$$

FIGURE 5.13 Arbitrary polygon with parallel lines that meet at infinity, form an exterior angle of 2π, and the vertex is mapped to infinity.

The fourth case is where two parallel lines have their vertex mapped to infinity ($b_n = \infty$) and have an external angle $\phi_n = 2\pi$. We can write (5.53) as

$$\int_{z[w_1]}^{z[w_2]} dz = A_1 \int_{w_1}^{w_2} \prod_{i \neq n} (w - b_i)^{-\phi_i/\pi} dw \qquad (5.85)$$

Let line 1 be the line between vertices a_{n-1} and a_n and line 2 be the line between vertices a_n and a_1 in the z-plane. Let line 1 go to infinity in the $+x$ direction at an angle θ_{n-1} (see Fig. 5.13). Let the distance between lines 1 and 2 be a_0. Integrating along a large semicircle centered at the origin of the w-plane allows writing $w = re^{j\theta}$ where $r \gg |b_i|$ for $b_i \neq b_n$ allowing the approximation

$$\left(re^{j\theta} - b_i\right)^{-\phi_i/\pi} \approx \left(re^{j\theta}\right)^{-\phi_i/\pi}\left(1 + \frac{b_i\phi_i}{\pi r}e^{-j\theta} + O\left[r^{-2}\right]\right) \qquad (5.86)$$

Since we are excluding $\phi_n = 2\pi$ from the integral, we have

$$\sum_{i \neq n} \phi_i = 2\pi - \phi_n = 0 \qquad (5.87)$$

Since the image of w_1 is on line 1 and the image of w_2 is on line 2, we can write

$$z[w_1] = a_{n-1} + r_1 e^{j\theta_{n-1}}, \quad z[w_2] = a_{n-1} + ja_0 e^{j\theta_{n-1}} - r_2 e^{j\theta_{n-1}} \qquad (5.88)$$

Substituting (5.86), (5.87), and (5.88) into (5.85) and performing the integral over z gives

$$- (r_1 + r_2 - ja_0)\, e^{j\theta_{n-1}}$$

$$\approx A_1 \int_0^\pi \left(1 + \frac{e^{-j\theta}}{\pi r} \left(\sum_{i \neq n} b_i \phi_i \right) + O\left[r^{-2} \right] \right) j r e^{j\theta}\, d\theta \qquad (5.89)$$

From (5.60), $\arg[A_1] = \theta_{n-1}$ allowing (5.89) to be written as

$$- r_1 - r_2 + ja_0$$

$$\approx |A_1| \int_0^\pi \left(1 + \frac{e^{-j\theta}}{\pi r} \left(\sum_{i \neq n} b_i \phi_i \right) + O\left[r^{-2} \right] \right) j r e^{j\theta}\, d\theta \qquad (5.90)$$

In the limit that $r \to \infty$, the imaginary part of (5.90) remains finite and results in

$$a_0 = |A_1| \sum_{i \neq n} b_i \phi_i \qquad (5.91)$$

or

$$a_0 e^{j\theta_{n-1}} = A_1 \sum_{i \neq n} b_i \phi_i \qquad (5.92)$$

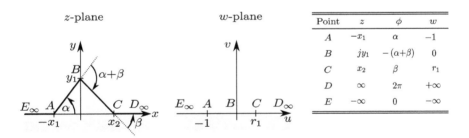

FIGURE 5.14 Transform defined in [7].

5.10.5 Example

As an example of using the boundary conditions for parallel lines that meet at infinity, consider the infinite polygon shown in Fig. 5.14. There are four vertices but we are only allowed to choose the value of three vertices. The remaining vertex must be determined by the constraints of the polygon. If we let

$$b_1 = -1, \quad b_2 = 0, \quad b_3 = r_1, \quad b_4 = \infty, \quad \phi_1 = \alpha, \quad \phi_3 = \beta \qquad (5.93)$$

then $\phi_2 = -(\alpha + \beta)$ and the resulting S-C transformation is

$$z = A_1 \int \frac{w^{\alpha+\beta}}{(w+1)^\alpha (w-r_1)^\beta} dw + B_1 \qquad (5.94)$$

It was found in [7] through numerical solutions that

$$r_1 = \alpha/\beta \qquad (5.95)$$

Over the years, several methods have been published to prove (5.95) [6, 1, 12]. The relationship given in (5.95) is easily obtained with the parallel lines boundary condition (5.92). From Fig. 5.14, $\theta_{n-1} = 0$ and $a_0 = 0$. Plugging these values into (5.92) gives

$$0 = A_1 \left(-\alpha + r_1\beta + 0\left(\alpha + \beta\right)\right) = A_1 \left(-\alpha + r_1\beta\right) \qquad (5.96)$$

Since A_1 cannot equal zero, the terms in parentheses of (5.96) must sum to zero

$$-\alpha + r_1\beta = 0 \quad \text{or} \quad r_1 = \alpha/\beta \qquad (5.97)$$

proving the relationship originally found numerically in [7].

5.11   POLYGONS WITH BOTH VERTICES AT INFINITY

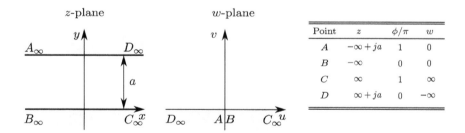

FIGURE 5.15   S-C transformation of an infinite strip of height a in the z-plane onto the upper half of the w-plane.

For our first example, we use the S-C transformation on a polygon consisting of two parallel horizontal lines separated by a distance a as shown in Fig. 5.15. In Section 4.14, we showed using the Riemann sphere that parallel lines can be viewed as meeting at infinity. Thus, we can imagine points A_∞ and B_∞ meet at infinity to close the left side of the polygon and points C_∞ and D_∞ meet at infinity to close the right side of the polygon. Then we have a polygon with two vertices. As shown in Fig. 5.15, let the real axis of z-plane coincide with the bottom side of the polygon and the line $z = ja$ coincide with the top side of the polygon. Vertices at infinity typically occupy two points in one of the planes. When this occurs, it is useful to label the vertex with two letters as shown in Fig. 5.15. For this polygon, each vertex has an external angle of π. Since there are only two vertices, there are only two b_i. We are free to choose the location of up to three b_i. We choose the vertex AB as $b_1 = 0$ and the vertex CD as $b_2 = \infty$. We also choose $z = ja$ when $w = -u_1$ to completely define the problem. This choice of b_i and $z[-u_1] = ja$ results in the top side of the polygon from D_∞ to the y-axis mapping to the negative real axis with $-\infty < u < -u_1$, the top side of the polygon from the y-axis to A_∞ mapping to the negative real axis with $-u_1 < u < 0$, the bottom side of the polygon from B_∞ to 0 mapping to the positive real axis with $0 < u < u_1$, and the bottom side of the polygon from 0 to C_∞ mapping to the positive real axis with $u_1 < u < \infty$ in the w-plane. The S-C integral (5.50) becomes

$$z = A_1 \int (1/w)\, dw + B_1 = A_1 \ln[w] + B_1 \qquad (5.98)$$

Since we have parallel lines that meet at infinity with an external angle of π and have their vertex mapped to infinity, we can use boundary condition (5.74) with $a_0 = a$ to obtain

$$A_1 = a/\pi \qquad (5.99)$$

Using (5.98) with $z[-u_1] = ja$ and A_1 obtained from (5.99) allows determining

$$B_1 = -(a/\pi) \ln[u_1] \qquad (5.100)$$

We now have the final transformation between the z and w planes of

$$z[w] = (a/\pi) \ln[w/u_1], \quad w[z] = u_1 \exp[\pi z/a] \qquad (5.101)$$

We are free to choose the value of u_1. Often $u_1 = 1$ is chosen to simplify the transformation equations. We can also define the w_1-plane as $w_1 = w/u_1$ resulting in

$$z[w_1] = (a/\pi) \ln[w_1], \quad w_1[z] = \exp[\pi z/a] \qquad (5.102)$$

Both are equivalent but the w_1-plane approach allows dimensional analysis to be used as a check of the transformation equations. Therefore, we use the w_1-plane approach.

To demonstrate how the relationship between z and w_1 is used to obtain the complex potential in one plane from the complex potential in a different plane, we let the polygon in the z-plane represent the parallel plate capacitor that was solved in Section 2.21. The complex potential for the top plate at potential V_1 and the bottom plate at potential V_0 was given by (2.152) and repeated here

$$\Phi[z] = -j(V_1 - V_0)(z/a) + V_0 \qquad (5.103)$$

The transformation (5.102) transforms the parallel plate capacitor into a capacitor where one plate occupies the negative real axis and the other plate occupies the positive real axis in the w_1-plane. The complex potential in the w_1-plane is obtained by replacing z in (5.103) with $z[w_1]$ from (5.102)

$$\Phi[w_1] = -j(V_1 - V_0)(\ln[w_1]/\pi) + V_0 \qquad (5.104)$$

This same problem of two plates occupying the real axis was solved in Section 3.2 using the complex potential of a line charge located at the origin. The complex potential for the upper half plane was given by (3.12). Comparing (5.104) with (3.12) we see that they are identical. The field map is given in Fig. 3.1 with the equipotential curves represented by the dashed lines and the lines of force curves represented by the solid lines.

Another example that uses the same polygon, is a line charge q_0 located at z_0 between two grounded plates. In the w_1-plane, the problem transforms into a line charge q_0 located at w_0 above a grounded plane that lies along the real axis. The complex potential in the w_1-plane is obtained using the methods of images as

$$\Phi[w_1] = \frac{q_0}{2\pi\epsilon} \ln\left[\frac{w_1 - w_0^*}{w_1 - w_0}\right] \qquad (5.105)$$

To transform the complex potential to the z-plane, we replace w_1 with $w_1[z]$ and w_0 with $w_1[z_0]$ from (5.101) to obtain

$$\begin{aligned}
\Phi[z] &= \frac{q_0}{2\pi\epsilon} \ln\left[\frac{e^{\pi z/a} - e^{\pi z_0^*/a}}{e^{\pi z/a} - e^{\pi z_0/a}}\right] \\
&= \frac{q_0}{2\pi\epsilon} \ln\left[\frac{\sinh\left[\frac{1}{2}\pi(z - z_0^*)/a\right]}{\sinh\left[\frac{1}{2}\pi(z - z_0)/a\right]}\right] + C_0
\end{aligned} \qquad (5.106)$$

where $C_0 = -j\pi\text{Im}[z_0]/a$. The field map for the z-plane and w_1-plane are shown in Fig. 5.16. The same problem (except the grounded planes

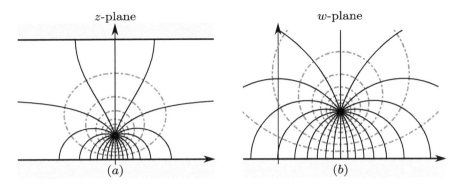

FIGURE 5.16 (a) Field map for a line charge located between two grounded planes. The distance between grounded planes is a and the line charge is located at $z_0 = 0.3a$. (b) Field map for a line charge located above a grounded plane. $\Delta V = \Delta \Psi = 0.05 q_0 / \epsilon$.

were vertical instead of horizontal) was solved in Section 3.14 using the method of images. Replacing z with jz and z_0 with jb in (5.106) and then simplifying shows that (5.106) and (3.127) differ only by a constant.

5.12 POLYGONS WITH ONE FINITE VERTEX AND ONE VERTEX AT INFINITY

In this section we investigate polygons with one finite vertex and one vertex at infinity in the z-plane. The interior angle, $\pi\alpha$, at the finite vertex is defined in terms of the exterior angle as

$$\pi\alpha = \pi - \phi \tag{5.107}$$

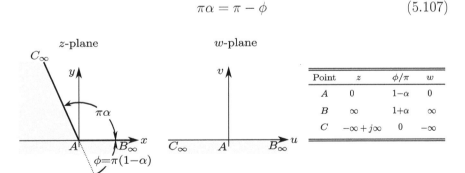

FIGURE 5.17 S-C transformation of the region formed by the intersection of two plane with angle $\pi\alpha$ between them onto the upper half of the w-plane.

We have already encountered a polygon with a single finite vertex back in Section 3.2 where we had two semi-infinite plates that met at the origin with an angle $\pi\alpha$ between them. Although we typically would not associate the region between two lines that intersect with a polygon, we can imagine that the end points of each line that go to infinity meet at infinity. To simplify the math, we define the finite vertex as the origin of the z-plane as shown in Fig. 5.17. No loss in generality occurs with this choice of vertex. The other vertex of the polygon is located at infinity. We have two b_i and we let $b_1 = 0$ correspond to vertex A and $b_2 = \infty$ correspond to vertex BC. The S-C integral (5.50) becomes

$$z = A_1 \int w^{\alpha-1} dw + B_1 = A_1 \left(1/\alpha\right) w^\alpha + B_1 \tag{5.108}$$

The boundary condition $w = 0$ when $z = 0$ requires $B_1 = 0$. To determine A_1, we need to specify one more corresponding point between the z-plane and w-plane. If we let $z = a$ correspond to $w = u_1$, then

$$A_1 = a\alpha u_1^{-\alpha} \tag{5.109}$$

Defining $w_1 = w/u_1$ gives the final transformation between the z and w_1 as

$$z[w_1] = aw_1^\alpha, \quad w_1[z] = (z/a)^{1/\alpha} \tag{5.110}$$

The polygon of Fig. 5.17 can represent a semi-infinite conducting plate with potential V_1 on side CA and a semi-infinite conducting plate with a potential V_0 on side AB. The complex potential in the z-plane is obtained by replacing w_1 in (5.104) with $w_1[z]$ in (5.110) giving

$$\Phi[z] = -j\left(V_1 - V_0\right) \ln[z/a]/(\pi\alpha) + V_0 \tag{5.111}$$

This same problem of two semi-infinite plates with an angle $\pi\alpha$ between them was solved in Section 3.2 using the complex potential of a line charge located at the origin. The complex potential was given by (3.11). Comparing (5.111) with (3.11) we see that they are identical if we let $a = 1$. The field map is given in Fig. 3.3.

Another example that uses the same polygon, is a line charge q_0 located at z_0 inside the polygon. In the w_1-plane, the problem transforms into a line charge q_0 located at w_0 above a grounded plane that lies along the real axis. The complex potential in the w_1-plane is given by (5.105). To obtain the complex potential in the z-plane, we replace w_1 with $w_1[z]$ and w_0 with $w_1[z_0]$ from (5.110) to obtain

$$\Phi[z] = \frac{q_0}{2\pi\epsilon} \ln \left[\frac{z^{1/\alpha} - \left(z_0^{1/\alpha}\right)^*}{z^{1/\alpha} - z_0^{1/\alpha}}\right] \tag{5.112}$$

In Section 3.9, a similar problem was solved using the method of images except the angle between the plates was restricted to π/m where m was an integer greater than one. In that case, the complex potential was also a finite sum of terms dependent on the value of m. With conformal mapping, there is no longer a restriction on angle and the complex potential is a single term. The field maps for $\alpha = 4/3$ (which cannot be obtained by the methods of Section 3.9) and $\alpha = 1/3$ are shown in Fig. 5.18.

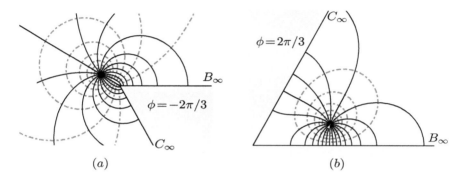

$$\phi = 2\pi/3$$

$$\phi = -2\pi/3$$

(a)   (b)

FIGURE 5.18 Field map of a line charge and two semi-infinite grounded plates that meet at the origin with an external angle of (a) $\phi = -2\pi/3$ with $z_0 = (0.5j)^{5/3}$ and (b) $\phi = 2\pi/3$ with $z_0 = 0.3e^{j\pi/12}$. $\Delta V = \Delta \Psi = 0.05 q_0/\epsilon$.

We can also obtain the complex potential due to a line charge q_0 at z_0 and a grounded semi-infinite plate by letting $\alpha = 2$ which folds the semi-infinite line CA back onto the positive real axis. The transformation between the z- and w_1-planes is

$$z[w_1] = aw_1^2, \quad w_1[z] = \sqrt{z/a} \tag{5.113}$$

and the complex potential is

$$\Phi[z] = \frac{q_0}{2\pi\epsilon} \ln \left[\frac{\sqrt{z} - \sqrt{z_0^*}}{\sqrt{z} - \sqrt{z_0}} \right] \tag{5.114}$$

The field map for a charge and a grounded semi-infinite plate is shown in Fig. 5.19.

We can convert the semi-infinite plate in the z-plane into a finite plate in the z_1-plane using complex inversion. Let the radius and center of the circle of inversion be $z_c = -L$ and $r_I = L$ where L is the length of the finite plate. We also perform a translation of L so that the left edge of the finite plate is at the origin in the z_1-plane. The transformation is

$$z_1[z] = \frac{L^2}{z + L}, \quad z[z_1] = \frac{L^2}{z_1} - L \tag{5.115}$$

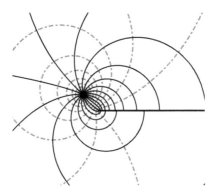

FIGURE 5.19 Field map of a line charge and a grounded semi-infinite plate. The plate lies on the positive real axis with the left edge at the origin. The line charge is located at $z_0 = -0.25L_1 + j0.25L_1$ where L_1 is the length of the finite plate in Fig. 5.20. $\Delta V = \Delta \Psi = 0.05q_0/\epsilon$.

The complex potential due to a line charge q_0 at $z_{1,0}$ and a finite grounded plate is obtained by replacing z with $z[z_1]$ and z_0 with $z[z_{1,0}]$ from (5.115) in (5.114) to obtain

$$\Phi[z_1] = \frac{q_0}{2\pi\epsilon} \ln \left[\frac{\sqrt{L - z_1} - \left(\sqrt{z_1 (L - z_{1,0})/z_{1,0}}\right)^*}{\sqrt{L - z_1} - \sqrt{z_1 (L - z_{1,0})/z_{1,0}}} \right] \qquad (5.116)$$

The field map for two different sized finite plates is shown in Fig. 5.20. In both cases, the line charge is located in the same location relative to the left edge of the plate as the semi-infinite plate. Let the plate length in Fig. 5.20 (a) be L_1. Then the charge is located at $z_0 = (1 + j)/L_1$ and the plate length in Fig. 5.20 (b) is $5L_1$. A comparison of the field lines of Fig. 5.20 (b) and Fig. 5.19 seem to indicate that semi-infinite plates may be good approximations for finite plates in certain cases. We quantify this statement in Chapter 6.

5.13 POLYGONS WITH ONE FINITE VERTEX AND TWO VERTICES AT INFINITY

5.13.1 General S-C Integral

A general polygon with one finite vertex and two vertices at infinity along with its mapping to the w-plane are shown in Fig. 5.21. The solution to the S-C integral for the general polygon involves non-elementary functions. Before solving the S-C integral for the general polygon, we first

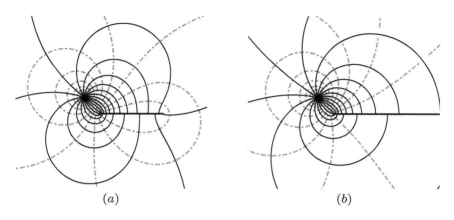

FIGURE 5.20 Field map of a line charge and a grounded plate of length (a) L_1 and (b) $5L_1$. The plate lies on the positive real axis with the left edge at the origin for both field maps. The line charge is located at $z_0 = -0.25L_1 + j0.25L_1$ in both cases and $\Delta V = \Delta \Psi = 0.05 q_0/\epsilon$.

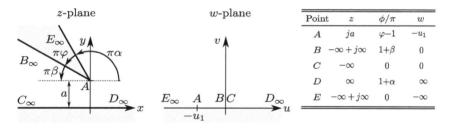

Point	z	ϕ/π	w
A	ja	$\varphi - 1$	$-u_1$
B	$-\infty + j\infty$	$1+\beta$	0
C	$-\infty$	0	0
D	∞	$1+\alpha$	∞
E	$-\infty + j\infty$	0	$-\infty$

FIGURE 5.21 S-C transformation of an arbitrary polygon with one finite vertex and two vertices at infinity in the z-plane onto the upper half of the w-plane.

determine and analyze the S-C mappings for the two polygons shown in Figs. 5.22 and 5.23 that can be expressed in terms of elementary functions. The mapping used for the general polygon is also applied to the two other polygons. This allows all three polygons to share the same S-C integral determined by the general polygon. To make A_1 a real constant, the portion CD in the z-plane is aligned with the real axis and corresponds to the portion of real axis in the w-plane that goes to ∞. The S-C transformation for the general polygon is

$$z = A_1 \int \frac{(w + u_1)^{1-\varphi}}{w^{1+\beta}} dw + B_1 \tag{5.117}$$

where φ and β have the restrictions

$$0 \le \varphi < 1, \quad 0 \le \beta \le 1 \tag{5.118}$$

5.13.2 Case 1: $\varphi = \beta = 0$

For the polygon shown in Fig. 5.22, $\varphi = \beta = 0$ and (5.117) reduces to

$$z = A_1 \int \frac{(w + u_1)}{w} dw + B_1 = A_1 (w + u_1 \ln[w]) + B_1 \qquad (5.119)$$

Parallel boundary condition (5.68) at vertex BC gives

$$A_1 = a/ (\pi u_1) \qquad (5.120)$$

The boundary condition $z = ja$ when $w = -u_1$ gives

$$B_1 = \frac{a}{\pi} (1 - \ln[u_1]) \qquad (5.121)$$

The final transformation is

$$z = \frac{a}{\pi} (1 + w_1 + \ln[w_1]) \qquad (5.122)$$

where $w_1 = w/u_1$. If we let the real axis in the z-plane be a grounded conductor and the semi-infinite line EAB be a conductor with a potential V_1, then we have a capacitor that can be used to model the fringing capacitance at the end of a capacitor plate. The complex potential in the w_1-plane is given by (5.104) with $V_0 = 0$. Since (5.122) cannot be inverted to obtain an expression for $w_1[z]$, we cannot obtain an expression for $\Phi[z]$. We can obtain an expression for $z[\Phi]$ by replacing w_1 with $w_1[\Phi]$ obtained from (5.122) as

$$w_1[\Phi] = \exp[j\pi\Phi/V_1] \qquad (5.123)$$

Substituting (5.123) for w_1 in (5.122) gives

$$z[\Phi] = \frac{a}{\pi} (1 + \exp[j\pi\Phi/V_1] + j (\pi\Phi/V_1)) \qquad (5.124)$$

Although we can obtain the field maps by numerically solving $z[\Phi]$ for constant V and constant Ψ, it is often easier to determine the lines of force and equipotential contours in the w_1-plane and then transform the contours from the w_1-plane to the z-plane using (5.122). The field map obtained using this method is shown in Fig. 5.22 (b).

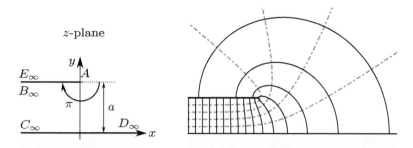

z-plane

FIGURE 5.22 Field map of a semi-infinite conductor and a parallel infinite conductor at different potentials. This configuration can be used to model the fringing fields at the edge of a parallel plate capacitor with thin contacts. $\Delta V = \Delta \Psi = 0.2V_1$.

5.13.3 Case 2: $\varphi = 1/2$, $\beta = 0$

For the polygon shown in Fig. 5.23, $\varphi = 1/2$, $\beta = 0$ and (5.117) reduces to

$$
\begin{aligned}
z &= A_1 \int \frac{\sqrt{w + u_1}}{w} dw + B_1 \\
&= A_1 2\sqrt{u_1} \left(\sqrt{w_1 + 1} - \tanh^{-1}\left[\sqrt{w_1 + 1}\right] \right) + B_1
\end{aligned}
\tag{5.125}
$$

where $w_1 = w/u_1$. Parallel boundary condition (5.68) at vertex BC gives

$$
A_1 = a/\left(\pi\sqrt{u_1}\right)
\tag{5.126}
$$

The boundary condition $z = ja$ when $w_1 = -1$ gives

$$
B_1 = ja
\tag{5.127}
$$

The final transformation is

$$
z = \frac{2a}{\pi}\left(\sqrt{w_1 + 1} - \tanh^{-1}\left[\sqrt{w_1 + 1}\right] \right) + ja
\tag{5.128}
$$

If we let the real axis in the z-plane be a grounded conductor and the line EAB be a conductor with a potential V_1, then we have a geometry that models the field lines of a metal corner near a grounded plane. Since an expression for $z[w]$ cannot be obtained, the field map in the z-plane is generated by first determining the field map in the w_1-plane and then (5.128) is used to transform the w_1 points into z points. The field map for metal corner near a grounded plane is shown in Fig. 5.23 (b).

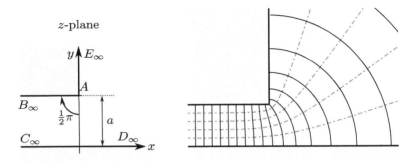

FIGURE 5.23 Field map for a conductor with a 90 degree bend above an infinite conductor at different potentials. A crowding of the lines of force near the corner of the conductor indicates a high density of surface charge in this region resulting in a high electric field. $\Delta V = \Delta \Psi = 0.2V_1$.

5.13.4 General Solution

Now we tackle the general polygon shown in Fig. 5.21 (a). From Appendix F, the identity

$$\int \frac{(w-b)^\gamma}{(w-a)^{1+\beta}} dw = \frac{(w-a)^{-\beta}}{(-\beta)(a-b)^{-\gamma}} {}_2F_1\left[-\gamma, -\beta, 1-\beta, \frac{w-a}{b-a}\right] \quad (5.129)$$

allows integrating (5.117) as

$$z = A_1 \frac{(w_1+1)^{2-\varphi}e^{-j\pi\beta}}{(\varphi-2)u_1^{\varphi+\beta-1}} {}_2F_1[1+\beta, 2-\varphi, 3-\varphi, w_1+1] + B_1 \quad (5.130)$$

where $w_1 = w/u_1$ and ${}_2F_1[\,]$ is Gauss's hypergeometric function. When $z = ja$, $w_1 = -1$ requiring $B_1 = ja$. Let δ be real, positive and $\delta \ll 1$. From Fig. 5.21, $z[\delta]$ is real. The series expansion of (5.130) near zero can be written in terms of δ as

$$z[\delta] = A_1 \left(\frac{u_1^{1-\varphi}\delta^{-\beta}}{-\beta} + \frac{\pi u_1^{1-\varphi-\beta} \cot[\pi\beta]\,\Gamma[2-\varphi]}{\Gamma[1+\beta]\,\Gamma[2-\varphi-\beta]} \right)$$
$$+ j \left(a - \frac{A_1 \pi u_1^{1-\varphi-\beta}\Gamma[2-\varphi]}{\Gamma[1+\beta]\,\Gamma[2-\varphi-\beta]} \right) \quad (5.131)$$

Since A_1 and $z[\delta]$ are real, the imaginary part of (5.131) must be zero requiring

$$A_1 = \frac{a\Gamma[1+\beta]\,\Gamma[2-\varphi-\beta]}{\pi u_1^{1-\varphi-\beta}\Gamma[2-\varphi]} \quad (5.132)$$

The final transformation is

$$\frac{z[w_1]}{a} = j + \left(\frac{\Gamma[1+\beta]\,\Gamma[2-\varphi-\beta]}{\Gamma[3-\varphi]} \left(\frac{(w_1+1)^{2-\varphi}}{-\pi e^{j\pi\beta}} \right) \right.$$
$$\left. \times\,_2F_1[1+\beta, 2-\varphi, 3-\varphi, w_1+1] \right)$$

(5.133)

The S-C transformation for other polygons that have one finite vertex and two infinite vertices can be found in the conformal mapping dictionary located at the book's website.

5.14 POLYGONS WITH TWO FINITE VERTICES AND ONE VERTEX AT INFINITY

5.14.1 General S-C Integral

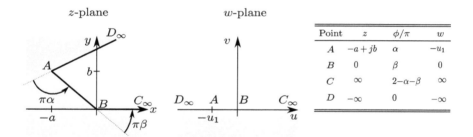

FIGURE 5.24 S-C transformation of an arbitrary polygon with two finite vertices and one vertex at infinity in the z-plane onto the upper half of the w-plane.

A general three vertices polygon with one vertex at infinity along with its mapping to the w-plane are shown in Fig. 5.24. The solution to the S-C integral for the general polygon involves non-elementary functions. Before solving the S-C integral for the general polygon, we first determine and analyze the S-C mappings for the three polygons in Fig. 5.25 and Fig. 5.26 that can be expressed in terms of elementary functions. The mapping used for the general polygon is also applied to the three other polygons. This allows all four polygons to share the same S-C integral determined by the general polygon. To make A_1 a real constant, the portion BC in the z-plane is aligned with the real axis and made to correspond to the portion of real axis in the w-plane that goes to ∞. The S-C transformation for the general polygon is

$$z = A_1 \int \frac{(w+u_1)^{-\alpha}}{w^\beta}\,dw + B_1$$

(5.134)

5.14.2 Case 1: $\alpha = \beta = 1/2$

For the polygon shown in Fig. 5.25 (a) known as a degenerate triangle, $\alpha = \beta = 1/2$ and (5.134) reduces to

$$z = A_1 \int \frac{1}{\sqrt{w}\sqrt{w + u_1}}\,dw + B_1 = 2A_1\sinh^{-1}[\sqrt{w_1}] + B_1 \qquad (5.135)$$

where $w_1 = w/u_1$. Parallel boundary condition (5.74) at vertex CD gives

$$A_1 = b/\pi \qquad (5.136)$$

The boundary condition $z = 0$ when $w = 0$ gives $B_1 = 0$. The final transformation is

$$z = (2b/\pi)\sinh^{-1}[\sqrt{w_1}] \qquad (5.137)$$

which can be inverted to give

$$w_1[z] = \sinh^2[\pi z/(2b)] \qquad (5.138)$$

If we let the real axis in the w_1-plane be a grounded conductor, then the potential due to a point charge at w_0 above the grounded plane is given by (5.105). To transform the complex potential to the z-plane, we replace w_1 with $w_1[z]$ and w_0 with $w_1[z_0]$ from (5.138) to obtain

$$\Phi[z] = \frac{q_0}{2\pi\epsilon}\ln\left[\frac{\sinh^2[\pi z/(2b)] - \sinh^2[\pi z_0^*/(2b)]}{\sinh^2[\pi z/(2b)] - \sinh^2[\pi z_0/(2b)]}\right] \qquad (5.139)$$

The field map of (5.139) is shown in Fig. 5.25 (a).

5.14.3 Case 2: $\alpha = -1/2,\ \beta = 1/2$

For the polygon shown in Fig. 5.25 (b) (an infinite horizontal plate with a step height b), $\alpha = -1/2$, $\beta = 1/2$, and (5.134) reduces to

$$\begin{aligned}
z &= A_1 \int \frac{\sqrt{w + u_1}}{\sqrt{w}}\,dw + B_1 \\
&= A_1 u_1\left(\sqrt{w_1}\sqrt{w_1 + 1} + \sinh^{-1}[\sqrt{w_1}]\right) + B_1
\end{aligned} \qquad (5.140)$$

where $w_1 = w/u_1$. Parallel boundary condition (5.92) at vertex CD gives

$$A_1 = 2b/(u_1\pi) \qquad (5.141)$$

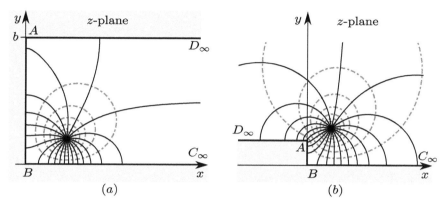

FIGURE 5.25 (a) Field map of a line charge inside a degenerate triangle with the walls grounded. The location of the charge is $z_0 = (b/3) + j\,(b/5)$. (b) Field map of a line charge above a grounded plane with a step in it. The location of the line charge is $z_0 = b + j\,(3b/2)$. For both field maps $\Delta V = \Delta \Psi = 0.05q_0/\epsilon$.

The boundary condition $z = 0$ when $w = 0$ gives $B_1 = 0$. The final transformation is

$$z = (2b/\pi)\left(\sqrt{w_1}\sqrt{w_1 + 1} + \sinh^{-1}[\sqrt{w_1}]\right) \qquad (5.142)$$

If we let the real axis in the w_1-plane be a grounded conductor, then the potential due to a point charge at w_0 above the grounded plane is given by (5.105). Since an expression for $z[w]$ cannot be obtained, the field map in the z-plane is generated by first determining the field map in the w_1-plane and then (5.142) is used to transform the w_1 points into z points. The field map for a line charge above a plate with a step in it is shown in Fig. 5.25 (b).

5.14.4 Case 3: $\beta = -1$

For the polygon shown in Fig. 5.26, α is left unspecified, $\beta = -1$, and (5.134) reduces to

$$\begin{aligned}
z &= A_1 \int \frac{w}{(w + u_1)^\alpha}\,dw + B_1 \\
&= -A_1 u_1^{2-\alpha}\frac{(w_1 + 1)^{1-\alpha}\,(1 - w_1\,(1 - \alpha))}{(2 - \alpha)\,(1 - \alpha)} + B_1
\end{aligned} \qquad (5.143)$$

where $w_1 = w/u_1$. The boundary condition $z[-1] = a$ gives $B_1 = a$. The boundary condition $z[0] = 0$ gives

$$A_1 = a\,(2 - \alpha)\,(1 - \alpha)\,u_1^{\alpha-2} \qquad (5.144)$$

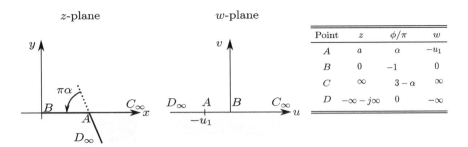

FIGURE 5.26 S-C transformation for Case 3: $\beta = -1$.

The final transformation is

$$z[w_1] = a - a(w_1 + 1)^{1-\alpha}(1 - (1 - \alpha)w_1) \qquad (5.145)$$

5.14.5 General Solution

Now we tackle the general polygon shown in Fig. 5.24. With the identity given in (5.129), we can integrate (5.134) in terms of $_2F_1[\ ]$ as

$$z = A_1 \frac{w_1^{1-\beta} u_1^{1-\alpha-\beta}}{1 - \beta} \, _2F_1[\alpha, 1 - \beta, 2 - \beta, -w_1] + B_1 \qquad (5.146)$$

where $w_1 = w/u_1$. From Fig. 5.24, the origin of the z-plane corresponds to the origin of the w-plane which results in $B_1 = 0$. Using the identity

$$_2F_1[a, b, c, 1] = \frac{\Gamma[c]\,\Gamma[c - a - b]}{\Gamma[c - a]\,\Gamma[c - b]} \qquad (5.147)$$

and boundary condition $w_1 = -1$ when $z = -a + jb$ gives

$$-a + jb = \frac{-A_1 e^{-j\pi\beta} u_1^{1-\alpha-\beta}}{1 - \beta} \frac{\Gamma[1 - \alpha]\,\Gamma[2 - \beta]}{\Gamma[2 - \alpha - \beta]} \qquad (5.148)$$

Since A_1 is real, we have

$$a = \frac{A_1 \cos[\pi\beta]\, u_1^{1-\alpha-\beta}}{1 - \beta} \frac{\Gamma[1 - \alpha]\,\Gamma[2 - \beta]}{\Gamma[2 - \alpha - \beta]}$$
$$b = \frac{A_1 \sin[\pi\beta]\, u_1^{1-\alpha-\beta}}{1 - \beta} \frac{\Gamma[1 - \alpha]\,\Gamma[2 - \beta]}{\Gamma[2 - \alpha - \beta]} \qquad (5.149)$$

Using the second equation in (5.149) to solve for A_1 gives

$$A_1 = \frac{b\Gamma[\beta]\,\Gamma[2 - \alpha - \beta]}{\Gamma[1 - \alpha]\,\pi u_1^{1-\alpha-\beta}} \qquad (5.150)$$

The final transformation for the general polygon is

$$z[w_1] = b \left(\frac{\Gamma[\beta] \, \Gamma[2 - \alpha - \beta] w_1^{1-\beta}}{\Gamma[1 - \alpha] \, \pi \, (1 - \beta)} \right) {}_2F_1[\alpha, 1 - \beta, 2 - \beta, -w_1] \quad (5.151)$$

The S-C transformations for other polygons that have two finite vertices and one vertex at infinity can be found in the conformal mapping dictionary located at the book's website.

5.15 POLYGONS WITH TWO FINITE VERTICES AND TWO VERTICES AT INFINITY

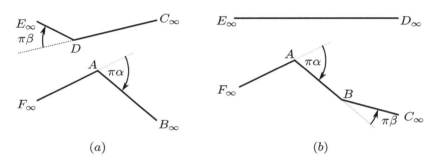

(a)   (b)

FIGURE 5.27 (a) An arbitrary polygon defined by two lines that meet at infinity with each line having a finite vertex. (b) An arbitrary polygon defined by two lines that meet at infinity with one line having two finite vertices and the other line having no finite vertices.

The final polygon with vertices at infinity that we solve in this chapter is a polygon with two finite vertices and two vertices at infinity. In general, there are two types of polygon configurations that fit this description. The first type consists of two separate lines each containing a finite vertex as shown in Fig. 5.27 (a). The second type consists of two separate lines as well, but with one line having no finite vertex and the other line has two finite vertices as shown in Fig. 5.27 (b). To minimize the complexity of the mathematics, we solve a specific case for each type shown in Fig. 5.27.

5.15.1 Case 1

The first case we solve consists of two semi-infinite plates that are coincident with the real axis. The first semi-infinite plate occupies the portion of the real axis from $x = -\infty$ to $x = -a$ and the second semi-infinite plate occupies the portion of the real axis from $x = a$ to $x = \infty$ (see Fig. 5.28). To envision this geometry as a closed geometry, we let points E_∞

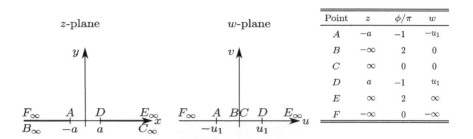

Point	z	ϕ/π	w
A	$-a$	-1	$-u_1$
B	$-\infty$	2	0
C	∞	0	0
D	a	-1	u_1
E	∞	2	∞
F	$-\infty$	0	$-\infty$

FIGURE 5.28 S-C transformation of a plate with a gap in the z-plane onto the upper half of the w-plane.

and F_∞ be connected by an arc at infinity and points B_∞ and C_∞ also be connected by an arc at infinity in the z-plane. The S-C integral is

$$z = A_1 \int \frac{w^2 - u_1^2}{w^2} dw + B_1 \tag{5.152}$$

Integrating (5.152) gives

$$z[w_1] = A_1 u_1 \left(w_1 + w_1^{-1} \right) + B_1 \tag{5.153}$$

where $w_1 = w/u_1$. The boundary conditions $z[\pm 1] = \pm a$ result in

$$A_1 = a/2u_1, \quad B_1 = 0 \tag{5.154}$$

The final transform is

$$z[w] = \frac{a}{2} \left(w_1 + w_1^{-1} \right), \quad w_1[z] = \frac{1}{a} \left(z + \sqrt{z^2 - a^2} \right) \tag{5.155}$$

The complex potential for a line charge q_0 located at z_0 and two grounded semi-infinite plates is obtained by replacing w with $w[z]$ and w_0 with $w[z_0]$ from (5.155) to obtain

$$\Phi[z] = \frac{q_0}{2\pi\epsilon} \ln \left[\frac{z + \sqrt{z^2 - a^2} - \left(z_0 + \sqrt{z_0^2 - a^2} \right)^*}{z + \sqrt{z^2 - a^2} - \left(z_0 + \sqrt{z_0^2 - a^2} \right)} \right] \tag{5.156}$$

A field map of (5.156) is shown in Fig. 5.29.

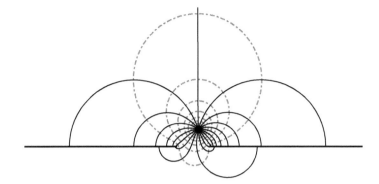

FIGURE 5.29 Field map of a line charge above a grounded plane with a gap. The plate lies on the real axis and is symmetric with respect to the imaginary axis. The line charge is located at $z_0 = 0.3a + ja$ where $2a$ is the length of the gap and $\Delta V = \Delta \Psi = 0.05 q_0 / \epsilon$.

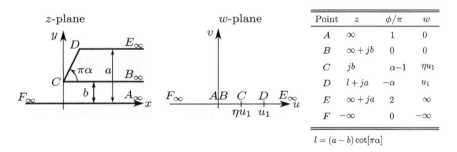

Point	z	ϕ/π	w
A	∞	1	0
B	$\infty + jb$	0	0
C	jb	$\alpha - 1$	ηu_1
D	$l + ja$	$-\alpha$	u_1
E	$\infty + ja$	2	∞
F	$-\infty$	0	$-\infty$

$l = (a - b) \cot[\pi \alpha]$

FIGURE 5.30 S-C transformation of semi-infinite plate of thickness $a - b$ above an infinite plate that coincides with the real axis. The edge of the semi-infinite plate has an angle $\pi \alpha$ as shown in the diagram.

5.15.2 Case 2

The second geometry to solve is shown in Fig. 5.30. The S-C integral is

$$z = A_1 \int \frac{(w - \eta u_1)^{1-\alpha}(w - u_1)^{\alpha}}{w} dw + B_1 \qquad (5.157)$$

with $0 < \eta < 1$. Using the substitution $w = u_1 + u^{-1}$, (5.157) becomes

$$z = -\frac{A_1}{u_1^{\alpha}(1 - \eta)^{\alpha - 1}} \int \frac{\left(u + (1 - \eta)^{-1} u_1^{-1}\right)^{1-\alpha}}{\left(u + u_1^{-1}\right) u^2} du + B_1 \qquad (5.158)$$

From appendix F, the identity

$$\int \frac{(w-b)^{\varphi-1}}{w^2\,(w-a)}dw = \frac{(w-b)^{\varphi}}{a^2\varphi}\frac{(a\varphi-a-b)}{b^2}\,_2F_1\!\left[1,\varphi,\varphi+1,\frac{b-w}{b}\right]$$
$$-\frac{(w-b)^{\varphi}}{a^2(a-b)}\,_2F_1\!\left[1,\varphi,\varphi+1,\frac{w-b}{a-b}\right]-\frac{(w-b)^{\varphi}}{abw} \tag{5.159}$$

allows integrating (5.158) as

$$z = B_1 + A_1 u_1 \frac{(w_1-\eta)^{2-\alpha}}{(w_1-1)^{1-\alpha}}$$
$$+\frac{A_1 u_1}{(2-\alpha)\,\eta}\left(\frac{w_1-\eta}{w_1-1}\right)^{2-\alpha}\,_2F_1\!\left[1,2-\alpha,3-\alpha,\frac{w_1-\eta}{(w_1-1)\,\eta}\right] \tag{5.160}$$
$$+A_1 u_1\left(\frac{w_1-\eta}{w_1-1}\right)^{2-\alpha}\frac{(\eta\alpha-\eta-\alpha)}{2-\alpha}\,_2F_1\!\left[1,2-\alpha,3-\alpha,\frac{w_1-\eta}{w_1-1}\right]$$

where $w_1 = w/u_1$. Using parallel lines boundary condition (5.68) for the vertex at AB gives

$$A_1 = \frac{b}{\eta^{1-\alpha}u_1\pi} \tag{5.161}$$

Using parallel lines boundary condition (5.92) for the vertex EF with $a_0 = -a$ (since the line going to infinity in the $-x$ direction is a below the line going to infinity in the $+x$ direction) gives

$$A_1 = \frac{a}{(\eta+\alpha\,(1-\eta))\,u_1\pi} \tag{5.162}$$

Combining (5.161) and (5.162) gives

$$\alpha\left(1-\frac{1}{\eta}\right) = 1 - \frac{1}{\eta^\alpha}\frac{a}{b} \tag{5.163}$$

For a given a/b and α, (5.163) can be can be solved numerically for η which then gives the value of A_1. Evaluating (5.160) with $w_1 = \eta$ gives

$$B_1 = jb \tag{5.164}$$

The final transformation is

$$\frac{z[w_1]}{b} = \frac{\eta^{\alpha-1}}{\pi}\frac{(w_1-\eta)^{2-\alpha}}{(w_1-1)^{1-\alpha}} + j$$
$$-\frac{a}{\pi b}\frac{1}{(2-\alpha)}\left(\frac{w_1-\eta}{w_1-1}\right)^{2-\alpha}\,_2F_1\!\left[1,2-\alpha,3-\alpha,\frac{w_1-\eta}{w_1-1}\right] \tag{5.165}$$
$$+\frac{\eta^{\alpha-2}}{\pi(2-\alpha)}\left(\frac{w_1-\eta}{w_1-1}\right)^{2-\alpha}\,_2F_1\!\left[1,2-\alpha,3-\alpha,\frac{w_1-\eta}{(w_1-1)\,\eta}\right]$$

This transformation allows visualizing the field lines at the edge of a parallel plate capacitor with the plates having finite thickness by letting the real axis of the z-plane be the bottom plate of the capacitor and the geometry defined by $BCDE$ be one end of the top plate. Let the top plate and ground plane have potentials V_0 and 0, respectively. Using (5.104), the complex potential in the w_1-plane is

$$\Phi[w_1] = j\,(V_0/\pi)\ln[w_1] + V_0 \tag{5.166}$$

A field map of (5.166) is shown in Fig. 5.31 with $\alpha = 1/3$.

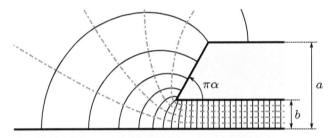

FIGURE 5.31   Field map of a conductor of finite thickness above an infinite conductor that lies along the real axis. The two conductors have a potential difference V_0. This configuration can be used to model the fringing fields of a parallel plate capacitor with contacts of finite thickness. The geometry has $a = 3b$ and $\alpha = 1/3$. $\Delta V = \Delta \Psi = 0.2V_0$.

5.15.3   Case 3

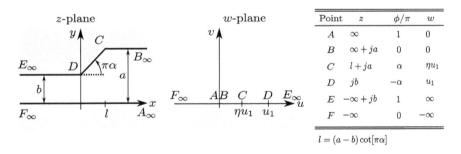

FIGURE 5.32   S-C transformation of two semi-infinite strips of different heights connected by a linear transition in the z-plane onto the upper half of the w-plane.

The final geometry to solve is shown in Fig. 5.32. The S-C integral is

$$z = A_1 \int \frac{1}{w_1}\left(\frac{w_1 - 1}{w_1 - \eta}\right)^{\alpha} dw_1 + B_1 \tag{5.167}$$

where $w_1 = w/u_1$. Using partial fractions, the integrand can be written as

$$\frac{1}{w_1}\left(\frac{w_1-1}{w_1-\eta}\right)^\alpha = \frac{(w_1-1)^{\alpha-1}}{(w_1-\eta)^\alpha} - \frac{(w_1-1)^{\alpha-1}}{w_1(w_1-\eta)^\alpha} \tag{5.168}$$

The first term can be integrated in terms of $_2F_1[\ ]$. Using the substitution $u = 1/w_1$, the integral of the second term becomes

$$-\int \frac{(w_1-1)^{\alpha-1}}{w_1(w_1-\eta)^\alpha}dw_1 = -\frac{1}{\eta^\alpha}\int \frac{(u-1)^{\alpha-1}}{(u-\eta^{-1})^\alpha}du \tag{5.169}$$

which can also be integrated in terms of $_2F_1[\ ]$. Performing the integration and simplifying gives

$$\begin{aligned}z = B_1 &+ \frac{A_1(w_1-\eta)^{1-\alpha}}{(1-\alpha)(\eta-1)^{1-\alpha}}{}_2F_1\left[1-\alpha,1-\alpha,2-\alpha,\frac{w_1-\eta}{1-\eta}\right]\\ &- \frac{A_1(w_1-\eta)^{1-\alpha}w_1^{\alpha-1}}{(1-\alpha)(\eta-1)^{1-\alpha}\eta^\alpha}{}_2F_1\left[1-\alpha,1-\alpha,2-\alpha,\frac{(w_1-\eta)}{w_1(1-\eta)}\right]\end{aligned} \tag{5.170}$$

Using parallel lines boundary condition (5.68) for the vertex at AB with $a_0 = a$ and $\varphi = 0$ gives

$$A_1 = -\frac{a}{\pi}\eta^\alpha \tag{5.171}$$

Using parallel lines boundary condition (5.74) for the vertex at EF with $a_0 = b$ and $\varphi = \pi$ gives

$$A_1 = -b/\pi \tag{5.172}$$

Combining (5.171) and (5.172) allows solving for η as

$$\eta = (b/a)^{1/\alpha} \tag{5.173}$$

Evaluating (5.170) with $w_1 = \eta$ gives

$$B_1 = l + ja \tag{5.174}$$

The transformation between z and w is

$$\begin{aligned}z = l + ja &+ \frac{b\pi^{-1}(w_1-\eta)^{1-\alpha}}{(\alpha-1)(\eta-1)^{1-\alpha}}{}_2F_1\left[1-\alpha,1-\alpha,2-\alpha,\frac{w_1-\eta}{1-\eta}\right]\\ &- \frac{b(w_1-\eta)^{1-\alpha}w_1^{\alpha-1}}{\pi(\alpha-1)(\eta-1)^{1-\alpha}\eta^\alpha}{}_2F_1\left[1-\alpha,1-\alpha,2-\alpha,\frac{(w_1-\eta)}{w_1(1-\eta)}\right]\end{aligned} \tag{5.175}$$

The boundaries in the z-plane can model two parallel plate capacitors of different plate spacing joined by a metal transition in the top plate. The capacitance due to the transition ($\underline{C}_{sw}$) is defined as

$$\underline{C}_{sw} = \underline{Q}_{sw}/\Delta V \qquad (5.176)$$

where $\underline{Q}_{sw}$ is the total charge on the sidewall and ΔV is the potential difference between the bottom and top plates. Let the bottom plate have a potential $-V_1$ and the top plate grounded. The complex potential in the w_1-plane is given by (5.104)

$$\Phi[w_1] = jV_1 \ln[w_1]/\pi \qquad (5.177)$$

The charge on the metal transition is easily obtained in the w_1-plane as

$$\underline{Q}_{sw} = \epsilon \left(\Psi[1] - \Psi[\eta] \right) = \frac{\epsilon V_1}{\pi \alpha} \ln[a/b] \qquad (5.178)$$

The capacitance of the transition is

$$\underline{C}_{sw} = \frac{\epsilon}{\pi \alpha} \ln[a/b] \qquad (5.179)$$

The field map for $\alpha = 1/2$ is shown in Fig. 2.15.

5.16   THE JOUKOWSKI TRANSFORMATION

The transformation of (5.155) is useful for more than just transforming two semi-infinite plates separated by a gap into two semi-infinite plates with no gap. Let w_1 in (5.155) represent a point on a circle centered at the origin of the w_1-plane, $w_1 = r_1 e^{j\theta_1}$ with $r_1 = |w|/u_1$. Solving for x and y gives

$$\frac{x}{(r_1^2 + 1)} = \frac{a}{2r_1} \cos[\theta_1], \quad \frac{y}{(r_1^2 - 1)} = \frac{a}{2r_1} \sin[\theta_1] \qquad (5.180)$$

Squaring both equations in (5.180) and then adding them gives

$$\frac{x^2}{(r_1^2 + 1)^2} + \frac{y^2}{(r_1^2 - 1)^2} = \frac{a^2}{4r_1^2} \qquad (5.181)$$

For $r_1 > 1$, (5.181) is the equation of an ellipse with major axis $2a_1$ and minor axis $2b_1$ of

$$2a_1 = \frac{a \left(r_1^2 + 1 \right)}{r_1}, \quad 2b_1 = \frac{a \left(r_1^2 - 1 \right)}{r_1} \qquad (5.182)$$

Thus, (5.155) transforms circular cylinders in the w_1-plane into elliptical cylinders in the z-plane as shown in Fig. 5.33. Solving for r_1 and a in terms of the major and minor axis lengths of the ellipse gives

$$r_1 = \frac{\sqrt{a_1 + b_1}}{\sqrt{a_1 - b_1}}, \quad a = \sqrt{a_1^2 - b_1^2} \tag{5.183}$$

When $r_1 = 1$, $b_1 = 0$ and the circular cylinder in the w-plane is transformed into a flat plate in the z-plane. This provides another method for obtaining Green's function for a grounded finite plate already obtained in Section 5.12. Note that the region $|w_1| < 1$ is not mapped to the z-plane. This complication prevents mapping circular concentric dielectric cylinders into elliptic concentric dielectric cylinders but has no impact on conductive cylinders.

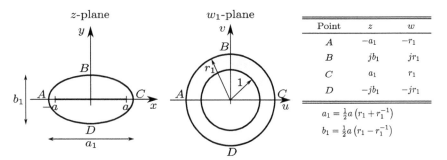

FIGURE 5.33 Conformal transformation of concentric circles in the w-plane onto concentric ellipses in the z-plane. The unit circle in the w-plane is mapped to a flat line between $-a < z < a$ in the z-plane.

From (3.91), the complex potential for a line charge q_0 located at w_0 and a grounded circular cylinder with radius r_c and its center at the origin of the w-plane is

$$\Phi[w, w_0] = \frac{q_0}{2\pi\epsilon} \ln\left[\frac{|w_0|}{r_c}\left(\frac{w - (r_c^2/w_0^*)}{w - w_0}\right)\right] \tag{5.184}$$

To transform (5.184) to the w_1-plane, we replace w with $u_1 w_1$ and r_c with $u_1 r_1$ to obtain

$$\Phi[w_1, w_{1,0}] = \frac{q_0}{2\pi\epsilon} \ln\left[\frac{|w_{1,0}|}{r_1}\left(\frac{w_1 - (r_1^2/w_{1,0}^*)}{w_1 - w_{1,0}}\right)\right] \tag{5.185}$$

The complex potential for a line charge q_0 located at z_0 and a grounded elliptical cylinder with major axis $2a_1$, minor axis $2b_1$, and its center at the origin of the z-plane is obtained by replacing w_1 with $w_1[z]$ and $w_{1,0}$ with $w_1[z_0]$ from (5.155) in (5.185) to give

$$\Phi[z] = \frac{q_0}{2\pi\epsilon} \ln\left[\frac{|f[z_0]|}{(a_1 + b_1)} \frac{(f[z_0])^* f[z] - (a_1 + b_1)^2}{(f[z_0])^* (f[z] - f[z_0])} \right] \qquad (5.186)$$

where

$$f[z] = aw_1[z] = z + \sqrt{z - (a_1^2 - b_1^2)^{1/2}} \sqrt{z + (a_1^2 - b_1^2)^{1/2}} \qquad (5.187)$$

A field map of (5.187) is shown in Fig. 5.34.

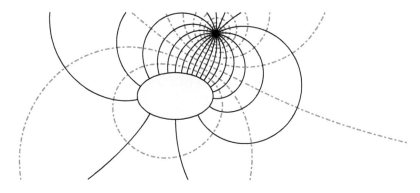

FIGURE 5.34 Field map of a line charge and a grounded ellipse. The ellipse is centered at the origin of the z-plane with its major axis coincident with the real axis. The location of the line charge is $z_0 = (4/3)a + j2a$ and $r_1 = 2$ for the transformed circle. $\Delta V = \Delta \Psi = 0.05 q_0/\epsilon$.

In the field of aerodynamics, the transform

$$z = w_1 + w_1^{-1} \qquad (5.188)$$

is known as the Joukowski transformation. It was originally used to transform circles into shapes that resemble the cross-section of an airfoil. To obtain the airfoil like cross-section, the circle is not centered at the origin of the w_1-plane, goes through $w_1 = 1$, and encloses $w_1 = -1$. Examples of how different shapes are produces by simply changing the location of the center of the circle in the w_1-plane is shown in Fig. 5.35.

5.17   POLYGONS WITH THREE FINITE VERTICES

Our first finite polygon to transform onto the upper half plane is the triangle. With only three vertices, we can specify the location of each

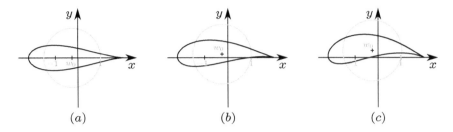

FIGURE 5.35 Evolution of an airfoil by transforming a circle using the Joukowski transformation. The center of the circle is (a) $w_0 = -0.2$, (b) $w_0 = -0.2 + j0.15$, and (c) $w_0 = -0.2 + j0.3$. The radius of each circle is 1.2. The w-plane is overlaid on the z-plane showing the circle, $w = \pm 1$ and w_0.

vertex in the w-plane. For mathematical simplicity, we let one vertex map to infinity. We setup our coordinate system in the z-plane so that one vertex is located at the origin and one side lies along the positive real axis as shown in Fig. 5.36. Let $\pi\alpha$, $\pi\beta$, and $\pi\gamma$ be the external angles of the triangle in the z-plane. Let the remaining two vertices map to $w = 0$ and $w = u_1$. The S-C integral is

$$z = A_1 \int \frac{(w - u_1)^{-\beta}}{w^\alpha} dw + B_1 \tag{5.189}$$

The integration can be performed in terms of $_2F_1[\;]$ as

$$z = A_1 \left(\frac{w_1^{1-\alpha} u_1^{1-\alpha-\beta}}{(1 - \alpha)\, e^{j\pi\beta}} \right) {}_2F_1[\beta, 1 - \alpha, 2 - \alpha, w_1] + B_1 \tag{5.190}$$

where $w_1 = w/u_1$. The boundary condition $z = 0$ when $w_1 = 0$ requires $B_1 = 0$. Requiring $w_1 = 1$ when $z = a$ gives

$$
\begin{aligned}
a &= \frac{A_1 u_1^{1-\alpha-\beta}}{(1 - \alpha)\, e^{j\pi\beta}} {}_2F_1[\beta, 1 - \alpha, 2 - \alpha, 1] \\
&= \frac{A_1 u_1^{1-\alpha-\beta}}{(1 - \alpha)\, e^{j\pi\beta}} \frac{\Gamma[2 - \alpha]\, \Gamma[1 - \beta]}{\Gamma[2 - \alpha - \beta]}
\end{aligned}
\tag{5.191}
$$

Solving for A_1 gives

$$A_1 = a \frac{(1 - \alpha)\, e^{j\pi\beta}}{u_1^{1-\alpha-\beta}} \frac{\Gamma[2 - \alpha - \beta]}{\Gamma[2 - \alpha]\, \Gamma[1 - \beta]} \tag{5.192}$$

The final transformation is

$$z = a \frac{w_1^{1-\alpha} \Gamma[2 - \alpha - \beta]}{\Gamma[2 - \alpha]\, \Gamma[1 - \beta]} {}_2F_1[\beta, 1 - \alpha, 2 - \alpha, w_1] \tag{5.193}$$

z-plane   w-plane

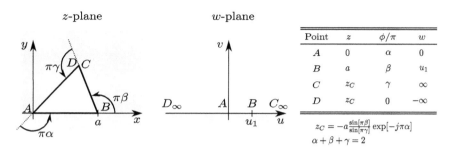

Point	z	ϕ/π	w
A	0	α	0
B	a	β	u_1
C	z_C	γ	∞
D	z_C	0	$-\infty$

$$z_C = -a\frac{\sin[\pi\beta]}{\sin[\pi\gamma]}\exp[-j\pi\alpha]$$
$$\alpha + \beta + \gamma = 2$$

FIGURE 5.36 S-C transformation of a triangle in the z-plane onto the upper half of the w-plane.

The field map of a line charge inside a grounded cylinder with a triangular cross-section is shown in Fig. 5.37 (a). The field map where two sides of the triangle are grounded and the third side has a potential V_1 is shown in Fig. 5.37 (b). In both cases, the values of w_1 that produce constant potential and lines of force are determined and then these w_1 values are converted to z values using (5.193) to produce the field maps in the z-plane.

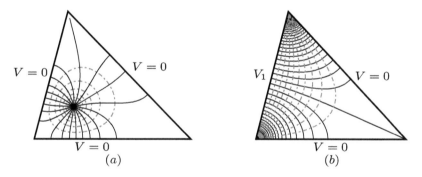

$V = 0$ $V = 0$

$V = 0$
(a)

V_1 $V = 0$

$V = 0$
(b)

FIGURE 5.37 (a) Field map of a line charge inside a grounded cylinder of triangular cross-section. The location of the line charge is $z_0 = 0.25a + j0.2a$ and $\Delta V = \Delta \Psi = 0.05q_0/\epsilon$. (b) Field map of a triangle with one side at potential V_1 and the other two sides grounded. $\Delta V = \Delta \Psi = 0.1V_1$. The bottom left vertex of each triangle coincides with the origin. Both triangles have $\alpha = 7/12$ and $\beta = 3/4$.

5.18 POLYGONS WITH FOUR FINITE VERTICES

A polygon with four finite vertices (a quadrilateral) is the final type of polygon we solve in this chapter. We limit our investigation to a rectangle, quadrilaterals with diagonal symmetry and quadrilaterals with at least on right angle.

5.18.1   General Rectangle Transformation

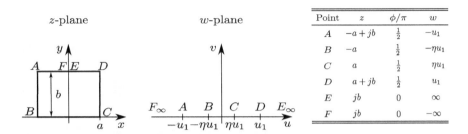

Point	z	ϕ/π	w
A	$-a+jb$	$\frac{1}{2}$	$-u_1$
B	$-a$	$\frac{1}{2}$	$-\eta u_1$
C	a	$\frac{1}{2}$	ηu_1
D	$a+jb$	$\frac{1}{2}$	u_1
E	jb	0	∞
F	jb	0	$-\infty$

FIGURE 5.38   S-C transformation of a rectangle in the z-plane onto the upper half of the w-plane.

Let the rectangle be symmetric about the imaginary axis with the bottom of the rectangle on the real axis (see Fig. 5.38). All the external angles of a rectangle are $\pi/2$. Due to symmetry, we can let the vertices in the z-plane map to $\pm\eta u_1$ and $\pm u_1$ in the w-plane. The S-C integral is

$$z = A_1 \int \frac{dw}{\sqrt{(w^2 - \eta^2 u_1^2)(w^2 - u_1^2)}} + B_1 \qquad (5.194)$$

Using the substitution $u = w/(\eta u_1)$ and $m = \eta^2$, (5.194) becomes

$$z = \frac{A_1}{u_1} \int \frac{du}{\sqrt{(1 - u^2)(1 - mu^2)}} + B_1 \qquad (5.195)$$

The integral of (5.195) is known as an incomplete elliptic integral of the first kind written as $F[\theta, m]$ where $\sin[\theta] = u$ (see appendix C). Elliptic integrals are built-in functions for many mathematical software packages such as Mathematica. Thus, z can be written in terms of $F[\theta, m]$ as

$$z = \frac{A_1}{u_1} F\left[\sin^{-1}[w_1/\eta], \eta^2\right] + B_1 \qquad (5.196)$$

where $w_1 = w/u_1$. Since $z = 0$ when $w_1 = 0$, $B_1 = 0$. The boundary condition $w_1 = \eta$ when $z = a$ gives

$$a = \frac{A_1}{u_1} F\left[\pi/2, \eta^2\right] \qquad (5.197)$$

$F[\pi/2, m]$ is known as the complete elliptic integral of the first kind and is written as $K[m]$. Solving for A_1 in terms of $K[m]$ gives

$$A_1 = u_1 a / K\left[\eta^2\right] \qquad (5.198)$$

Replacing A_1 from (5.198) in (5.196) gives

$$z = \frac{a}{K[\eta^2]} F\left[\sin^{-1}[w_1/\eta], \eta^2\right] \tag{5.199}$$

In order to completely define the mapping, we still need to determine η in terms of known parameters from the rectangle being mapped. The boundary condition $w_1 = 1$ when $z = a + jb$ gives

$$a + jb = \frac{a}{K[\eta^2]} F\left[\sin^{-1}[1/\eta], \eta^2\right] \tag{5.200}$$

From appendix C, we have the identity

$$F\left[\sin^{-1}[1/\sqrt{m}], m\right] = K[m] + jK[1 - m] \tag{5.201}$$

which leads to the relationship

$$\frac{b}{a} = \frac{K[1 - \eta^2]}{K[\eta^2]} \tag{5.202}$$

Thus, for a given rectangle aspect ratio (width to height ratio), η^2 can be determined by solving (5.202) and the mapping is fully determined.

5.18.2 Two Equal Finite Coplanar Plates

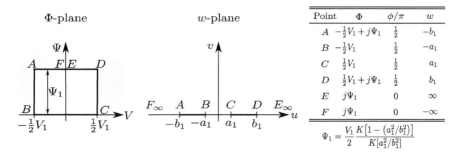

FIGURE 5.39 S-C transformation of the rectangle formed in the complex potential plane by tracing the boundary of two equal coplanar conductors at different potentials in the w-plane.

The mapping of a rectangle onto the upper half-plane is used for determining capacitance between coplanar plates. Let the region on the real axis defined by $-b_1 \le w \le -a_1$ and $a_1 \le w \le b_1$ be conductors with potentials of $\pm \frac{1}{2} V_1$ (see Fig. 5.39). A map of the complex potential along the real axis in the w-plane produces a rectangle in the complex potential plane of width V_1. The complex potential is obtained by setting $a = \frac{1}{2}V_1$, $b_1 = u_1$, $\eta = a_1/b_1$, and $w_1 = w/b_1$ in (5.199)

$$\Phi[w] = \frac{V_1}{2K[a_1^2/b_1^2]} F\left[\sin^{-1}[w/a_1], a_1^2/b_1^2\right] \qquad (5.203)$$

All parameters are known and the complex potential is fully defined. The capacitance between the coplanar plates is

$$\underline{C} = \epsilon \frac{\mathrm{Im}[\Phi[b_1]]}{V_1} = \epsilon \frac{K\left[1 - (a_1^2/b_1^2)\right]}{2K[a_1^2/b_1^2]} \qquad (5.204)$$

This is the capacitance for only the upper half plane. Due to symmetry, the lower half plane contributes the same amount of capacitance. Therefore, the capacitance of the full w-plane is twice that of (5.204).

5.18.3 Coplanar Center Conductor Between Grounds

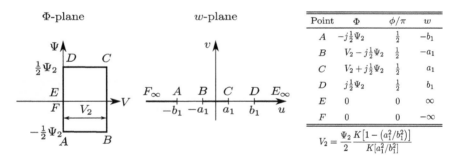

Point	Φ	ϕ/π	w
A	$-j\frac{1}{2}\Psi_2$	$\frac{1}{2}$	$-b_1$
B	$V_2 - j\frac{1}{2}\Psi_2$	$\frac{1}{2}$	$-a_1$
C	$V_2 + j\frac{1}{2}\Psi_2$	$\frac{1}{2}$	a_1
D	$j\frac{1}{2}\Psi_2$	$\frac{1}{2}$	b_1
E	0	0	∞
F	0	0	$-\infty$

$$V_2 = \frac{\Psi_2}{2} \frac{K\left[1 - (a_1^2/b_1^2)\right]}{K[a_1^2/b_1^2]}$$

FIGURE 5.40 S-C transformation of the rectangle formed in the complex potential plane by tracing the real axis of the w-plane that consists of two ground planes that occupy the real axis for $|w| > b_1$ and a center conductor at a higher potential that occupies the real axis for $|w| < a_1$.

The complex potential

$$\Phi[w] = j\frac{V_1}{2K[a_1^2/b_1^2]} F\left[\sin^{-1}[w/a_1], a_1^2/b_1^2\right] + \Psi_1 \qquad (5.205)$$

is obtained by multiplying (5.203) by j (which transforms the constant Ψ contours into constant V contours and constant V contours into constant Ψ contours) and then adding Ψ_1. Thus, regions along the real axis in the w-plane located between $-\infty \le w \le -b_1$ and $b_1 \le w \le \infty$ are now grounded conductors and the region located between $-a_1 \le w \le a_1$ is now a conductor with potential Ψ_1 (see Fig. 5.40). To help minimize confusion, let $V_2 = \Psi_1$ so that (5.205) can be written as

$$\Phi[w] = V_2 + j\frac{V_2}{K[1 - (a_1^2/b_1^2)]}F\left[\sin^{-1}[w/a_1], a_1^2/b_1^2\right] \qquad (5.206)$$

Note that the value of Ψ_1 given in Fig. 5.39 was used to simplify (5.206). The capacitance between the center conductor and the two grounds is

$$\underline{C} = \epsilon\frac{\Psi_2}{V_2} = \epsilon\frac{2K\left[a_1^2/b_1^2\right]}{K[1 - (a_1^2/b_1^2)]} \qquad (5.207)$$

This is the capacitance for only the upper half plane. Due to symmetry, the lower half plane contributes the same amount of capacitance. Therefore, the capacitance of the full w-plane is twice that of (5.207).

A field map comparison of (5.203) and (5.206) is shown in Fig. 5.41. Portions of the real axis that are not conductors in Fig. 5.41 (a) are replace with conductors in Fig. 5.41 (b) and the original conductors are removed. Using the method of curvilinear squares, the capacitance due to the upper half plane of Fig. 5.41 (a) is $\sim 0.5\epsilon$ and (b) is $\sim 2\epsilon$. For both field maps, $a_1^2/b_1^2 = 1/2$ which gives capacitances of 0.5ϵ and 2ϵ for Figures 5.41 (a) and (b), respectively. Thus, excellent agreement is obtained from the method of curvilinear squares.

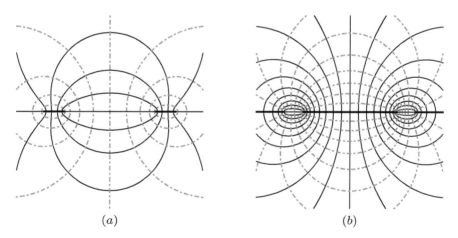

(a)   (b)

FIGURE 5.41 A field map comparison of (a) two symmetric coplanar plates and (b) two semi-infinite grounds with a center conductor. Portions of the real axis that are not conductors in Fig. 5.41(a) are replaced with conductors in Fig. 5.41(b) and the original conductors are removed. In both figures $a_1^2/b_1^2 = 1/2$ and $\Delta V = \Delta \Psi = 0.1V_1 = 0.1V_2$.

5.18.4 Coplanar Finite Plate and Semi-infinite Plate

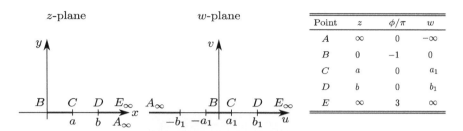

FIGURE 5.42   S-C transformation of the z-plane onto the upper half of the w-plane.

We can also obtain the complex potential in the z-plane of a finite plate of width a at a potential V_2 and a grounded coplanar semi-infinite plate located a distance $b - a$ away by letting the z-plane represent a polygon with one finite vertex of external angle $-\pi$ located at the origin (see Fig. 5.42). From Section 5.12, the transform is

$$z = (A_1/2)\, w^2 + B_1 \qquad (5.208)$$

Since $z = 0$ when $w = 0$, $B_1 = 0$. The boundary condition $z = a$ when $w = a_1$ gives

$$A_1 = 2a/a_1^2 \qquad (5.209)$$

The final transformation is

$$z/a = (w/a_1)^2 \qquad (5.210)$$

To obtain the complex potential in the z-plane, we still need to determine a_1/b_1 in terms of a and b. When $z = b$, $w = b_1$ giving

$$b/a = b_1^2/a_1^2 \qquad (5.211)$$

The complex potential in the z-plane is

$$\Phi[z] = V_2 + j \frac{V_2}{K[1 - (a/b)]} F\left[\sin^{-1}\left[\sqrt{z/a}\right], a/b\right] \qquad (5.212)$$

The capacitance between the finite plate and semi-infinite plate is

$$\underline{C} = \epsilon \frac{\Psi_2}{V_2} = \epsilon \frac{2K[a/b]}{K[1 - (a/b)]} \qquad (5.213)$$

This is the capacitance for the full plane. A field map for (5.212) is shown in Fig. 5.43.

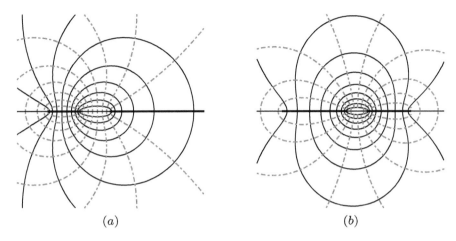

FIGURE 5.43 (a) Field map of a finite plate that is coplanar with a semi-infinite plate and $b = 2a$. (b) Field map of two finite unequal coplanar plates with $w_1 = 3g$ and $w_2 = 2g$. For both field maps, $\Delta V = \Delta \Psi = 0.1V_2$.

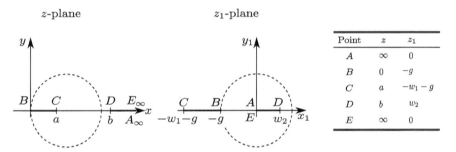

FIGURE 5.44 Complex inversion transformation of a finite plate and semi-infinite plate in the z-plane into two finite plates in the w-plane.

5.18.5 Two Coplanar Unequal Finite Plates

The complex potential of two coplanar plates of unequal width can be obtained by using complex inversion on the z-plane. Let the center of the circle of inversion in the z-plane be on the real axis between the two plates ($a < z_c < b$) and let the radius equal z_c so that the circle of inversion goes through the origin (see Fig. 5.44). We also add a translation of r_I so that the center of the circle of inversion is the origin in the transformed plane which we label the z_1-plane. The transformation is

$$z_1 = \frac{r_I^2}{z - z_c} = \frac{r_I^2}{z - r_I} \tag{5.214}$$

Let the two finite coplanar plates in the z_1-plane have widths w_1 and w_2 and let g be the distance between the plates. The boundary condition $z_1 = -g$ when $z = 0$ gives

$$r_I = g \tag{5.215}$$

The boundary conditions $z_1 = -w_1 - g$ when $z = a$ and $z_1 = w_2$ when $z = b$ give

$$a = \frac{g w_1}{w_1 + g}, \quad b = \frac{g\,(g + w_2)}{w_2} \tag{5.216}$$

The final transformation is

$$z_1 = \frac{g^2}{z - g}, \quad z = \frac{g\,(g + z_1)}{z_1} \tag{5.217}$$

The complex potential in the z_1-plane is now fully determined as

$$\Phi[z_1] = V_2 + j \frac{V_2}{K[1 - m_1]} F[\theta_1, m_1] \tag{5.218}$$

where

$$\theta_1 = \sin^{-1}\left[\sqrt{\frac{(g + w_1)\,(g + z_1)}{z_1 w_1}}\right], \quad m_1 = \frac{w_1 w_2}{(g + w_1)\,(g + w_2)} \tag{5.219}$$

The capacitance between unequal finite width coplanar plates is

$$C = \epsilon \frac{\Psi_2}{V_2} = \epsilon \frac{2K[m_1]}{K[1 - m_1]} \tag{5.220}$$

where m_1 is given by (5.219) and this is the capacitance for the full plane. A field map of (5.218) is shown in Fig. 5.43 (b).

5.18.6 Quadrilateral with Reflection Symmetry

Next we develop the transformation of a quadrilateral with reflection symmetry with respect to the diagonal line onto the upper half plane. A quadrilateral with diagonal symmetry can be viewed of being composed of a triangle and its mirror image about the line of symmetry as shown in Fig. 5.45. Therefore, a quadrilateral with diagonal symmetry is completely described by two angles and the length of one side of the triangle. We setup our coordinate system in the z-plane so that diagonally symmetric quadrilateral lies in the first quadrant with one vertex at the origin, one side along the real axis, and the line of symmetry

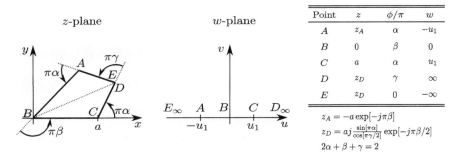

Point	z	ϕ/π	w
A	z_A	α	$-u_1$
B	0	β	0
C	a	α	u_1
D	z_D	γ	∞
E	z_D	0	$-\infty$

$z_A = -a\exp[-j\pi\beta]$

$z_D = aj\frac{\sin[\pi\alpha]}{\cos[\pi\gamma/2]}\exp[-j\pi\beta/2]$

$2\alpha + \beta + \gamma = 2$

FIGURE 5.45 S-C transformation of a quadrilateral with reflection symmetry with respect to the diagonal line in the z-plane onto the upper half of the w-plane.

going through the origin (see Fig. 5.45). Taking advantage of symmetry in the problem, we map the vertices that lie on the line of symmetry in the z-plane to the origin and $\pm\infty$. The other two vertices must then be mapped to $w = \pm u_1$. The S-C integral is

$$z = A_1 \int \frac{dw}{(w^2 - u_1^2)^\alpha w^\beta} + B_1 \tag{5.221}$$

With the substitution $u = w^2$, (5.221) is easily transformed into

$$z = A_1 \int \frac{(u - u_1^2)^{-\alpha}}{2u^{(1+\beta)/2}} du + B_1 \tag{5.222}$$

The integration can now be done in terms of Gauss's hypergeometric function as

$$z = \frac{A_1 w_1^{1-\beta} a^{1-\beta-2\alpha}}{(1-\beta)\, e^{j\pi\alpha}}\,{}_2F_1\left[\alpha, \tfrac{1}{2}(1-\beta), \tfrac{1}{2}(3-\beta), w_1^2\right] + B_1 \tag{5.223}$$

Since $z = 0$ when $w_1 = 0$, $B_1 = 0$. The boundary condition $w_1 = 1$ when $z = a$ gives

$$a = \frac{A_1 a^{1-\beta-2\alpha}}{(1-\beta)\, e^{j\pi\alpha}}\,{}_2F_1\left[\alpha, \tfrac{1}{2}(1-\beta), \tfrac{1}{2}(3-\beta), 1\right] \tag{5.224}$$

Solving for A_1 and using (5.135) gives

$$A_1 = \frac{(1-\beta)\, e^{j\pi\alpha}}{a^{-\beta-2\alpha}}\,\frac{\Gamma\left[\tfrac{1}{2}(3-2\alpha-\beta)\right]}{\Gamma[1-\alpha]\,\Gamma\left[\tfrac{1}{2}(3-\beta)\right]} \tag{5.225}$$

The final transformation is

$$z = \frac{w_1^{1-\beta} a \Gamma\left[\frac{1}{2}\left(3 - 2\alpha - \beta\right)\right]}{\Gamma[1-\alpha]\,\Gamma\left[\frac{1}{2}\left(3-\beta\right)\right]} \,_2F_1\left[\alpha, \tfrac{1}{2}\left(1-\beta\right), \tfrac{1}{2}\left(3-\beta\right), w_1^2\right] \quad (5.226)$$

If we let one pair of opposite sides of the diagonally symmetric quadrilateral correspond to conductors with a potential difference of V_1 and the remaining opposite sides correspond to lines of forces, it maps into a finite plate of width 1 and a semi-infinite plate with a spacing of 1 between them in the w_1-plane. Since the capacitance is obtained by only considering the upper half plane, the capacitance for this configuration is half of (5.213) with $a = 1$ and $b = 2$

$$\underline{C} = \epsilon \frac{K[1/2]}{K[1/2]} = \epsilon \quad (5.227)$$

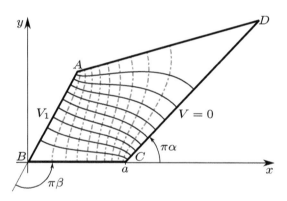

FIGURE 5.46 Field map of a diagonal symmetric quadrilateral. Side AB has a potential V_1, side CD is grounded and $\Delta V = \Delta \Psi = 0.1V_1$. The remaining sides have constant lines of force boundary conditions. The quadrilateral has $a = 1$, $\alpha = 1/4$, and $\beta = 2/3$.

A field map for a diagonally symmetric quadrilateral with side AB at potential V_1, side CD grounded and the remaining sides with constant lines of force boundary conditions is shown in Fig. 5.46. In general, forcing opposite sides of a quadrilateral to be constant lines of force is difficult. Thus, the result of (5.227) is of limited use. On the other hand, insulators exist that allow negligible current flow and a zero current boundary is easily achieved. Translating (5.227) into a resistor theorem that states a quadrilateral with diagonal symmetry made from ohmic material has a resistance between opposite side equal to the sheet resistance of the ohmic material [9] leads to a more useful result.

5.18.7   Quadrilateral with One Right Angle

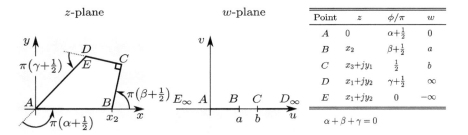

FIGURE 5.47   S-C transformation of a polygon with four finite vertices where one of the vertices has a right angle in the z-plane onto the upper half of the w-plane.

Next, we demonstrate how the S-C transformation can be used to determine the capacitance of a quadrilateral with at least one right angle corner, but otherwise arbitrary [10]. Let the z-plane contain the quadrilateral with the coordinate system setup as shown in Fig. 5.47. Let the sides AB and CD be conductors of different potential and the remaining two sides (AE and BC) represent constant lines of force contours. The S-C differential equation between the z and w-planes is

$$\frac{dz}{dw} = \frac{A_1}{(w-a)^\beta w^\alpha \sqrt{w(w-a)(w-b)}} \tag{5.228}$$

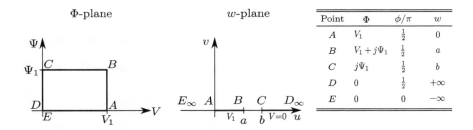

FIGURE 5.48   S-C transformation of a rectangle formed in the complex potential plane by tracing the real axis of the w-plane consisting of a finite plate and semi-infinite plate at different potentials.

Mapping the complex potential values along the boundary of the quadrilateral produces a rectangle in the complex potential plane which can be mapped into the w-plane as shown in Fig. 5.48. The S-C differential equation between the Φ and w-planes is

$$\frac{d\Phi}{dw} = \frac{A_2}{\sqrt{w(w-a)(w-b)}} \tag{5.229}$$

The relationship between the z and Φ-planes is

$$z = \int \frac{dz}{dw}\frac{dw}{d\Phi}d\Phi + B_1 = \frac{A_1}{A_2}\int \frac{d\Phi}{(w-a)^\beta w^\alpha} + B_1 \qquad (5.230)$$

A comparison of the w-plane in Fig. 5.48 with the z-plane of Fig. 5.42 shows they are the same. Therefore, the complex potential for Fig. 5.48 in the w-plane is given by (5.212) with z replaced with w, Ψ_2 replaced with $2\Psi_1$ (we are only considering the upper half plane) and V_2 replaced with V_1

$$\Phi[w] = j\frac{\Psi_1}{K[m]}F\left[\sin^{-1}\left[\sqrt{w/a}\right], m\right] + V_1 \qquad (5.231)$$

where $m = a/b$ and

$$\Psi_1 = V_1 K[m]/K[1-m] \qquad (5.232)$$

With the Jacobi elliptic function sn[ ], we can now express w in terms of Φ as

$$w = a\,\text{sn}^2[j(1-(\Phi/V_1))K[1-m], m] \qquad (5.233)$$

Defining a new variable

$$\Lambda = j(V_1 - \Phi)K[1-m]/V_1 \qquad (5.234)$$

we can express z as an integral function in the Λ-plane as

$$\int_0^z dz = z = A_3 \int_0^{\Lambda[z]} \frac{d\Lambda}{\text{cn}^{2\beta}[\Lambda, m]\,\text{sn}^{2\alpha}[\Lambda, m]} \qquad (5.235)$$

where A_3 is a constant that does not depend on Λ and the identity

$$\text{cn}^2[\Lambda, m] = 1 - \text{sn}^2[\Lambda, m] \qquad (5.236)$$

was used to simplify the expression.

The capacitance of the upper half-plane is half of (5.213) and only depends on m. To determine m, we need to eliminate A_3 which can be done by taking the ratio of the integral along side AE to the integral along side AB. The integral along side AE is

$$\int_0^{z_E} dz = z_E = A_3 \int_0^{jK[1-m]} \frac{d\Lambda}{\text{cn}^{2\beta}[\Lambda, m]\,\text{sn}^{2\alpha}[\Lambda, m]} \qquad (5.237)$$

Using the substitution $\Lambda = ju$ and the identities

$$\text{sn}[ju, m] = j\frac{\text{sn}[u, 1-m]}{\text{cn}[u, 1-m]}, \quad \text{cn}[ju, m] = \frac{1}{\text{cn}[u, 1-m]} \qquad (5.238)$$

allows re-writing (5.237) as

$$
\begin{aligned}
z_E &= A_3 \int_0^{K[1-m]} \frac{j\,du}{\mathrm{cn}^{2\beta}[ju,m]\,\mathrm{sn}^{2\alpha}[ju,m]} \\
&= A_3 \int_0^{K[1-m]} \frac{\mathrm{cn}^{2(\alpha+\beta)}[u,1-m]\,du}{j^{2\alpha-1}\mathrm{sn}^{2\alpha}[u,1-m]}
\end{aligned}
\tag{5.239}
$$

Using the substitution $t = \mathrm{sn}[u, 1-m]$ and the integral form of Gauss's hypergeometric function

$$
\frac{\Gamma[b]\,\Gamma[c-b]}{\Gamma[c]}\,_2F_1[a,b,c,z] = \int_0^1 t^{b-1}(1-t)^{c-b-1}(1-zt)^{-a}\,dt
\tag{5.240}
$$

allows expressing (5.239) as

$$
\begin{aligned}
z_E &= x_1 + jy_2 \\
&= A_3 \frac{\Gamma[\tfrac12 - \alpha]\,\Gamma[\tfrac12 + \alpha + \beta]}{2j^{2\alpha-1}\Gamma[1+\beta]}\,_2F_1[\tfrac12, \tfrac12 - \alpha, 1+\beta, 1-m]
\end{aligned}
\tag{5.241}
$$

The integral along side AB is

$$
\int_0^{z_B} dz = z_B = A_3 \int_0^{K[m]} \frac{d\Lambda}{\mathrm{cn}^{2\beta}[\Lambda,m]\,\mathrm{sn}^{2\alpha}[\Lambda,m]}
\tag{5.242}
$$

Using the substitution $t = \mathrm{sn}[\Lambda, m]$ and the integral form of Gauss's hypergeometric function allows expressing (5.242) as

$$
z_B = x_2 = A_3 \frac{\Gamma[\tfrac12 - \alpha]\,\Gamma[\tfrac12 - \beta]}{2\Gamma[1-\alpha-\beta]}\,_2F_1[\tfrac12, \tfrac12 - \alpha, 1-\alpha-\beta, m]
\tag{5.243}
$$

The ratio of (5.243) to (5.241) gives

$$
\begin{aligned}
&\frac{x_2}{\sqrt{x_1^2 + y_2^2}} \\
&= \frac{\Gamma[\tfrac12 - \beta]\,\Gamma[1+\beta]}{\Gamma[1-\alpha-\beta]\,\Gamma[\tfrac12 + \alpha + \beta]}\,\frac{_2F_1[\tfrac12, \tfrac12 - \alpha, 1-\alpha-\beta, m]}{_2F_1[\tfrac12, \tfrac12 - \alpha, 1+\beta, 1-m]}
\end{aligned}
\tag{5.244}
$$

The only unknown in (5.244) is m which can be determined using numerical methods. Once m is known, the capacitance can be calculated as

$$
\underline{C} = \epsilon\Psi_1/V_1 = \epsilon K[m]\,/K[1-m]
\tag{5.245}
$$

228 **■** 2D Electrostatic Fields: A Complex Variable Approach

Again, forcing the boundary of a quadrilateral to be a line of force in general is difficult and this result has more practical applications as a resistor theorem. In terms of a resistor theorem, (5.245) states the resistance between opposite sides of a quadrilateral with at least one right angle made from ohmic material is

$$R = R_{sh} K[1 - m] / K[m] \tag{5.246}$$

where R_{sh} is the sheet resistance of the ohmic material and m is dependent only on the geometry of the quadrilateral through (5.244). See Section 7.2 on how to translate capacitance results into resistance results.

5.18.8 Line Charge and Two Finite Coplanar Grounded Plates

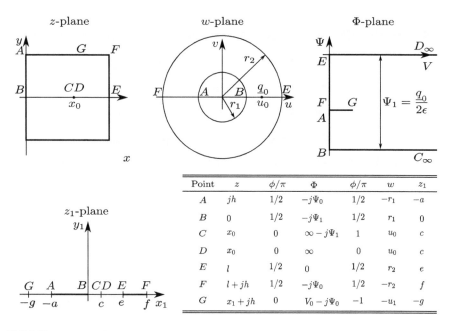

Point	z	ϕ/π	Φ	ϕ/π	w	z_1
A	jh	$1/2$	$-j\Psi_0$	$1/2$	$-r_1$	$-a$
B	0	$1/2$	$-j\Psi_1$	$1/2$	r_1	0
C	x_0	0	$\infty - j\Psi_1$	1	u_0	c
D	x_0	0	∞	0	u_0	c
E	l	$1/2$	0	$1/2$	r_2	e
F	$l+jh$	$1/2$	$-j\Psi_0$	$1/2$	$-r_2$	f
G	x_1+jh	0	$V_0 - j\Psi_0$	-1	$-u_1$	$-g$

FIGURE 5.49 In the z, w, and z_1-planes, the segments of the boundary between AB and EF are grounded conductors and the segments of the boundary between FA, BC, and DE are lines of force. All the boundaries then map to the same figure shown in the Φ-plane.

As shown in Fig. 5.49, tracing out the complex potential along the boundary for a line charge between two coaxial grounded conductive cylinders (already solved in Section 3.17), a rectangular enclosure where two opposite sides are grounded conductors and the other two sides are constant flux contours with a line charge that lies midway between the constant flux boundary contours, or two finite grounded plates and a line charge that all lie on the same plane produces the same contour in the complex potential plane. If the complex potential is known in any of these planes, then finding the complex potential in any of the other planes reduces to finding the conformal mapping between the planes. Although the complex potential is already known for a line charge between two coaxial grounded circular cylinders in the w-plane, it is given in terms of an infinite series. It is shown that the complex potential in the z and z_1-planes can be obtained in terms of elliptic integrals which then allows expressing the complex potential in the w-plane in terms of elliptic integrals instead of an infinite series. We first solve for the complex potential for a line charge in the z_1 plane with coplanar plates. Then we use conformal mapping to obtain the complex potential for the concentric grounded cylinders in the w-plane and rectangular geometry in z-plane of Fig. 5.49.

The S-C integral to map the geometry in the Φ-plane onto the upper half of the z_1-plane is

$$\Phi = A_1 \int \frac{(z_1 + g)\, dz_1}{\sqrt{z_1 + a}\sqrt{z_1}\sqrt{z_1 - e}\sqrt{z_1 - f}\,(z_1 - c)} + B_1 \qquad (5.247)$$

Using the substitutions

$$z_1 = \frac{(t^2 - 1)\, ef}{t^2 e - f}, \quad dz_1 = \frac{2e\,(e - f)\, ft}{(t^2 e - f)^2} dt \qquad (5.248)$$

converts (5.247) into

$$\Phi = \frac{2A_1}{\sqrt{(a + e)f}} \frac{(f - e)(g + c)}{(e - c)(f - c)} \int \frac{dt}{(1 - nt^2)\sqrt{1 - t^2}\sqrt{1 - mt^2}}$$

$$+ \frac{2A_1}{\sqrt{(a + e)f}} \frac{(g + f)}{(f - c)} \int \frac{dt}{\sqrt{1 - t^2}\sqrt{1 - mt^2}} + B_1 \qquad (5.249)$$

with

$$m = \frac{e\,(f + a)}{f\,(a + e)}, \quad n = \frac{e\,(f - c)}{f\,(e - c)} \qquad (5.250)$$

The first integral has the form of an incomplete elliptic integral of the first kind. The second integral has the form of an incomplete elliptic integral of the third kind written as $\Pi[n, \theta_1, m]$ where $\sin[\theta_1] = t$ (see

Appendix C). Thus, we can write (5.249) as

$$\Phi = \frac{2A_1}{\sqrt{(a+e)\,f}}\frac{(g+f)}{(f-c)}F[\theta_1,m]$$
$$+ \frac{2A_1}{\sqrt{(a+e)\,f}}\frac{(f-e)\,(g+c)}{(e-c)\,(f-c)}\Pi[n,\theta_1,m] + B_1 \tag{5.251}$$

where

$$\sin[\theta_1] = \sqrt{\frac{f\,(e-z_1)}{e\,(f-z_1)}} \tag{5.252}$$

Parallel boundary condition (5.68) at vertex CD gives

$$A_1 = \frac{q_0}{2\pi\epsilon}\frac{\sqrt{(e-c)\,(f-c)\,(c+a)\,c}}{(c+g)} \tag{5.253}$$

The boundary condition $\Phi[e] = 0$ gives $B_1 = 0$. The boundary condition $\mathrm{Im}[\Phi[g]] = 0$ gives

$$\left(\frac{f+g}{c+g}\right) = -\left(\frac{f-e}{e-c}\right)\frac{\Pi[n,m]}{K[m]} \tag{5.254}$$

Plugging (5.253) and (5.254) into (5.251) gives

$$\Phi = \frac{q_0}{\pi\epsilon}\left(\frac{f-e}{e-c}\right)\frac{\sqrt{c\,(e-c)\,(a+c)}}{\sqrt{f\,(f-c)\,(a+e)}}$$
$$\times\left(\Pi[n,\theta_1,m] - \frac{\Pi[n,m]}{K[m]}F[\theta_1,m]\right) \tag{5.255}$$

To obtain the complex potential in the z-plane, we must first determine $z_1[z]$. The S-C integral to map the geometry $ABEF$ in the z-plane onto the upper half of the z_1-plane is

$$z = A_2\int\frac{dz_1}{\sqrt{z_1+a}\sqrt{z_1}\sqrt{z_1-e}\sqrt{z_1-f}} + B_2 \tag{5.256}$$

This is just a special case of the integral given in (5.247) with $c = -g$. Setting $c = -g$ in (5.251) gives

$$z = \frac{2A_2}{\sqrt{(a+e)\,f}}F[\theta_1,m] + B_2 \tag{5.257}$$

The boundary condition $z = l$ when $z_1 = e$ gives $B_2 = l$. The boundary condition $z = jh$ when $z_1 = -a$ gives

$$A_2 = -\frac{l\sqrt{(a+e)\,f}}{2K[m]} \tag{5.258}$$

and

$$\frac{K[m]}{K[1-m]} = \frac{l}{h} \tag{5.259}$$

The final transformations between the z and z_1-planes are

$$z[z_1] = l\left(\frac{K[m] - F[\theta_1, m]}{K[m]}\right) \tag{5.260}$$

and

$$z_1[z] = \frac{f\left(1 - \text{sn}^2[K[m](l-z)/l, m]\right)}{(f/e) - \text{sn}^2[K[m](l-z)/l, m]} \tag{5.261}$$

To calculate the complex potential in the z-plane, z_1 in (5.255) is replaced with $z_1[z]$ from (5.261) to obtain

$$\Phi = \frac{q_0}{\pi\epsilon}\left(\frac{f-e}{e-c}\right)\frac{\sqrt{c(e-c)(a+c)}}{\sqrt{f(f-c)(a+e)}}$$
$$\times\left(\Pi[n, \phi_1, m] - \Pi[n, m]\left(\frac{l-z}{l}\right)\right) \tag{5.262}$$

where

$$\sin[\phi_1] = \text{sn}[K[m](l-z)/l, m] \tag{5.263}$$

The expressions for m and n in (5.250) allow writing a and c as

$$a = \frac{ef(1-m)}{mf-e}, \quad c = \frac{ef(1-n)}{e-nf} \tag{5.264}$$

Substituting a and c from (5.264) into (5.262) gives the complex potential in the z-plane as

$$\Phi = \frac{q_0}{\pi\epsilon}\sqrt{\frac{(1-n)(m-n)}{n}}\left(\Pi[n, \phi_1, m] - \Pi[n, m]\left(\frac{l-z}{l}\right)\right) \tag{5.265}$$

where m is obtained by solving (5.259) and n is determined by the boundary condition $z = x_0$ when $z_1 = c$ as

$$n = 1/\text{sn}^2[K[m](l-x_0)/l, m] \tag{5.266}$$

To obtain the complex potential in the w-plane, we first need to determine $z[w]$. This is done with the logarithm transformation (4.45)

$$z[w] = \ln[w/r_1] \tag{5.267}$$

The transform (5.267) maps $w = -r_2$ to $z = l + jh$. Thus

$$l = \ln[r_2/r_1], \quad h = \pi \tag{5.268}$$

Replacing z with $z[w]$ and using (5.268) in (5.265) gives the complex potential in the w-plane as

$$\Phi = \frac{q_0}{\pi\epsilon} \sqrt{\frac{(1-n)(m-n)}{n}} \left(\Pi[n, \varphi_1, m] - \Pi[n, m] \frac{\ln[r_2/w]}{\ln[r_2/r_1]} \right) \tag{5.269}$$

where

$$\sin[\varphi_1] = \mathrm{sn}\left[K[m] \frac{\ln[r_2/w]}{\ln[r_2/r_1]}, m\right], \quad \frac{1}{n} = \mathrm{sn}^2\left[K[m]\frac{\ln[r_2/u_0]}{\ln[r_2/r_1]}, m\right] \tag{5.270}$$

The value of m is determined by

$$\frac{\ln[r_2/r_1]}{\pi} = \frac{K[m]}{K[1-m]} \tag{5.271}$$

We now have the complex potential due to a line charge located between two grounded conductors in terms of elliptic functions instead of an infinite series. Since elliptic functions are typically built-in functions for most mathematical software programs, their use may be easier than an infinite series that requires the knowledge of how many terms to use. Field maps for the z and z_1-planes are shown in Fig. 5.50. The field map for the w-plane was given in Fig. 3.32.

EXERCISES

5.1 The z_i-planes are mapped to the same w-plane shown in Fig. 5.51. Determine the phase of the complex constant A_1 in front of the S-C integral for each mapping.

5.2 Four circular conductive cylinders of equal diameter d are place in contact with each other creating the bounded region shown in Fig. 5.52. Infinitesimal insulation is inserted to allow cylinder 1 to be raised to a potential V_0. The other three cylinders are grounded. What is total induced charge on cylinder 3?

5.3 A conductive hollow circular cylinder has a radius r_1 and its center at the origin of the z-plane. Infinitesimal insulation located at the boundary of each quadrant allow the potential in quadrants 1, 2, 3 and 4 to be held at V_1, 0, $-V_1$, and 0, respectively. Determine the complex potential inside the cylinder.

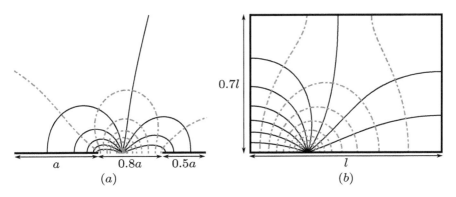

FIGURE 5.50 (a) Field map of a line charge between two finite coplanar plates. The left plate has a width of a and the right plate has a width of $0.5a$. The gap between the plates is $0.8a$. The line charge is located $0.3a$ from the edge of the left plate. $\Delta V = \Delta \Psi = 0.05 q_o/\epsilon$. (b) Field map of a line charge between two finite vertical plates with a distance l between plates and plate height of $0.7l$. The horizontal lines connecting the top and bottom of the two plates are lines of constant force. The line charge is located $0.3a$ from the left plate. $\Delta V = \Delta \Psi = 0.05 q_o/\epsilon$.

5.4 a. Show the region bounded by the real axis of the z-plane and a semi-infinite plate with zero thickness that makes an angle $\pi \alpha$ with the horizontal and is a distance a above the real axis can be mapped to the upper half of the w-plane with

$$ z = \frac{a}{\sin[\pi \alpha]} \left(w_1^\alpha \left(1 - \alpha\right) - \alpha w_1^{\alpha-1} - \cos[\pi\alpha] \right) $$

where $w_1 = w/u_1$, the semi-infinite plate is mapped to the negative real axis, and the original real axis is mapped to the positive real axis.

   b. Determine the complex potential and draw a field map in the z-plane when the semi-infinite plate is conductive and has a potential V_1 and the real axis is grounded.

5.5 a. Derive the transformation that maps the region $y > 0$ bounded by the real axis and the portion of the imaginary axis $0 \leq y \leq a$.

   b. The real axis and the portion of the imaginary axis $0 \leq y \leq a$ are grounded and serve as the boundary to an electric field that is uniform far from the origin of the z-plane. Let the uniform field be described by $\mathcal{E} = j|\mathcal{E}_0|$. What is the complex potential in the upper half of the z-plane. Plot the field map in the z-plane.

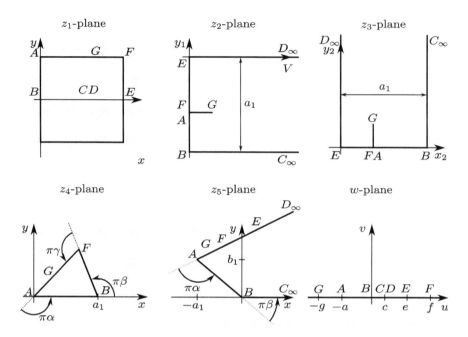

FIGURE 5.51 Each z_i-plane is mapped to the w-plane.

5.6 a. Derive the conformal transformation that transforms half the interior of a circle onto the upper half plane.

   b. A cylinder in the shape of a half a circle with radius r_1 has the straightline portion of the cylinder on the real axis symmetric with respect to the imaginary axis of the z-plane. If all of the cylinder is grounded except for the portion for $|x| < a < r_1$ which as at potential V_1, show that complex potential inside the cylinder is

$$\Phi = \frac{jV_1}{\pi} \ln \left[\left(\frac{r_1 - z}{r_1 + z} \right)^2 - \left(\frac{r_1 + a}{r_1 - a} \right)^2 \right]$$
$$- \frac{jV_1}{\pi} \ln \left[\left(\frac{r_1 - z}{r_1 + z} \right)^2 - \left(\frac{r_1 - a}{r_1 + a} \right)^2 \right]$$

5.7 a. What is the transformation that transforms the region bounded by $y > 0$ for $|x| \geq r_1$ and $y > \sqrt{r_1^2 - x^2}$ for $|x| \leq r_1$ onto the upper half plane.

   b. The boundary $y > 0$ for $|x| \geq r_1$ and $y > \sqrt{r_1^2 - x^2}$ for $|x| \leq r_1$

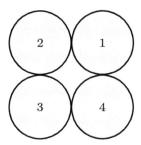

FIGURE 5.52 Four circular cylinder of equal diameter in contact with each other.

is a conductor at potential zero and serves as the boundary to an electric field that is uniform far from the origin of the z-plane. Let the uniform field be described by $\mathcal{E} = j|\mathcal{E}_0|$. What is the complex potential in the upper half of the z-plane. Plot the field map in the z-plane.

5.8 a. Show the region bounded by the positive imaginary axis, the positive real axis, the horizontal line $x + jb$ with $x > a$ and the vertical line $a + jy$ with $y > b$ can be mapped to the upper half of the w-plane with

$$\frac{z}{a} = \frac{j2}{\pi}\left(\sinh^{-1}\left[\frac{\sqrt{w_1 - 1}}{\sqrt{1 + \eta}}\right] - \frac{1}{\sqrt{\eta}}\tan^{-1}\left[\frac{\sqrt{w_1 - 1}\sqrt{\eta}}{\sqrt{w_1 + \eta}}\right]\right) + 1 + j\frac{b}{a}$$

b. Determine the complex potential and draw a field map in the z-plane when the positive x-axis and y-axis are at V_0 and the remaining boundary is grounded.

c. For the bounded region between $0 \le y \le y_1$ and $0 \le x \le x_1$, show the capacitance capacitance can be approximated as

$$\frac{C}{\epsilon} = \left(\frac{x_1 - a}{b}\right) + \left(\frac{y_1 - b}{a}\right) + \frac{2}{\pi}\ln\left[\frac{1 + \eta}{4\sqrt{\eta}}\right]$$
$$+ \frac{2}{\pi}\left(\sqrt{\eta}\sin^{-1}\left[\frac{1}{\sqrt{1 + \eta}}\right] + \frac{1}{\sqrt{\eta}}\tan^{-1}[\sqrt{\eta}]\right)$$

5.9 Conductor 1 occupies the region $y < a$ and $x < -b$ and conductor 2 occupies the region $y < 0$ and $x > b$. Determine the complex potential if conductor 2 is grounded and conductor 1 has a potential V_0.

5.10 One model for an electron gun is a wire inside a round conductive cylinder with a slot in it [4]. Long cylinders can be treated in 2D. If the wire is modeled as a line charge q_0 at the center of the conductive cylinder, the conductive cylinder has a radius r_1, is at zero potential and the slot forms a central angle $2\theta_0$, show that the complex potential is

$$\Phi[z] = \frac{q_0}{2\pi\epsilon} \ln \left[\frac{\left(\frac{r_1-z}{r_1+z}\right) + \sqrt{\left(\frac{r_1-z}{r_1+z}\right)^2 + \tan\left[\frac{1}{2}\theta_0\right]^2} + \sec\left[\frac{1}{2}\theta_0\right] + 1}{\left(\frac{r_1-z}{r_1+z}\right) + \sqrt{\left(\frac{r_1-z}{r_1+z}\right)^2 + \tan\left[\frac{1}{2}\theta_0\right]^2} - \sec\left[\frac{1}{2}\theta_0\right] - 1} \right]$$

5.11 Assume the end of a parallel plate capacitor can be modeled by the semi-infinite geometry shown in Fig. 5.30. Assuming a capacitor plate thickness t and sidewall angle $\pi\alpha$, determine the capacitance due only to the metal on the sidewall of the capacitor plate.

5.12 Two semi-infinite parallel plate capacitors share the same bottom plate and the top plates are connected by a linear conductive transition. Capacitor 1 has a distance a between its plates, capacitor 2 has a distance $b < a$ between its plates, and the transition angle between top plates is $0 \le \pi\alpha \le \pi/2$ with respect to the horizontal.

a. Assume the bottom plate has a voltage V_0 and the top plate is grounded. Create a field map of the region near the transition for $a/b = 3$ and two transition angles $\alpha = 1/2$ and $\alpha = 1/4$.

b. Far from the transition in either direction, the field maps show field lines that resemble the field lines of a parallel plate capacitor. Assuming the transition between top plates starts at $x = 0$ and capacitor 2 is to the left of the transition, the capacitance of the region between $-x_2 \le x \le x_1$ can be approximated as the sum of the two parallel plate capacitances and the transition capacitance

$$\frac{C}{\epsilon} \approx \frac{x_1 - l}{a} + \frac{x_2}{b} + \frac{1}{\pi\alpha} \ln[a/b]$$

Assuming the plate width is much larger than the plate separation for both capacitors and $a/b = 3$, generate a plot of the true capacitance for the region defined by $-5b \le x \le 5a + l$ with transition angles between $\pi/10$ and $\pi/2$. What is the maximum % error over this range between the true capacitance and the approximate capacitance given above. Where is the error largest and why?

c. When $\alpha = p_1/q_1$ is a rational fraction, an even more accurate expression for capacitance in terms of elementary functions is possible. If $p_1 < q_1$, show the more accurate capacitance equation consists of the capacitance equation given in part b of the problem plus the additional terms

$$= -\frac{1}{\pi}\left(\frac{q_1}{p_1}\ln\left[\frac{a}{b}\right] + \left(\frac{b}{a} + \frac{a}{b}\right)\ln\left[1 - \eta^{1/q_1}\right] - 2\ln\left[\frac{1-\eta}{q_1\sqrt{\eta}}\right]\right)$$

$$-\frac{1}{\pi}\sum_{k=1}^{q_1-1}e^{2\pi jkp_1/q_1}\left(\frac{b}{a}\ln\left[1 - \eta^{1/q_1}e^{2\pi jk/q_1}\right] - \ln\left[1 - e^{2\pi jk/q_1}\right]\right)$$

$$-\frac{1}{\pi}\sum_{k=1}^{q_1-1}e^{-2\pi jkp_1/q_1}\left(\frac{a}{b}\ln\left[1 - \eta^{1/q_1}e^{2\pi jk/q_1}\right] - \ln\left[1 - e^{2\pi jk/q_1}\right]\right)$$

The identity [5]

$$B[z, p/q, 0] = -\sum_{k=0}^{q-1}e^{-2k\pi jp/q}\ln\left[1 - z^{1/q}e^{2k\pi j/q}\right]$$

when $0 < p/q < 1$ may be useful.

5.13 The negative real axis is a grounded conductor of zero thickness. A conducting plate of zero thickness is located $z = a + jy$ where $-h \leq y \leq h$ and has a potential V_1. Determine the capacitance between the finite plate and the semi-infinite plate.

5.14 A horizontal plate with thickness $2c$ and width $2b$ has a voltage V_1 and is located midway between two infinite ground planes. The distance between ground planes is $2a$. Calculate the capacitance between the plate and ground planes.

5.15 A plate of zero thickness and width $2a$ has a voltage V_1. It is located at the center of a hollow grounded conductive circular cylinder of radius $r_1 > a$. Calculate the capacitance between the plate and cylinder.

REFERENCES

[1] A. Brown. Proof of a conformal mapping relationship. *Bulletin of the Australian Mathematical Society*, 10(1):91–94, February 1974.

[2] E. T. Copson. *An Introduction to the Functions of a Complex Variable*. Oxford University Press, 1935.

[3] Tobin A. Driscoll and Lloyd N. Trefethen. *Schwarz-Christoffel Mapping*. Cambridge University Press, 2002.

[4] Thornton C. Fry. Two problems in potential theory. *The American Mathematical Monthly*, 39(4):199–209, apr 1932.

[5] J. L. González-Santander. A note on some reduction formulas for the incomplete beta function and the lerch transcedent, 2020.

[6] O. F. Hughes. A simplification of the schwarz–christoffel formula for symmetric quadrilateral transformation. *SIAM Journal on Mathematical Analysis*, 6(2):258–261, apr 1975.

[7] O.F. Hughes. A useful relationship in the conformal mapping of quadrilaterals. *Bulletin of the Australian Mathematical Society*, 9(1):99–104, August 1973.

[8] J. D. Jackson. A curious and useful theorem in two-dimensional electrostatics. *American Journal of Physics*, 67(2):107–115, February 1999.

[9] Charles H. Lees. On the electrical resistance between opposite sides of a quadrilateral one diagonal of which bisects the other at right angles. *Memoirs and proceedings of the Manchester literary & philosophical society*, 44(1):1–3, 1899.

[10] H. Fletcher Moulton. Current flow in rectangular conductors. *Proceedings of the London Mathematical Society*, s2-3(1):104–110, January 1905.

[11] W. F. Osgood. On the existence of the green's function for the most general simply connected plane region. *Transactions of the American Mathematical Society*, 1(3):310, July 1900.

[12] S. Richardson. An identity arising in a problem of conformal mapping. *SIAM Review*, 31(3):484–485, September 1989.

[13] A. M. Thompson and D. G. Lampard. A new theorem in electrostatics and its application to calculable standards of capacitance. *Nature*, 177(4515):888–888, May 1956.

[14] L. J. van der Pauw. A method of measuring the resistivity and hall coefficient of lamellae of arbitrary shape. *Philips Technical Review*, 20:220–224, 1958.

CHAPTER 6

Case Studies with Conformal Mapping

In this chapter, we apply the tools developed in previous chapters to several engineering problems. The case studies chosen are meant to answer several questions that often arise when applying conformal mapping to problems. For example,

— When is an infinite geometry a good approximation for a finite boundary?

— Why use conformal mapping to solve problems when finite element simulators are available?

— When should conformal mapping not be used?

Although accuracy of the models produced by conformal mapping is important, physical insight gained by an analytic solutions is often just as important.

6.1 PARALLEL PLATE CAPACITOR

6.1.1 Introduction

The parallel plate capacitor has been studied for many years [9, 8, 10, 1, 13]. Despite the simple geometry, an analytic solution that takes into account the plate thickness is relatively new [13]. The analytic solution is based on Fourier series and requires solving six matrix equations with the matrix size determined by the number of terms in the Fourier series to be used. Despite the existence of the analytic solution, we use finite

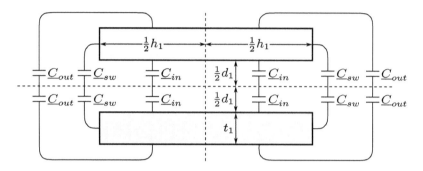

FIGURE 6.1 Parallel plate capacitor consisting of two conductors of thickness t_1 and width h_1 with a distance d_1 between plates. The total capacitance is composed of three components: the inside capacitance ($\underline{C}_{in}$), the side wall capacitance ($\underline{C}_{sw}$), and the outside capacitance ($\underline{C}_{out}$).

element simulation to determine the amount of error in the conformal mapping model to be developed. In order to determine a suitable model that accounts for the plate thickness, we first investigate a finite and semi-infinite model with zero plate thickness. Based on this analysis, we then develop an accurate semi-infinite model that take into account plate thickness.

The parallel plate capacitor for this study is symmetric, have a plate width h_1, a plate separation d_1, and a plate thickness t_1 as shown in Fig. 6.1. The total capacitance is composed of the three capacitances shown in Fig. 6.1. The inside capacitance $\underline{C}_{in}$ is associated with the charge located on the surfaces of the plates that face each other. The sidewall capacitance $\underline{C}_{sw}$ is associated with the charge on the sidewalls of the plates. The outside capacitance $\underline{C}_{out}$ is associated with the charge on the backside of the plates. The horizontal line of symmetry between the top and bottom plates is an equipotential curve and can be replace with a conductor of zero thickness. The vertical line of symmetry is a constant line of force. Thus we only need to consider $1/4$ of the structure in order to determine the field map and capacitances. Note $\underline{C}_{in}$, $\underline{C}_{sw}$, and $\underline{C}_{out}$ determined by considering only $1/4$ of the structure have the same value as the total inside capacitance, sidewall capacitance and outside capacitance that would be calculated for the full structure.

Figure 6.2 shows the field maps of two parallel plate capacitors generated with a finite element simulator. Due to symmetry, only one quarter of the capacitor is shown in each case. In one case, the capacitor plates have a thickness greater than the plate-to-plate distance. In the other case, the capacitor plates are thinner than the plate-to-plate distance.

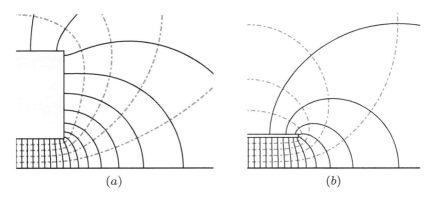

FIGURE 6.2 Field maps for a parallel plate capacitor with (a) thick plates and (b) thin plates. If plate to plate distance is larger (smaller) than the plate thickness, the plate is considered thin (thick). Due to symmetry, only 1/4 of the structure is shown. $\Delta V = \Delta \Psi = 0.1V_1$ where V_1 is the voltage difference between the capacitor plates.

Several key features should be noted. First, between the plates and far away from the edge of the plates, the field lines are essentially the same as the field lines of the infinite parallel plate capacitor. Second, near the edge of the plates, the lines of force cease to be straight lines. Third, depending on the thickness of the plates, the sidewall of the plates can contribute more capacitance than the backside of the plates. Based on these observations, we can conclude that the infinite parallel plate capacitor approximation $\underline{C} = \epsilon/d_1$ underestimates the capacitance of a real parallel plate capacitor as does any capacitor model that ignores the thickness of the plates.

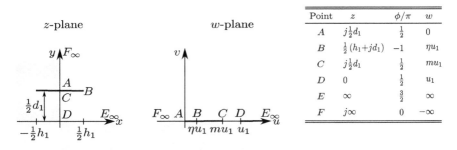

Point	z	ϕ/π	w
A	$j\frac{1}{2}d_1$	$\frac{1}{2}$	0
B	$\frac{1}{2}(h_1+jd_1)$	-1	ηu_1
C	$j\frac{1}{2}d_1$	$\frac{1}{2}$	$m u_1$
D	0	$\frac{1}{2}$	u_1
E	∞	$\frac{3}{2}$	∞
F	$j\infty$	0	$-\infty$

FIGURE 6.3 S-C transformation of a finite parallel plate capacitor with plates of zero thickness in the z-plane onto the upper half of the w-plane. Due to symmetry, only 1/4 of the structure is transformed.

6.1.2 Zero Thickness, Finite Plate Width Model

Let the capacitor plates have finite width, but zero thickness as shown in Fig. 6.3. Due to symmetry, we only need to consider $1/4$ of the structure labeled A through F in Fig. 6.3 which gives the S-C transformation

$$z = A_1 \int \frac{(w - \eta u_1)\, dw}{\sqrt{w}\sqrt{w - mu_1}\sqrt{w - u_1}} + B_1 \tag{6.1}$$

Using the substitution $u = \sqrt{w/(mu_1)}$ allows writing (6.1) as

$$z = -2A_1\sqrt{u_1} \int \left(\frac{(1-\eta)}{\sqrt{1-u^2}\sqrt{1-mu^2}} - \frac{\sqrt{1-mu^2}}{\sqrt{1-u^2}} \right) du + B_1 \tag{6.2}$$

The first term of integral (6.2) has the form of an elliptic integral of the first kind. The second term of integral (6.2) is also an elliptic integral known as an elliptic integral of the second kind defined as

$$E\left[\sin^{-1}[w], m\right] = \int_0^w \frac{\sqrt{1 - mw_1^2}}{\sqrt{1 - w_1^2}}\, dw_1 \tag{6.3}$$

Performing the integration of (6.2) gives

$$z = 2A_1\sqrt{u_1}\left(E[\theta_1[w_1], m] - (1-\eta)\, F[\theta_1[w_1], m]\right) + B_1 \tag{6.4}$$

where $w_1 = w/u_1$ and

$$\theta_1[w_1] = \sin^{-1}\left[\sqrt{w_1/m}\right] \tag{6.5}$$

The boundary condition $z = j\frac{1}{2}d_1$ when $w_1 = 0$ gives $B_1 = j\frac{1}{2}d_1$. The boundary condition $z = 0$ when $w_1 = 1$ gives

$$2A_1\sqrt{u_1}\left(E\left[\sin^{-1}\left[\frac{1}{\sqrt{m}}\right], m\right] - (1-\eta)F\left[\sin^{-1}\left[\frac{1}{\sqrt{m}}\right], m\right]\right) = -j\frac{d_1}{2} \tag{6.6}$$

Using the identities

$$F\left[\sin^{-1}[1/\sqrt{m}], m\right] = K[m] + jK[m']$$
$$E\left[\sin^{-1}[1/\sqrt{m}], m\right] = E[m] + j\left(K[m'] - E[m']\right) \tag{6.7}$$

where $m' = 1 - m$ allows writing (6.6) as

$$2A_1\sqrt{u_1}\left(E[m] - (1-\eta)\,K[m] - j\left(E[m'] - \eta K[m']\right)\right) + j\frac{1}{2}d_1 = 0 \tag{6.8}$$

Both the real and imaginary parts of (6.8) must be zero. Since A_1 is a real number, setting the real part of (6.8) equal to zero gives

$$\eta = 1 - (E[m]/K[m]) \tag{6.9}$$

Equating the imaginary part of (6.8) to zero combined with (6.9) gives

$$A_1 = \frac{d_1}{4\sqrt{u_1}} \left(\frac{K[m]}{(E[m] - K[m]) K[m'] + K[m] E[m']} \right) \tag{6.10}$$

Legendre's relation

$$(E[m] - K[m]) K[m'] + K[m] E[m'] = \pi/2 \tag{6.11}$$

allows writing A_1 as

$$A_1 = \frac{d_1 K[m]}{2\pi\sqrt{u_1}} \tag{6.12}$$

The final relationship between z and w_1 is

$$z = \frac{d_1}{\pi} \left(K[m] E[\theta_1[w_1], m] - E[m] F[\theta_1[w_1], m] \right) + j\tfrac{1}{2}d_1 \tag{6.13}$$

To determine m, we use the boundary condition $z = \frac{1}{2}(h_1 + jd_1)$ when $w_1 = \eta$ which gives

$$\frac{\pi h_1}{2d_1} = K[m] E[\theta_1[\eta], m] - E[m] F[\theta_1[\eta], m] \tag{6.14}$$

where

$$\theta_1[\eta] = \sin^{-1}\left[\sqrt{\frac{K[m] - E[m]}{mK[m]}} \right] \tag{6.15}$$

If we let the top plate have a potential $\frac{1}{2}V_1$ and the real axis in the z-plane have a potential of zero, then the complex potential for the w_1-plane is given by (5.212) with $a = m$, $b = 1$ and $V_2 = \frac{1}{2}V_1$

$$\Phi[w_1] = j\frac{V_1}{2K[m']} F[\theta_1[w_1], m] + \tfrac{1}{2}V_1 \tag{6.16}$$

To transform the complex potential from the w_1-plane to the z-plane, we replace $\theta_1[w_1]$ in (6.13) with

$$\theta_1[\Phi] = \sin^{-1}\left[\mathrm{sn}\left[\frac{(2\Phi - V_1) K[m']}{jV_1}, m \right] \right] \tag{6.17}$$

The capacitance for this structure is half of (5.213)

$$\underline{C} = \epsilon K[m]/K[m'] \tag{6.18}$$

For a given h_1/d_1, m is determined by (6.14) then the capacitance can be calculated with (6.18). Clearly this approximation does not take into account the capacitance due to the finite thickness of the capacitor plates and underestimates the true capacitance. For comparison to other approximations, it is useful to break (6.18) into $\underline{C}_{in}$ and $\underline{C}_{out}$. The capacitance $\underline{C}_{out}$ is the flux between points A and B divided by the potential difference $V_1/2$

$$\underline{C}_{out} = \epsilon \frac{(\Psi[\eta] - \Psi[0])}{V_1/2} = \frac{\epsilon}{K[m']} F\left[\sin^{-1}\left[\sqrt{\frac{K[m] - E[m]}{mK[m]}}, m\right]\right] \tag{6.19}$$

and the capacitance $\underline{C}_{in}$ is the flux between points B and C divided by the potential difference $V_1/2$

$$\underline{C}_{in} = \epsilon \frac{(\Psi[m] - \Psi[\eta])}{V_1/2} = \epsilon \frac{K[m]}{K[m']} - \underline{C}_{out} \tag{6.20}$$

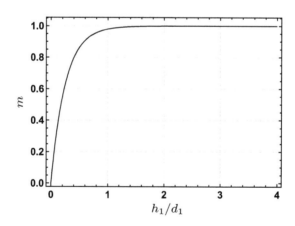

FIGURE 6.4 The dependence of m on the ratio of capacitor width (h_1) to plate separation distance (d_1). The value of m becomes very close to 1 for values of $h_1 > 1$ requiring high precision numerical analysis to obtain accurate values for the capacitance (6.18).

A plot of m vs. h_1/d_1 (Fig. 6.4) shows m approaches 1 quickly. When $h_1/d_1 > 4$, the first six digits for m are already 9. Thus, care should be taken when solving for m if large h_1/d_1 values are used.

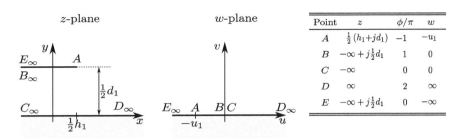

FIGURE 6.5 S-C transformation of a semi-infinite parallel plate capacitor of zero plate thickness in the z-plane onto the upper half of the w-plane.

6.1.3 Zero Thickness, Semi-infinite Plate Width Model

Next we investigate a semi-infinite plate above an infinite grounded plate as shown in Fig. 6.5. It is instructive to understand the difference between this approximation and the finite width zero thickness model in accuracy and mathematical complexity. We have already solved the S-C transformation for Fig. 6.5 back in Section 5.13 with the right edge of the top plate at $z = ja$. We want to approximate a capacitor that has a plate width h_1 and spacing between plates of d_1. Therefore, we let $a = d_1/2$ in (5.123) and translate the origin to place the right edge of the top plate at $(h_1 + jd_1)/2$. The transformation including this translation is

$$2\pi z = d_1 (1 + w_1 + \ln[w_1]) + \pi h_1 \qquad (6.21)$$

To determine an expression for the capacitance, we assume the charge on both sides of the top plate between $z = jd_1/2$ and $z = (h_1 + jd_1)/2$ is the same as the charge on the half plate of the true capacitor. The charge is obtained by first determining the location on the real axis in the w_1-plane where

$$\pi jd_1 = d_1 (1 + w_1 + \ln[w_1]) + \pi h_1 \qquad (6.22)$$

Then the flux in the w_1-plane is calculated allowing the capacitance to be determined.

There are two solutions for (6.22). The first solution $-u_a$ is located between -1 and 0. The second solution $-u_b$ is less than -1. Although numerical methods can be used to solve (6.22), we use Taylor polynomial approximations to obtain closed form approximations for $-u_a$ and $-u_b$. Replacing w_1 with $-u_a$ in (6.22) and simplifying gives

$$u_a = \exp[u_a - \alpha_0] \approx e^{-\alpha_0} (1 + u_a) \qquad (6.23)$$

where

$$\alpha_0 = 1 + \pi \left(h_1/d_1 \right) \tag{6.24}$$

Solving for u_a in (6.23) gives

$$u_a \approx \frac{e^{-\alpha_0}}{1 - e^{-\alpha_0}} \tag{6.25}$$

Replacing w_1 with $-u_b$ in (6.22) and rearranging gives

$$u_b = \alpha_0 + \ln[u_b] \tag{6.26}$$

Letting $u_b = \alpha_0 + \delta$ with $\delta/\alpha_0 < 1$ allows rewriting (6.26) as

$$\delta = \ln[\alpha_0 + \delta] \approx \ln[\alpha_0] + (\delta/\alpha_0) \tag{6.27}$$

Solving for δ gives

$$\delta = \alpha_0 \ln[\alpha_0] / (\alpha_0 - 1) \tag{6.28}$$

and

$$u_b = \frac{\alpha_0 \left(\alpha_0 - 1 + \ln[\alpha_0] \right)}{\alpha_0 - 1} \tag{6.29}$$

If we let the top plate have a potential of $V_1/2$, then using (5.104), the complex potential in the w_1-plane is

$$\Phi[w_1] = -jV_1\ln[w_1] / (2\pi) \tag{6.30}$$

Letting $\Phi[-u_a] = \frac{1}{2}V_1 + j\Psi_1$ gives

$$\Phi[-u_a] = \tfrac{1}{2}V_1 + j\Psi_1 = \tfrac{1}{2}V_1 - j\tfrac{1}{2}V_1 \left(\ln[u_a]/\pi \right) \tag{6.31}$$

Plugging (6.25) into (6.31) and solving for Ψ_1 gives

$$\Psi_1 = \tfrac{1}{2} \left(V_1/\pi \right) \left(\alpha_0 + \ln\left[1 - e^{-\alpha_0}\right] \right) \tag{6.32}$$

Letting $\Phi[-u_b] = \frac{1}{2}V_1 + j\Psi_2$ gives

$$\Phi[-u_b] = \tfrac{1}{2}V_1 + j\Psi_2 = \tfrac{1}{2}V_1 - j\tfrac{1}{2}V_1 \left(\ln[u_b]/\pi \right) \tag{6.33}$$

Plugging (6.29) into (6.33) and solving for Ψ_2 gives

$$\Psi_2 = -\frac{V_1}{2\pi} \ln\left[\frac{\alpha_0 \left(\alpha_0 - 1 + \ln[\alpha_0] \right)}{\alpha_0 - 1} \right] \tag{6.34}$$

Before calculating the total capacitance, we calculate $\underline{C}_{in}$ and $\underline{C}_{out}$ requiring Ψ at the edge of the plate. Letting $\Phi = \frac{1}{2}V_1 + j\Psi_3$ when $w_1 = -1$ gives

$$\Psi_3 = 0 \qquad (6.35)$$

With Ψ_3, we can calculate $\underline{C}_{in}$ as

$$\underline{C}_{in} = \frac{\epsilon\left(\Psi_1 - \Psi_3\right)}{\frac{1}{2}V_1} = \epsilon\frac{h_1}{d_1} + \frac{\epsilon}{\pi}\left(1 + \ln\left[1 - e^{-\alpha_0}\right]\right) \qquad (6.36)$$

and $\underline{C}_{out}$ as

$$\underline{C}_{out} = \epsilon\left(\Psi_3 - \Psi_2\right) / \left(\tfrac{1}{2}V_1\right) = \frac{\epsilon}{\pi}\ln\left[\frac{\alpha_0\left(\alpha_0 - 1 + \ln[\alpha_0]\right)}{\alpha_0 - 1}\right] \qquad (6.37)$$

The total capacitance per unit length $\underline{C}$ is $\underline{C}_{in} + \underline{C}_{out}$

$$\underline{C} = \epsilon\frac{h_1}{d_1} + \frac{\epsilon}{\pi}\left(1 + \ln\left[\frac{\alpha_0\left(1 - e^{-\alpha_0}\right)\left(\alpha_0 - 1 + \ln[\alpha_0]\right)}{\alpha_0 - 1}\right]\right) \qquad (6.38)$$

The first term in (6.38) is the infinite parallel plate capacitance approximation. The second term is due to the fringing capacitance.

6.1.4 Comparison of Zero Plate Thickness Models

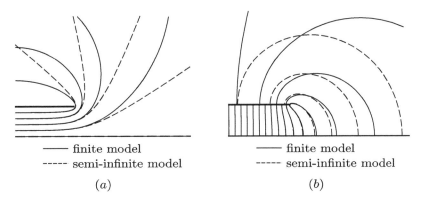

——— finite model
- - - - semi-infinite model

(a)

——— finite model
- - - - semi-infinite model

(b)

FIGURE 6.6 Comparison of the finite plate capacitor model and the semi-infinite plate capacitor model. Both models have zero plate thickness. (a) Constant potential contour comparison and (b) lines of force comparison. The semi-infinite model underestimates the amount of charge on the backside of the plate.

In Fig. 6.6, a comparison of the equipotential curves and constant lines of force for one quarter of the capacitor with $h_1/d_1 = 2$ is shown for the two capacitor models of zero thickness. In Fig. 6.6, the solid lines are for the finite plate model and the dashed lines are for the semi-infinite plate model. The field maps align well for the region located below the top capacitor plate. In the region outside the top plate, the field maps begin to show differences. The lines of force contours that terminate on the backside of the top plate for the finite plate model have less distance between them than the lines of force contours for the semi-infinite plate model. This means the charge density is higher for the finite plate than the semi-infinite plate. We can compute the charge density on the backside of the top plate for both models using (2.144). The charge density for the finite plate model is

$$\sigma = -\epsilon \frac{V_1}{d_1} \frac{\pi}{2K[m']} \frac{1}{(K[m](w_1 - 1) + E[m])}, \quad 0 < w_1 < \eta \qquad (6.39)$$

The charge density for the semi-infinite plate model is

$$\sigma = -\epsilon \frac{V_1}{d_1} \frac{1}{(w_1 + 1)}, \quad -u_b < w_1 < -1 \qquad (6.40)$$

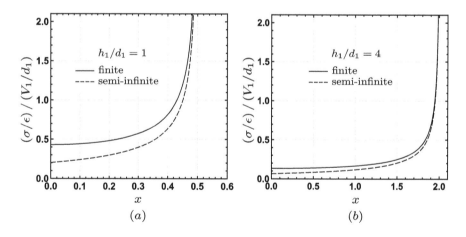

FIGURE 6.7 Comparison of the charge density on the backside of top plate for the finite plate capacitor model and the semi-infinite plate capacitor model. Both models have zero plate thickness. (a) $h_1/d_1 = 1$ and (b) $h_1/d_1 = 4$.

Comparisons of the charge density on the backside of the top plate for $h_1/d_1 = 1$ and $h_1/d_1 = 4$ are shown in Fig. 6.7. Consistent with the field map comparison, the charge density for the finite plate model is larger than the semi-infinite plate model. A larger charge density results in a larger $\underline{C}_{out}$ for the finite plate model compared to the semi-infinite plate model. The larger charge density over the entire backside of the finite plate also shows our initial assumption of equal charge for the same plate width in the two models was incorrect.

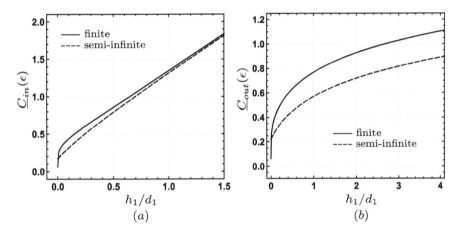

FIGURE 6.8 Comparison of (a) the inside capacitance $\underline{C}_{in}$ and (b) the outside capacitance $\underline{C}_{out}$ for the finite plate capacitor model and the semi-infinite plate capacitor model. Both models have zero plate thickness. Due to the under estimation of the charge on the backside of the top contact, the semi-infinite model has a smaller $\underline{C}_{out}$ than the finite model.

A comparison of $\underline{C}_{in}$ and $\underline{C}_{out}$ is shown in Fig. 6.8 for the two models. For $h_1/d_1 > 1$, excellent agreement is observed for $\underline{C}_{in}$ and the difference in $\underline{C}_{out}$ accounts for almost all the difference in capacitance between the two models. Again, this is easily predicted from the field map comparison. One item not obvious from the field map comparison or charge density comparisons is the almost constant difference in $\underline{C}_{out}$ for $h_1/d_1 > 1$. This is not a result of inaccuracy from the series expansion expression for $\underline{C}_{out}$, but instead a consequence of assuming a semi-infinite plate and a finite plate have the same amount of charge over the same region. The almost constant difference in $\underline{C}_{out}$ and excellent agreement for $\underline{C}_{in}$ results in an almost constant difference in the total capacitance for $h_1/d_1 > 1$ shown in Fig. 6.9. If the difference of 0.225ϵ is added to semi-infinite model, the accuracy is better than 1 % for $h_1/d_1 > 1$. This correction offset is expected to hold for all models that use a semi-infinite capacitor plate to approximate a finite capacitor plate combined with the assumption that both plates have the same charge for the same capacitor plate width.

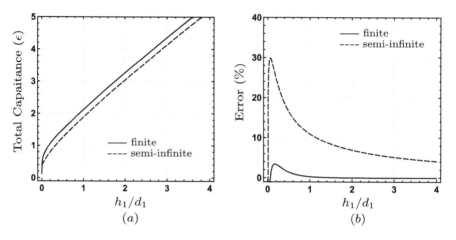

FIGURE 6.9 Comparison of (a) total capacitance and (b) relative error for the semi-infinite plate model. The dashed error line has no correction. The solid error line is the remaining error after adding the 0.225ϵ correction to the capacitance calculated by the semi-infinite plate model.

6.1.5 Finite Thickness, Semi-infinite Plate Width Model

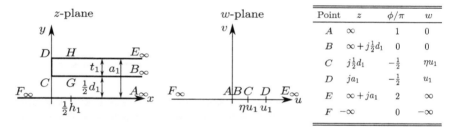

Point	z	ϕ/π	w
A	∞	1	0
B	$\infty + j\frac{1}{2}d_1$	0	0
C	$j\frac{1}{2}d_1$	$-\frac{1}{2}$	ηu_1
D	ja_1	$-\frac{1}{2}$	u_1
E	$\infty + ja_1$	2	∞
F	$-\infty$	0	$-\infty$

FIGURE 6.10 S-C transformation of a semi-infinite geometry that models the end of a parallel plate capacitor where the plates have a finite thickness t_1 in the z-plane onto the upper half of the w-plane.

Building on the zero thickness semi-infinite plate approximation, we develop the semi-infinite capacitor plate with finite thickness model shown in Fig. 6.10. From Section 5.15, the S-C transformation for Fig. 6.10 is given by (5.165) with $\alpha = 1/2$, $b = d_1/2$, and $a = t_1 + (d_1/2)$

$$z = \frac{d_1}{\pi}\left(\frac{\sqrt{w_1 - 1}\sqrt{w_1 - \eta}}{2\sqrt{\eta}} + \ln\left[\frac{\sqrt{\eta(w_1 - 1)} + \sqrt{w_1 - \eta}}{\sqrt{(1 - \eta)w_1}}\right]\right.$$
$$\left. + \left(\frac{1 + \eta}{2\sqrt{\eta}}\right)\ln\left[\frac{\sqrt{w_1 - 1} - \sqrt{w_1 - \eta}}{\sqrt{\eta - 1}}\right]\right)$$

$$(6.41)$$

For $\alpha = 1/2$, an analytic expression for η can be obtained as

$$\eta = \left(\alpha_1 - \sqrt{\alpha_1^2 - 1}\right)^2, \quad \alpha_1 = 1 + \frac{2t_1}{d_1} = \frac{1+\eta}{2\sqrt{\eta}} \tag{6.42}$$

To determine an expression for the capacitance, we assume the charge on the semi-infinite top plate located between points G and H is the same as the charge on the real capacitor's half plate. If we let the potential on the top plate be $V = 0$ and the real axis have a potential $V = \frac{1}{2}V_1$, then from (5.104) the relationship between w_1 and the complex potential in the w_1-plane is

$$w_1 = \exp[j2\pi\Phi/V_1], \quad \Phi = -jV_1\ln[w_1]/(2\pi) \tag{6.43}$$

We can express z as a function of Φ by replacing w_1 in (6.41) with (6.43). When $z = \frac{1}{2}(h_1 + jd_1)$, we have $w_1 = \exp[-2\pi\Psi_2/V_1] \ll \eta, 1$ allowing (6.41) to be approximated as

$$h_1 \approx \frac{2d_1}{\pi}\left(\ln\left[\frac{2\sqrt{\eta}}{\sqrt{1-\eta}}\right] + \left(\frac{1+\eta}{2\sqrt{\eta}}\right)\ln\left[\frac{1-\sqrt{\eta}}{\sqrt{1-\eta}}\right] + \pi\frac{\Psi_2}{V_1} - \frac{1}{2}\right) \tag{6.44}$$

Letting $\Phi = j\Psi_1$ when $w_1 = \eta$ gives

$$\eta = \exp[-2\Psi_1\pi/V_1] \tag{6.45}$$

Combining (6.44) and (6.45) gives

$$\frac{C_{in}}{\epsilon} = \frac{(\Psi_2 - \Psi_1)}{\frac{1}{2}V_1} = \frac{h_1}{d_1} + \frac{1}{\pi} - \frac{1}{\pi}\ln\left[\frac{4}{1-\eta}\right] - \frac{(1+\eta)}{\pi\sqrt{\eta}}\ln\left[\frac{1-\sqrt{\eta}}{\sqrt{1-\eta}}\right] \tag{6.46}$$

The first term of (6.46) is the ideal parallel plate capacitor approximation. The remaining terms are due to distortions in the field lines caused by the finite edge of the capacitor (labeled fringing capacitance previously).

The sidewall capacitance is easily determined since the location of the start ($w_1 = \eta$) and end ($w_1 = 1$) of the sidewall is known in the w_1-plane. The flux due to the sidewall alone is obtained from (6.43) as

$$\epsilon\Delta\Psi_{sw} = \left(\frac{1}{2}V_1\epsilon/\pi\right)(\ln[1] - \ln[\eta]) \tag{6.47}$$

which gives the sidewall capacitance

$$\underline{C}_{sw} = -(2\epsilon/\pi)\ln\left[\alpha_1 - \sqrt{\alpha_1{}^2 - 1}\right] = -(\epsilon/\pi)\ln[\eta] \tag{6.48}$$

Note that (6.48) predicts a sidewall capacitance that is independent of h_1 and is only depends on t_1/d_1.

When $z = \frac{1}{2}(h_1 + j(d_1 + 2t_1))$, let $w_1 = \delta^{-1} = \exp[-2\pi\Psi_3/V_1]$ which leads to

$$
h_1 = \frac{d_1}{\pi}\left(\left(\frac{1+\eta}{\sqrt{\eta}}\right)\ln\left[\frac{\sqrt{1-\eta\delta}-\sqrt{1-\delta}}{\sqrt{\delta}\sqrt{1-\eta}}\right]\right.
$$
$$
\left. +\frac{\sqrt{1-\delta}\sqrt{1-\eta\delta}}{\delta\sqrt{\eta}} + 2\ln\left[\frac{\sqrt{\eta(1-\delta)}+\sqrt{1-\eta\delta}}{\sqrt{1-\eta}}\right]\right)
\tag{6.49}
$$

In order to obtain a closed form expression for Ψ_3/V_1, engineering judgment must be used. From Fig. 6.2, the contribution of $\underline{C}_{out}$ to the total capacitance is largest when the capacitor plate is thin. Therefore, it is more important to have an accurate value for $\underline{C}_{out}$ when the capacitor plate is thin than when it is thick. A series expansion of (6.49) with the condition $\delta^{-1} \gg \eta, 1$ and terms in δ^n where $n \geq 1$ neglected gives

$$
h_1 \approx \frac{d_1}{\pi\sqrt{\eta}}\left(\left(\frac{1+\eta}{2}\right)\left(\ln\left[\frac{1-\eta}{4}\right]-1\right)\right.
$$
$$
\left. +\delta^{-1} + \left(\frac{1+\eta}{2}\right)\ln[\delta] + 2\sqrt{\eta}\ln\left[\frac{1+\sqrt{\eta}}{\sqrt{1-\eta}}\right]\right)
\tag{6.50}
$$

There are two issues with the series expansion of (6.50). First, there is no justification to ignore the $\ln[\delta]$ term since the magnitude of this term increases as δ decreases. Unfortunately, keeping this term prevents an analytic expression for δ from being obtained. Second, there is no reason for the series expansion to be accurate for thin capacitor plates since it was developed assuming $\delta^{-1} \gg \eta$. From (6.42), η increases as the capacitor plate thickness decreases which may impact the $\delta^{-1} \gg \eta$ assumption. Both of these issues can be addressed by adding a term x to the series expansion of (6.50). The value of x is determined through the following requirement. In the limit $t_1 \to 0$, the new series expansion should give the equivalent location for $w_1 = \delta^{-1}$ as the semi-infinite plate of zero thickness approximation of $w_1 = u_b$ where u_b is given by (6.29). From (6.42), $\eta \to 1$ in the limit that $t_1 \to 0$. Thus, adding x to right side of (6.50), taking the limit as $t_1 \to 0$ and solving for x gives

$$
\lim_{t_1\to 0}[x] = -\frac{d_1}{\pi}\left(\left(\frac{\alpha_0}{\alpha_0-1}\right)\ln[\alpha_0]+\ln[\delta]\right)
\tag{6.51}
$$

where α_0 is given by (6.24). Since (6.51) defines the limiting value of x, there is some flexibility on the final expression of x. To eliminate the $\ln[\delta]$ term from the original series expansion, we let

$$x = -\frac{d_1}{\pi\sqrt{\eta}}\left(\left(\frac{\sqrt{\eta}\alpha_0}{\alpha_0 - 1}\right)\ln[\alpha_0] + \left(\frac{1+\eta}{2}\right)\ln[\delta]\right) \tag{6.52}$$

which has the limiting value given by (6.51). Adding (6.52) to (6.50) and with $\delta^{-1} = \exp[-2\pi\Psi_3/V_1]$ allows solving for $2\Psi_3/V_1$ as

$$\begin{aligned}
\frac{2\Psi_3}{V_1} = -\frac{1}{\pi}\ln\Bigg[&\sqrt{\eta}\left(\frac{1+\eta}{2\sqrt{\eta}}\left(1 - \ln\left[\frac{1-\eta}{4}\right]\right)\right) \\
&- 2\ln\left[\frac{1+\sqrt{\eta}}{\sqrt{1-\eta}}\right] + \frac{\alpha_0\ln[\alpha_0]}{\alpha_0 - 1} + \pi\frac{h_1}{d_1}\Bigg)\Bigg]
\end{aligned} \tag{6.53}$$

Letting $\Phi = j\Psi_4$ when $w_1 = 1$ gives

$$1 = \exp[-2\pi\Psi_4/V_1] \tag{6.54}$$

requiring $\Psi_4 = 0$. The outside capacitance is

$$\begin{aligned}
\frac{C_{out}}{\epsilon} = \frac{(\Psi_4 - \Psi_3)}{\frac{1}{2}V_1} = \frac{1}{\pi}\ln\Bigg[&\sqrt{\eta}\left(\frac{1+\eta}{2\sqrt{\eta}}\left(1 - \ln\left[\frac{1-\eta}{4}\right]\right)\right) \\
&- 2\ln\left[\frac{1+\sqrt{\eta}}{\sqrt{1-\eta}}\right] + \frac{\alpha_0\ln[\alpha_0]}{\alpha_0 - 1} + \pi\frac{h_1}{d_1}\Bigg)\Bigg]
\end{aligned} \tag{6.55}$$

The total capacitance is

$$C_{total} = C_{in} + C_{sw} + C_{out} \tag{6.56}$$

6.1.6 Comparison of Finite Plate Thickness Model and Finite Element Method (FEM) Simulations

In Fig. 6.11, a comparison of the equipotential curves and constant lines of force for one quarter of the capacitor with $h_1/d_1 = 4$ and $t_1/d_1 = 1.5$ is shown for the semi-infinite model with plate thickness greater than zero (dashed lines) and a finite element simulation (solid lines). Similar features are present in this comparison as were seen in the comparison of the zero thickness models. The field maps align well for the region located below the top capacitor plate and along most of the sidewall of the plate. This should result in good agreement between C_{in} and C_{sw} calculated by the semi-infinite model and the finite element simulation. The lines of force contours that terminate on the backside of the top plate for the finite element simulation have less distance between them

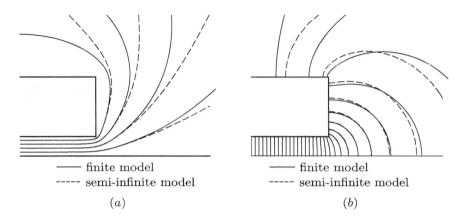

——— finite model   ——— finite model

- - - - semi-infinite model   - - - - semi-infinite model

(a)   (b)

FIGURE 6.11 Field map comparison of semi-infinite model (dashed lines) and finite element simulation (solid lines) with $h_1/d_1 = 4$ and $t_1/d_1 = 1.5$. (a) Equipotential curves (b) lines of force curves. Semi-infinite model again underestimates the charge on the backside of the plate.

than the lines of force contours for the semi-infinite plate model. This means that the charge density is higher for the finite element simulation than the semi-infinite plate model and $\underline{C}_{out}$ is lower for the semi-infinite model compared to the finite element simulation.

Error analysis for $\underline{C}_{in}$, $\underline{C}_{sw}$, and $\underline{C}_{out}$ from the semi-infinite model compared to finite element simulations is shown in Fig. 6.12. The error value of each curve is relative to the total capacitance. This means the total error is the sum of the three error values for a given h_1/d_1 value in Fig. 6.12. As predicted by the field map comparison, excellent agreement is observed for $\underline{C}_{in}$ and good agreement is observed in $\underline{C}_{sw}$ for $h_1/d_1 > 4$. Although the semi-infinite models predicted a sidewall capacitance that was independent of h_1, Fig. 6.12 shows $\underline{C}_{sw}$ has some dependence on h_1 for $h_1/d_1 < 1$ but is almost constant for $h_1/d_1 > 2$. For $h_1/d_1 > 2$, the largest difference is observed in $\underline{C}_{out}$ and accounts for almost all the difference in capacitance between the semi-infinite model and the finite element simulations. This again is a consequence of assuming a semi-infinite plate and a finite plate have the same amount of charge over the same region. The almost constant difference in $\underline{C}_{out}$ results in an almost constant difference in the total capacitance for $h_1/d_1 > 1$ shown in Fig. 6.13.

Based on the previous zero thickness model analysis, the accuracy of the semi-infinite model should improve by adding 0.225ϵ to it. Associating this offset with $\underline{C}_{out}$ shifts the C_{out} error curve in Fig. 6.12 up so that

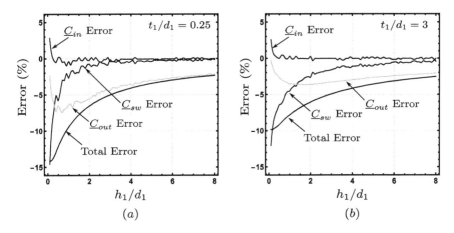

FIGURE 6.12 Error analysis for semi-infinite model compared to finite element simulations for the three capacitors that contribute to the total capacitance. The total error is the sum of the individual errors. (a) For a thin capacitor plate ($t_1/d_1 = 0.25$) and (b) for a thick capacitor plate ($t_1/d_1 = 3$).

it is approximately zero for large h_1/d_1 values. Since the shape of the C_{out} error curve is almost opposite to the shape of the C_{sw} error curve, the two errors can be made to almost cancel each other for $h_1/d_1 < 2$. Error analysis performed with and without the correction of 0.225ϵ is shown in Fig. 6.14. With the correction, the semi-infinite model agrees with the finite element simulations to better than 1 % for $h_1/d_1 > 1$. Thus, an accurate model using a semi-infinite geometry and conformal mapping has been developed.

6.1.7 Key Findings and Summary

Let us know summarize the path to an accurate conformal mapping model for the parallel plate capacitor and the key findings of this section. First, conformal mapping was used to produce a semi-infinite model of zero plate thickness and a finite model of zero plate thickness. This led to the key finding that in some cases finite geometries can be well modeled by infinite geometries. Additionally, a capacitance correction offset of 0.225ϵ was found to compensate for the charge difference between the semi-infinite and finite models. Building on this understanding, conformal mapping was used to develop a semi-infinite model of plate thickness greater than zero. Using engineering judgment to force an accurate C_{out} for thin capacitor plates and adding the 0.225ϵ correction resulted in a

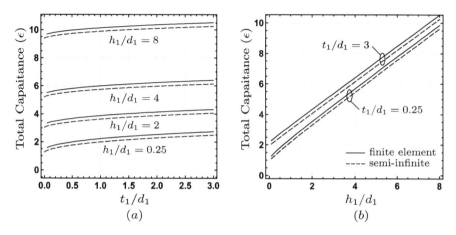

FIGURE 6.13 Comparison between the semi-infinite model and finite element simulations for the total capacitance as a function of (a) plate thickness to plate separation (t_1/d_1) and (b) plate width to plate separation (h_1/d_1). An almost constant difference between the two models is seen in both comparisons.

closed form highly accurate ($< 1\%$ error) parallel plate capacitor model with finite plate thickness.

Closed form solutions are highly desired as they allow determining the key variables and the natural unit of the length for the problem. In the parallel plate capacitor with finite plate thickness case, h_1/d_1 and t_1/d_1 are the key variables that determine the capacitance and the plate to plate distance d_1 is the natural unit of length. Thus, the definition of a thick capacitor plate must depend on d_1. This results is easily obtained due to the closed form solution, but may not be apparent at first with finite element solutions. The key to obtaining the closed form solution was the reduced mathematical complexity of the semi-infinite model. This is seen in the other case studies in this chapter. Typically, the trade-off for using a semi-infinite geometry to approximate a finite geometry is the geometric region over which the model is valid/accurate. The insight into the physics of the problem obtained by the closed form solution is often well worth the trade-off in accuracy.

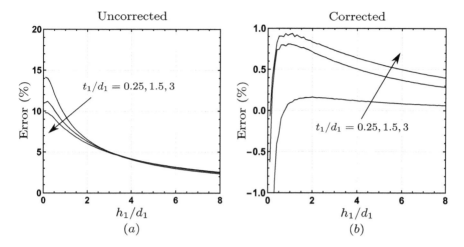

FIGURE 6.14 (a) Uncorrected error between the semi-infinite model and finite difference simulations as a function of plate width to plate separation (h_1/d_1). (b) Remaining error after adding 0.225ϵ to the semi-infinite model. After correction, the error is less than 1 % for $h_1/d_1 > 1$.

6.2 CHARACTERISTIC IMPEDANCE OF LOSSLESS TRANS-MISSION LINES

6.2.1 Introduction

Transmission lines are often used to send dynamic electrical signals from a signal source to another desired point in the circuit. In order to design the circuit for maximum power transfer, the circuit designer must know the characteristic impedance of the transmission line or design the transmission lines for a specific characteristic impedance. When the loss of the transmission line can be neglected, the characteristic impedance of the transmission line has the simple form [7]

$$Z_0 = \frac{1}{v\underline{C}} \tag{6.57}$$

where v is the velocity of the signal and $\underline{C}$ is the capacitance per unit length of the transmission line. When the medium surrounding the transmission line conductors has a uniform relative permittivity ϵ_r, the velocity is

$$v = c/\sqrt{\epsilon_r} \tag{6.58}$$

where c is the speed of light in vacuum. Thus, determining the characteristic impedance of a lossless transmission line that consists of parallel conductors surrounded by material of uniform permittivity simply requires determining the capacitance per unit length of the transmission line. For these cases, conformal mapping can often be used to find $\underline{C}$.

6.2.2 Small Diameter Wire Approximation

Let a line charge q_0 exist at $z = z_0$. If there is a region $|z - z_0| \leq r_0$ that consists of only uniform permittivity dielectric, then there exist a region $|z - z_0| < r_1 \leq r_0$ where the equipotential contours are approximately circular independent of the environment for the region $|z - z_0| > r_0$. The value of r_1 is determined by the environment in the region $|z - z_0| > r_0$. This leads to the ability to approximate wires of small (to be defined later) diameter by an equipotential contour near a line charge. For example, the complex potential of the region between a wire with circular cross-section of radius r_1 and center a distance d above a ground plane can be approximated by the complex potential of a line charge q_0 located at $z = jd$ above a ground plane

$$\Phi = \frac{q_0}{2\pi\epsilon} \ln\left[\frac{z + jd}{z - jd}\right] \tag{6.59}$$

The potential along the perimeter of the wire for this approximation is plotted in Fig. 6.15 and seen to vary with position. Typically, the average value (or suitable approximation to the average value) is used as the potential value for the wire. For this example, we take the value of the potential at $z = jd + r_1$ (which is approximately the average potential value) as the potential on the wire. One over capacitance can then be computed as

$$\frac{1}{\underline{C}} = \frac{V[jd + r_1]}{q_0} = \frac{1}{2\pi\epsilon}\text{Re}\left[\ln\left[\frac{r_1 + j2d}{r_1}\right]\right] \approx \frac{1}{2\pi\epsilon}\ln\left[\frac{\sqrt{4 + (r_1/d)^2}}{r_1/d}\right] \tag{6.60}$$

and the characteristic impedance of the transmission line in vacuum is

$$Z_0 = \frac{1}{2c\pi\epsilon_0}\ln\left[\frac{2 - (r_1/d)}{r_1/d}\right] \tag{6.61}$$

The true capacitance for this case was obtained in Section 3.3.3 giving

$$\frac{1}{\underline{C}_{true}} = \frac{1}{2\pi\epsilon}\cosh^{-1}[d/r_1] \tag{6.62}$$

A plot of relative error $(\underline{C} - \underline{C}_{true})/\underline{C}_{true}$ is shown in Fig. 6.15 (c). The definition of small depends on two items. The first item is the ratio of

r_1/d and not the absolute value of r_1. The second item is the amount of relative error that is tolerable. Based on this example, a rule of thumb is r_1 should be about 3× smaller than the distance to the closest change in environment to the line charge. A change in environment could be a boundary (as in this example) or even another line charge.

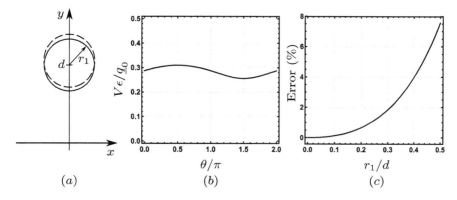

(a)   (b)   (c)

FIGURE 6.15 (a) Comparison of equipotential contour based on the line charge approximation (dashed line) and equipotential contour for a wire of radius $r_1 = d/3$. The equipotential contour for the line charge has the value $V[d - jr_1]$. (b) The voltage due to the line charge along the perimeter of the wire. (c) Relative error in the capacitance as a function of r_1/d. As r_1/d decreases, the error becomes smaller. The relative error is less than 2.5 % when $r_1/d < 1/3$.

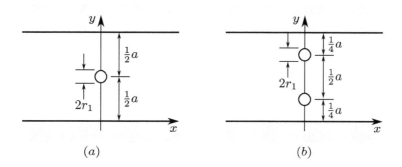

(a)   (b)

FIGURE 6.16 (a) Single wire located between two ground planes. (b) Two wires located between two ground planes.

As a second example, consider a small diameter wire located midway between two infinite horizontal ground planes with a separation distance a as shown in Fig. 6.16 (a). The complex potential can be approximated by the complex potential of a line charge located midway between two ground planes

$$\Phi = \frac{q_0}{2\pi\epsilon} \ln \left[\frac{\sinh \left[\frac{1}{2}\pi \left(z - z_0^* \right) /a \right]}{\sinh \left[\frac{1}{2}\pi \left(z - z_0 \right) /a \right]} \right] \tag{6.63}$$

Let the wire have a radius r_1 and its center at $z = ja/2$. The average value of the potential on the perimeter of the wire occurs at $z \approx j\frac{1}{2}a + e^{j\pi/4}r_1$. One over capacitance between the wire and the ground planes is

$$\frac{1}{C} = \frac{V\left[j\frac{1}{2}a + e^{j\pi/4}r_1 \right]}{q_0} \tag{6.64}$$

$$= \frac{1}{2\pi\epsilon} \mathrm{Re} \left[\ln \left[\coth \left[\frac{e^{j\pi 4}\pi r_1}{2a} \right] \right] \right] \approx \frac{1}{2\pi\epsilon} \ln \left[\frac{2a}{\pi r_1} \right]$$

and the characteristic impedance of the transmission line is

$$Z_0 = \frac{\sqrt{\epsilon_r}}{2\pi c\epsilon} \ln \left[\frac{2a}{\pi r_1} \right] \tag{6.65}$$

The final example is for a pair of wires between ground planes where the wires carry opposite currents (see Fig. 6.16 (b)). Let a be the distance between ground planes and let the bottom ground plane lie along the real axis. Let the center of the positive current wire be $z = ja/4$ and the center of the other wire be $z = j3a/4$. The characteristic impedance depends on the capacitance between the wires. Approximating the complex potential by two equal magnitude but opposite polarity line charges between ground planes gives

$$\Phi = \frac{q_0}{2\pi\epsilon} \ln \left[\frac{\sinh \left[\frac{1}{2}\pi \left((z/a) + \frac{1}{4}j \right) \right] \sinh \left[\frac{1}{2}\pi \left((z/a) - \frac{3}{4}j \right) \right]}{\sinh \left[\frac{1}{2}\pi \left((z/a) - \frac{1}{4}j \right) \right] \sinh \left[\frac{1}{2}\pi \left((z/a) + \frac{3}{4}j \right) \right]} \right] \tag{6.66}$$

The average value of the potential on the perimeter of each wire occurs at $z \approx z_0 + r_1 e^{j\pi/4}$ where z_0 is the center of the wire. One over capacitance between the two wires is

$$\frac{1}{C} = \frac{2V\left[j\frac{1}{4}a + r_1 e^{j\pi/4} \right]}{q_0} \tag{6.67}$$

$$= \frac{1}{\pi\epsilon} \mathrm{Re} \left[\ln \left[\coth \left[e^{j\pi/4}\frac{\pi r_1}{a} \right] \right] \right] \approx \frac{1}{\pi\epsilon} \ln \left[\frac{a}{\pi r_1} \right]$$

and the characteristic impedance of the transmission line is

$$Z_0 = \frac{\sqrt{\epsilon_r}}{\pi c\epsilon} \ln \left[\frac{a}{\pi r_1} \right] \tag{6.68}$$

Additional examples and an alternative conformal mapping approach for obtaining the capacitance in these examples can be found in [6].

6.2.3 Key Findings and Summary

Although this section focused on determining the characteristic impedance of lossless transmission lines, the true message is in the use of line charges to model circular conductors. The fact that the equipotential contours near a line charge surrounded by uniform dielectric are approximately circular even in the presence of other charged bodies and boundaries can be exploited for other applications as well. Lossless transmission lines were used due to their mathematical simplicity and the large amount of literature on transmission line models generated by conformal mapping techniques (which are not just of historical interest). These publications often provide different methods of attack that can be translated to other applications. All approximations have trade-offs and one approximation may be more useful than another. Additional line charges can be introduced to reduce the variation of potential on the perimeter of the wire [15, 16, 11], but if additional accuracy is the goal, finite element simulations would be more useful compared to this refinement.

6.3 CHARGE IMAGING ON INFINITE PLATE

Before solving problems where fixed charge is part of the boundary conditions, it is useful to understand how charge images on a plate held at a constant potential. Consider the case where a positive line charge q_0 is located at a distance y_0 above an infinite grounded plate that lies along the real axis. Without any loss in generality, we can let the location of the line charge be $z_0 = jy_0$. The complex potential is obtained using the method of images as

$$\Phi[z] = \frac{q_0}{2\pi\epsilon} \ln \left[\frac{z - z_0^*}{z - z_0} \right] \qquad (6.69)$$

The positive line charge induces an equal amount of negative charge on the infinite grounded plate. The induced charge density can be calculated using (2.144) as

$$\sigma = \epsilon \text{Im} \left[\mathcal{E}^* \frac{dz}{|dz|} \right] = -\frac{q_0}{\pi} \left(\frac{y_0}{x^2 + y_0^2} \right) \qquad (6.70)$$

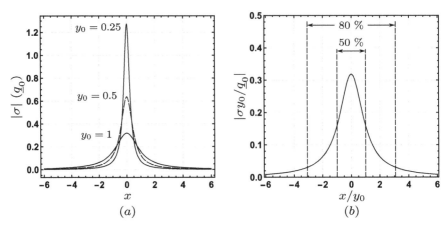

FIGURE 6.17 (a) Induced charge density on an infinite grounded conductor due to a line charge located at $z_0 = jy_0$. Note that the charge distribution becomes more spread as the line charge moves further away from the plate (y_0 becomes larger). (b) Dividing the induced charge distribution by $\underline{q_0}/y_0$ results in a universal curve that describes the induced charge distribution on the plate. 50 % of the charge is induced between $\pm y_0$ and 80 % of the charge is induced between $\pm 3.1 y_0$.

A plot of $|\sigma|$ for several y_0 values is shown in Fig. 6.17(a). The peak charge density occurs at $x = 0$. This is the location on the plate closest to the line charge. As the charge moves away from the plate (y_0 increases), the induced charge is more spread out on the plate and the magnitude of the peak charge density goes down. If instead of plotting $|\sigma|$, we plot the magnitude of the normalized charge density $|\sigma_n|$ defined as

$$\sigma_n = \sigma / \left(\underline{q_0}/y_0 \right) = -\frac{1}{\pi} \left(\frac{1}{x_n^2 + 1} \right) \tag{6.71}$$

where $x_n = x/y_0$, then all the curves previously plotted become the single curve shown in Fig. 6.17(b). This means the induced charge distribution on an infinite plate due to a single line located at z_0 can be fully described by a universal curve. Normalized variables often allow the relationships between variables to be most compactly described. To demonstrate the benefit and usage of a universal curve, we analyze the induced charge density universal curve in detail. First, the x_n-axis is in unit of y_0 and the $|\sigma_n|$-axis is in unit of $\underline{q_0}/y_0$. Thus any x_n value can be converted to its x value as $x = y_0 x_n$. Similarly, any value on the curve is multiplied by $\underline{q_0}/y_0$ to obtain the $|\sigma|$ value at that point. The peak σ_n value is $-1/\pi$ which has a σ value of $-q_0/(\pi y_0)$. Next, integration of σ_n from $x_{n,1}$ to $x_{n,2}$ gives

$$\int_{x_{n,1}}^{x_{n,2}} \sigma_n dx_n = \frac{1}{q_0} \int_{x_1}^{x_2} \sigma dx = \frac{\tan^{-1}[x_{n,1}] - \tan^{-1}[x_{n,2}]}{\pi} \tag{6.72}$$

Thus, the value obtained by integrating σ_n over an x_n-range should be multiplied by q_0 to obtain the value of σ integrated over the equivalent x-range. With (6.72), we can quantify how charge is imaged onto the infinite plate . Since $\tan^{-1}[\pm 1] = \pm \pi/4$, half of the total charge is induced on the plate between $-y_0 < x < y_0$. The region $-3.1y_0 < x < 3.1y_0$ has about 80 % of the charge induced on it. Based on this analysis, the definition of long and short plates depends on the minimum distance the charge is from the plate. For example, the minimum length for a long plate could be defined as the same length that satisfies 80 % of the charge on an infinite plate $(6.2y_0)$. If the charge is 1 unit away from the plate, the plate must be at least 6.2 unit long to be considered a long plate. If the charge is 0.1 unit away from the plate, then the plate only needs to be 0.62 unit long to be considered a long plate.

Using the complex potential for a grounded finite plate of length L and a charge of q_0 located at $z_{1,0}$ given by (5.116) allows comparing field maps of a finite plate to an infinite plate. Figure 6.18 shows the field map (solid lines) for a charge located at $z_0 = jy_0$ with a grounded finite plate of length $6.2y_0$ that is coincident with the real axis and symmetric about the imaginary axis compared to the field map (dashed lines) of an infinite grounded plate with the same charge located y_0 above it. The spacing between adjacent contours (dashed or solid) is $\Delta V = \Delta \Psi = 0.1q_0/\epsilon$. The lines of force comparison shows 60 % of the induced charge closest to the line charge is almost the same for the finite plate and infinite plate. Near the edges of the finite plate, the distance between the solid lines of force curves decreases compared to the dashed lines of force curves. This indicates that the charge density is higher toward the ends of the finite plate compared to the infinite plate. For the case where a charge q_0 is located y_0 above the center of a finite plate of length l, the normalized charge density on the side of the plate that faces the charge is

$$\sigma_n = \sigma/(q_0/y_0) = -\frac{2}{\pi \left(\sqrt{l_n^2 + 4} - \sqrt{l_n^2 - 4x_n^2}\right)\sqrt{l_n^2 - 4x_n^2}} \tag{6.73}$$

where

$$l_n = l/y_0, \quad x_n = x/y_0 \tag{6.74}$$

A comparison of the charge density for the infinite plate (6.71) and the

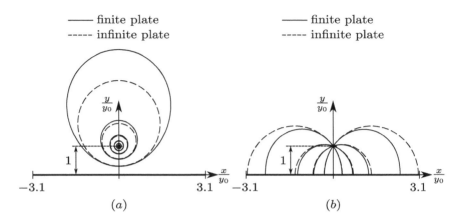

FIGURE 6.18 Comparison of (a) constant potential contours and (b) constant lines of force for a finite plate of length $6.2y_0$ and an infinite plate. Good agreement is obtained for the electrostatic potential in the region between the line charge and the plate. Good agreement is obtained for the lines of force away from the edges of the finite plate.

charge density of a finite plate (6.74) for two lengths is shown in Fig. 6.19. Two main conclusions can be drawn from Fig. 6.19. First, the induced charge density of the infinite plate is always lower than the induced charge density of the finite plate. Second, the edge of a plate that is $5y_0$ or more away from charge has little impact on the distribution of induced charge density due to the charge. Thus, infinite plates should form good approximations for such cases.

Although infinite charge densities are observed at the ends of the plate in Fig. 6.19, it should be noted that this does not mean infinite charge at the edges. The infinite charge density is due to the fact that charge can exist on the edges of the plate, but the edges have no area. Therefore, finite charge density divided by zero area results in infinite charge density. This also explains the infinite charge density at all corners with negative external angles. We are now ready to tackle problems that consist of boundary conditions that include fixed charges.

6.4 FIELD PLATES

6.4.1 Introduction

In the field of high power radio frequency (RF) field effect transistors (FETs), large electric fields can lead to catastrophic failure or early degradation in performance. One approach to mitigate the creation of

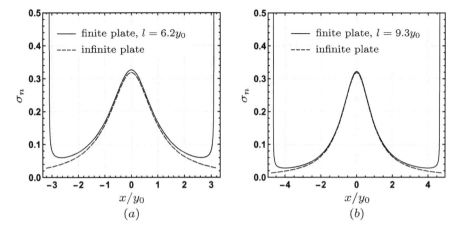

FIGURE 6.19 Comparison of induced charge density on an infinite plate and a finite plate of length (a) $6.2y_0$ and (b) $9.3y_0$. Good agreement is obtained for $|x/y_0| < 2$ for both plates, but as the plate becomes longer, the infinite plate becomes an even better approximation.

large electric fields in FETs is to use a field plate. The cross-section of the gated region of a typical single channel high electron mobility transistor with and without a field plate is shown in Fig. 6.20. The gate to charge distance is d_1 and the field plate to charge distance is d_2. The transition from gate to field plate has an angle $\pi\alpha$. The two main design parameters for the field plate are transition angle $\pi\alpha$ and the distance from the field plate to the fixed charge d_2. As will be shown, the field plate allows a more uniform induced charge on the gate/field plate metal which lowers the peak electric field. The objective of this section is to use conformal mapping to create a model that provides insight on how $\pi\alpha$ and d_2 impact the peak electric field in the region bounded on top by the metal, on bottom by the sheet charge, on the left by line EF and on the right by line AB as shown in Fig. 6.20.

The highest electric field during transistor operation occurs when the transistor is in its off-state. This means no current is flowing and we have effectively an electrostatic condition. To obtain a geometry that can be conformally mapped, the region ABEF in Fig. 6.20 (b) is approximated by the geometry in Fig. 6.21 (a) that places A at $z = \infty$, B at $z = \infty + jd_2$, E at $z = -\infty + jd_1$ and F at $z = -\infty$. Due to these approximations, the model is only valid for long gate/field plate designs. A quantitative definition of a long gate/field plate can be found [3] as satisfying $l/d > 6$ where l is the length of the plate and d is the distance

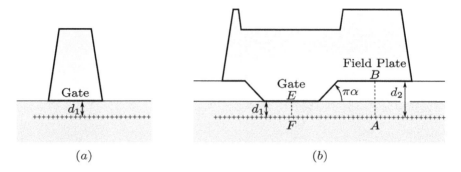

FIGURE 6.20 Cross-section of a field effect transistor (a) without and (b) with a field plate.

from plate to fixed charge. We also assume the region between the metal and fixed charge has a uniform permittivity ϵ_u. To obtain the electric field in the z-plane, we use conformal mapping to flatten the plate and place the fixed charge a constant distance from the plate in the w_2-plane. The complex potential for a unit charge above an infinite grounded plane is easily obtained by the method of images. Superposition can then be used to obtain the total complex potential and electric field due to all the charges.

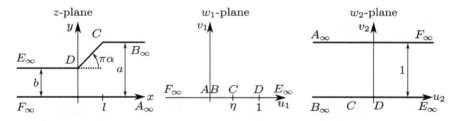

FIGURE 6.21 Conformal mapping of gate/field plate geometry into a flat plate with charge a constant distance away from the gate/field plate.

Before we determine expressions for the complex potential and electric fields, it is useful to determine the limiting behavior in certain regions of the model. For example, far away from the gate/field plate transition (in either direction), the model reduces to a uniform sheet of charge a fixed distance from a grounded conducting plate. These are equivalent conditions to the infinite parallel plate capacitor. Therefore, the potential far to the left or right of the transition should approach (2.149) with σ replace with $-\sigma_z$ since the plate now has negative charge induced on it

$$\Phi[z] - \Phi[0] = j\sigma_z z/\epsilon_u \tag{6.75}$$

The electrostatic potential drop from gate to the location of fixed charge in the region far away from the transition is

$$V_{p,g} = \text{Re}[\Phi[0] - \Phi[-x_1 + jd_1]] = \sigma_z d_1/\epsilon_u \tag{6.76}$$

and is called the pinch-off voltage of the gate. The electrostatic potential drop from field plate to the location of fixed charge in the region far away from the transition is

$$V_{p,fp} = \text{Re}[\Phi[0] - \Phi[x_1 + jd_2]] = \sigma_z d_2/\epsilon_u \tag{6.77}$$

and is called the pinch-off voltage of the field plate. These two values can be used as checks for the potential equation. Based on the same arguments, the electric field far away from the transition is a constant value in the positive y-direction

$$\mathcal{E}[w_2] = j\sigma_z/\epsilon_u \tag{6.78}$$

In the parallel plate capacitor problem, uniform charge is present on the plate because it is a flat plate. This implies that any changes in geometry to the gate/field plate that result in less deviation from a flat plate should result in a more uniform induced charge distribution on the gate/field plate. In other words, the closer d_2/d_1 is to 1 or the smaller α is, the more uniform the induced charge distribution on the gate/field plate. A more uniform charge distribution should result in lower peak electric field.

6.4.2 Conformal Mapping to a Flat Plate

Two conformal transformations are needed to obtain the flat plate along the real axis of the w_2-plane with the fixed charge a constant distance from the flat plate. The first transformation places the gate/field plate metal on the positive real axis and the fixed charge on the negative real axis of the w_1-plane. This is the same transformation given by (5.175) with $\pi\alpha$ as the transition angle between gate and field plate, $b = d_1$, $a = d_2$, and $\eta = (d_1/d_2)^{1/\alpha}$

$$z = l + jd_2 + \frac{d_1(w_1 - \eta)^{1-\alpha}}{\pi(\alpha - 1)(\eta - 1)^{1-\alpha}} {}_2F_1\left[1 - \alpha, 1 - \alpha, 2 - \alpha, \frac{\eta - w_1}{\eta - 1}\right]$$

$$- \frac{d_1(w_1 - \eta)^{1-\alpha}w_1^{\alpha-1}}{\pi(\alpha - 1)(\eta - 1)^{1-\alpha}\eta^\alpha} {}_2F_1\left[1 - \alpha, 1 - \alpha, 2 - \alpha, \frac{(\eta - w_1)}{w_1(\eta - 1)}\right] \tag{6.79}$$

where

$$l = (d_2 - d_1) \cot[\pi\alpha] \tag{6.80}$$

The second transformation places the gate/field plate metal along the entire real axis and the charge a distance 1 above the real axis in the w_2-plane

$$w_1 = \exp[\pi w_2] \tag{6.81}$$

6.4.3 Complex Potential Analysis

The complex potential in the w_2-plane due to a unit charge located at $w_0 = u_0 + j$ is

$$\Phi_1[w_2, w_0] = \frac{1}{2\pi\epsilon_u} \ln\left[\frac{w_2 - w_0^*}{w_2 - w_0}\right] \tag{6.82}$$

The complex potential (Φ) at any point w_2 due to the entire charge distribution is obtained by superposition

$$\Phi[w_2] = \int_{-\infty}^{\infty} \sigma_{w_2}[u_0]\, \Phi_1[w_2, w_0]\, du_0 \tag{6.83}$$

where $\sigma_{w_2}[u_0]$ is the charge density in the w_2-plane. From (4.8), the charge density in the w_2-plane is

$$\sigma_{w_2}[u_0] = \sigma_z \left|\frac{dz}{dw_2}\right|_{w_2 = u_0 + j} = \sigma_z d_1 \left(\frac{e^{\pi u_0} + 1}{e^{\pi u_0} + \eta}\right)^{\alpha} \tag{6.84}$$

Plugging (6.82) and (6.84) into (6.83) gives

$$\Phi[w_2] = \frac{\sigma_z d_1}{\epsilon_u} \left(\frac{1}{2\pi} \int_{-\infty}^{\infty} \left(\frac{e^{\pi u_0} + 1}{e^{\pi u_0} + \eta}\right)^{\alpha} \ln\left[\frac{w_2 - u_0 + j}{w_2 - u_0 - j}\right] du_0\right) \tag{6.85}$$

Unfortunately, a closed form solution to the integral of (6.85) has not been found and numerical integration must be employed.

With $w_2 = u_2 + jv_2$, the real part of (6.85) gives the electrostatic potential ($V = \mathrm{Re}[\Phi]$) as

$$V[w_2] = \frac{\sigma_z d_1}{4\pi\epsilon_u} \int_{-\infty}^{\infty} \left(\frac{e^{\pi u_0} + 1}{e^{\pi u_0} + \eta}\right)^{\alpha} \ln\left[\frac{(u_2 - u_0)^2 + (v_2 + 1)^2}{(u_2 - u_0)^2 + (v_2 - 1)^2}\right] du_0 \tag{6.86}$$

To help with convergence issues that arise with the numerical integration of (6.86), the following transformation is used. Let

$$f_1[u_0] = \left(\frac{e^{\pi u_0} + 1}{e^{\pi u_0} + \eta}\right)^{\alpha} \tag{6.87}$$

and

$$K_1[u_0] = \frac{1}{2} \int_{-\infty}^{u_0} \ln\left[\frac{(u_2 - x)^2 + (v_2 + 1)^2}{(u_2 - x)^2 + (v_2 - 1)^2}\right] dx = \pi v_2 + f_3[u_0] \quad (6.88)$$

where

$$f_3[u_0] = (1 - v_2) \tan^{-1}\left[\frac{u_2 - u_0}{1 - v_2}\right] - (v_2 + 1) \tan^{-1}\left[\frac{u_2 - u_0}{v_2 + 1}\right]$$
$$+ \left(\frac{u_0 - u_2}{2}\right) \ln\left[\frac{(u_2 - u_0)^2 + (v_2 + 1)^2}{(u_2 - u_0)^2 + (v_2 - 1)^2}\right] \quad (6.89)$$

Then the normalized electrostatic potential V_n can be written as

$$V_n = \frac{V\epsilon_u}{\sigma_z d_1} = \frac{1}{2\pi} \int_{-\infty}^{\infty} f_1[u_0] \left(\frac{dK_1[u_0]}{du_0}\right) du_0 \quad (6.90)$$

Integrating (6.90) by parts gives

$$V_n = \frac{1}{2\pi} \left(f_1[u_0] K_1[u_0]\big|_{-\infty}^{\infty} + \int_{-\infty}^{\infty} K_1[u_0] \left(-\frac{df_1[u_0]}{du_0}\right) du_0 \right) \quad (6.91)$$

where

$$-\frac{df_1[u_0]}{du_0} = -f_1'[u_0] = \frac{\pi\alpha e^{\pi u_0}(1 - \eta)}{(e^{\pi u_0} + \eta)^{1+\alpha}(e^{\pi u_0} + 1)^{1-\alpha}} \quad (6.92)$$

Since $K[-\infty] = 0$, $f_1[\infty] = 1$, and $K[\infty] = 2\pi v_2$, (6.90) becomes

$$V_n = \frac{1}{2} v_2 \left(\eta^{-\alpha} + 1\right) + \frac{1}{2\pi} \int_{-\infty}^{\infty} f_3[u_0] \left(-\frac{df_1[u_0]}{du_0}\right) du_0 \quad (6.93)$$

A plot of $-f_1'[u_0]$ for various values of α with $d_2/d_1 = 6$ is shown in Fig. 6.22 (a). For $u_0 > 0$, $e^{\pi u_0}$ becomes large quickly and $-f_1'[u_0] \approx 0$ for $u_0 > 5$. For $u_0 < 0$, $-f_1'[u_0]$ has a maximum value and then quickly becomes ≈ 0. The value of u_0 where $-f_1'[u_0]$ is a maximum is

$$u_{0,p} = \frac{1}{\pi} \ln\left[\frac{1}{2}\left(\alpha(\eta - 1) + \sqrt{4\eta + \alpha^2(\eta - 1)^2}\right)\right] \quad (6.94)$$

A plot of $-f_1'[5]$ and $-f_1'[u_{0,p} - 6]$ as a function of α is shown in Fig. 6.22 (b). The value of $-f_1'$ for these two conditions is less than 10^{-6} for all values of α. Additionally, it can be shown that $|f_3[u_0]| < \pi$. Based on this analysis, (6.93) can be approximated as

$$V_n \approx \frac{1}{2} v_2 \left(\eta^{-\alpha} + 1\right) + \frac{1}{2\pi} \int_{u_{0,p}-6}^{5} f_3[u_0] \left(-\frac{df_1[u_0]}{du_0}\right) du_0 \quad (6.95)$$

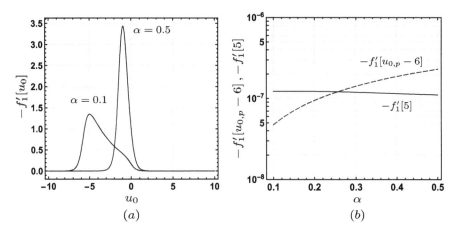

FIGURE 6.22 (a) $f_1'[u_0]$ for two different values of α. In both cases f_1' has a peak and then goes to zero over a very short range. (b) The value of f_1' evaluated at $u_0 = u_{0,p} - 6$ and $u_0 = 5$. For all reasonable values of α, f_1' is basically zero at these two values of u_0. This sets the limits of integration for (6.93).

The normalized electrostatic potential along the line of charge for various values of α with $d_2/d_1 = 6$ is shown in Fig. 6.23 (a). As α decreases, the distance required to go from the pinch-off voltage of the gate to the pinch-off voltage of the field plate increases. Since the average electric field is $\Delta V/\Delta l$, this means that the average electric field in the x-direction must decrease as α decreases. The electrostatic potential along the line of charge for various values of d_2/d_1 with $\alpha = 1/6$ is shown in Fig. 6.23 (b). As d_2/d_1 increases, V_n increases significantly for all $x/d_1 > 0$, but is approximately the same for all $x/d_1 < 0$. Thus, the average electric field in the x-direction increases as d_2/d_1 increases.

6.4.4 Electric Field Analysis

We now turn our attention to computing the electric field. In general, the electric field can be determined from the complex potential as

$$\mathcal{E}[w_2] = \left(-\frac{d\Phi}{dz}\right)^* = \int_{-\infty}^{\infty} \sigma_{w_2}[u_0] \left(-\frac{d\Phi_1[w_2, w_0]}{dz}\right)^* du_0 \qquad (6.96)$$

Plugging (6.84) and

$$\left(-\frac{d\Phi_1}{dz}\right)^* = \left(-\frac{d\Phi_1}{dw_2}\frac{dw_2}{dz}\right)^* = \frac{2j}{(w_2^* - u_0)^2 + 1}\frac{1}{d_1}\left(\frac{e^{\pi w_2^*} - \eta}{e^{\pi w_2^*} - 1}\right)^\alpha \qquad (6.97)$$

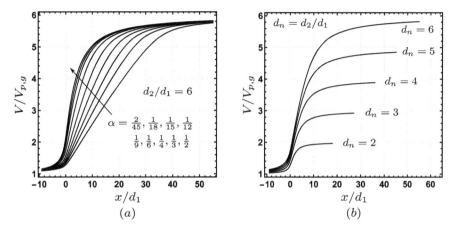

FIGURE 6.23 Electrostatic potential along the sheet of charge in the z-plane (a) as a function of gate to field plate transition angle α with $d_2/d_1 = 6$ and (b) as a function of $d_2/d_1 = d_n$ with $\alpha = 1/6$.

into (6.96) gives

$$\mathcal{E}_n = \frac{\mathcal{E}[w_2]}{\sigma/\epsilon_u} = \frac{j}{\pi}\left(\frac{e^{\pi w_2^*} - \eta}{e^{\pi w_2^*} - 1}\right)^{\alpha} \int_{-\infty}^{\infty}\left(\frac{e^{\pi u_0} + 1}{e^{\pi u_0} + \eta}\right)^{\alpha} \frac{1}{(w_2^* - u_0)^2 + 1}\,du_0 \quad (6.98)$$

where $\mathcal{E}_n$ is the normalized electric field. Similar to the normalized potential, $\mathcal{E}_n$ only depends on d_2/d_1, α, and w_2. A closed form solution to (6.98) has not been found and numerical integration is used for evaluation. As with the normalized potential integral, we transform the normalized electric field integral to improve convergence. With

$$K_2[u_0] = \int_{-\infty}^{u_0} \frac{dx}{(w_2^* - x)^2 + 1} = \frac{\pi}{2} + \tan^{-1}[u_0 - w_2^*] \quad (6.99)$$

the normalized electric field can be written as

$$\mathcal{E}_n = \frac{j}{\pi}\left(\frac{e^{\pi w_2^*} - \eta}{e^{\pi w_2^*} - 1}\right)^{\alpha} \int_{-\infty}^{\infty} f_1[u_0] \frac{dK_2[u_0]}{du_0}\,du_0 \quad (6.100)$$

Integrating (6.100) by parts gives

$$\begin{aligned}
\mathcal{E}_n = &\frac{j}{2}\left(\frac{e^{\pi w_2^*} - \eta}{e^{\pi w_2^*} - 1}\right)^{\alpha}(\eta^{-\alpha} + 1) \\
&+ \frac{j}{\pi}\left(\frac{e^{\pi w_2^*} - \eta}{e^{\pi w_2^*} - 1}\right)^{\alpha} \int_{-\infty}^{\infty} \tan^{-1}[u_0 - w_2^*]\left(-\frac{df_1[u_0]}{du_0}\right)du_0
\end{aligned} \quad (6.101)$$

Based on previous arguments, $-f_1'[u_0] \approx 0$ for $u_0 > 5$ and $u_0 < u_{0,p} - 6$ allowing the approximation

$$\mathcal{E}_n = \frac{j}{2}\left(\frac{e^{\pi w_2^*} - \eta}{e^{\pi w_2^*} - 1}\right)^\alpha (\eta^{-\alpha} + 1)$$

$$+ \frac{j}{\pi}\left(\frac{e^{\pi w_2^*} - \eta}{e^{\pi w_2^*} - 1}\right)^\alpha \int_{u_{0,p}-6}^{5} \tan^{-1}[u_0 - w_2^*]\left(-\frac{df_1[u_0]}{du_0}\right)du_0 \tag{6.102}$$

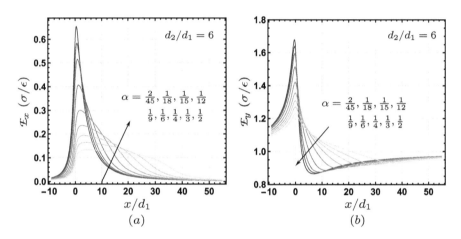

FIGURE 6.24 (a) x and (b) y components of the electric field along the sheet of charge in the z-plane as a function of α with $d_2/d_1 = 6$.

The x-component and y-component of the electric field along the line of charge $(y = 0)$ for various values of α with $d_2/d_1 = 6$ is shown in Fig. 6.24. Several items we expected to see are indeed present in Fig. 6.24. Far away from the transition, $\mathcal{E}_x$ approaches zero and $\mathcal{E}_y$ approaches σ_z/ϵ_u. The peak electric field for $\mathcal{E}_x$ decreases as α decreases. The x-component and y-component of the electric field along the line of charge $(y = 0)$ for various values of d_2/d_1 with $\alpha = 1/6$ is shown in Fig. 6.25. Note that the peak electric field decreases as d_2/d_1 is decreases. As discussed previously, the closer d_2/d_1 is to 1 or the smaller α is, the gate/field plate have less deviation from a flat plate. If the plate were flat, then charge density on the plate would be a constant and the electric field would be a constant. The model captures all of this behavior.

6.4.5 Induced Charge Density Analysis

Constant lines of force contours allows visualizing how the field plate influences where the charge in the channel images on the gate/field plate.

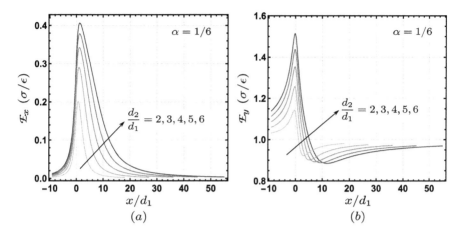

FIGURE 6.25 (a) x and (b) y components of the electric field along the sheet of charge in the z-plane as a function of d_2/d_1 with $\alpha = 1/6$.

To determine the line of force contours, we need to evaluate the imaginary part of the (6.85). All values of Ψ are referenced to $\Psi[0]$. In other words, we calculate

$$\Delta\Psi[w_2] = \Psi[w_2] - \Psi[0] \tag{6.103}$$

The imaginary part of (6.85) gives

$$\Delta\Psi_n[w_2] = \frac{1}{2\pi} \int_{-\infty}^{\infty} \left(\frac{e^{\pi u_0} + 1}{e^{\pi u_0} + \eta}\right)^\alpha \mathrm{Im}\left[\ln\left[\frac{w_2 - u_0 + j}{w_2 - u_0 - j}\right]\right] du_0$$
$$- \frac{1}{2\pi} \int_{-\infty}^{\infty} \left(\frac{e^{\pi u_0} + 1}{e^{\pi u_0} + \eta}\right)^\alpha \mathrm{Im}\left[\ln\left[\frac{-u_0 + j}{-u_0 - j}\right]\right] du_0 \tag{6.104}$$

where $\Psi_n = \mathrm{Im}[\Phi]/(\sigma_z d_1/\epsilon_u)$. A closed form solution to (6.104) has not been found and numerical integration is used for evaluation. As with the normalized potential integral, we transform the Ψ_n integral to improve convergence. The identity

$$\tan^{-1}\left[x^{-1}\right] - \tan^{-1}\left[y^{-1}\right] = \tan^{-1}[y] - \tan^{-1}[x] \tag{6.105}$$

allows writing

$$\mathrm{Im}\left[\ln\left[\frac{w_2 - u_0 + j}{w_2 - u_0 - j}\right] - \ln\left[\frac{-u_0 + j}{-u_0 - j}\right]\right] =$$
$$- \tan^{-1}\left[\frac{u_2 - u_0}{1 - v_2}\right] - \tan^{-1}\left[\frac{u_2 - u_0}{1 + v_2}\right] - 2\tan^{-1}[u_0] \tag{6.106}$$

With

$$K_3[u_0] = -\int_{-\infty}^{u_0} 2\tan^{-1}[x]\,dx$$
$$-\int_{-\infty}^{u_0}\left(\tan^{-1}\left[\frac{u_2-x}{1-v_2}\right]+\tan^{-1}\left[\frac{u_2-x}{1+v_2}\right]\right)dx \qquad (6.107)$$

the Ψ_n integral can be written as

$$\Psi_n = \frac{1}{2\pi}\int_{-\infty}^{\infty} f_1[u_0]\frac{dK_3[u_0]}{du_0}\,du_0 \qquad (6.108)$$

where f_1 is given by (6.87). Integrating (6.108) by parts gives

$$\Delta\Psi_n[w_2] = -\tfrac{1}{2}u_2\left(1+\eta^{-\alpha}\right)+\frac{1}{2\pi}\int_{-\infty}^{\infty} f_4[u_0]\left(-\frac{df_1[u_0]}{du_0}\right)du_0 \quad (6.109)$$

where

$$f_4[u_0] = (u_2-u_0)\left(\tan^{-1}\left[\frac{u_2-u_0}{1-v_2}\right]+\tan^{-1}\left[\frac{u_2-u_0}{1+v_2}\right]\right)$$
$$-2u_0\tan^{-1}[u_0]+\frac{v_2}{2}\ln\left[\frac{(1-v_2)^2+(u_2-u_0)^2}{(1+v_2)^2+(u_2-u_0)^2}\right]$$
$$+\ln\left[(1+u_0^2)\frac{\left((1-v_2)^2+(u_2-u_0)^2\right)^{-1/2}}{\left((1+v_2)^2+(u_2-u_0)^2\right)^{1/2}}\right] \qquad (6.110)$$

Based on previous arguments, $-f_1'[u_0]\approx 0$ for $u_0>5$ and $u_0<u_{0,p}-6$ allowing the approximation

$$\Delta\Psi_n[w_2]\approx -\tfrac{1}{2}u_2\left(1+\eta^{-\alpha}\right)+\frac{1}{2\pi}\int_{u_{0,p}-6}^{5} f_4[u_0]\left(-\frac{df_1[u_0]}{du_0}\right)du_0 \quad (6.111)$$

Constant lines of force contours near the gate edge for two field plate configurations are shown in Fig. 6.26. For both cases, $d_2/d_1 = 6$ and $\Delta\Psi_n = 1$. For Fig. 6.26 (a), $\alpha = 1/2$ and the field lines are concentrated near the gate edge. The field line spacing increases along the sidewall in the direction away from the gate edge. This indicates that the amount of charge induced on the sidewall of the gate decreases in the direction away from the gate edge. The locations labeled z_1 and z_2 are $d_1/2$ away from the edge of the gate. The distance between two field lines along the gate $\Delta\Psi_n$ is proportional to the charge between these field lines.

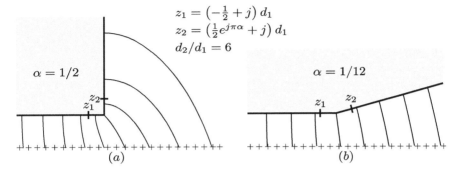

$$z_1 = \left(-\tfrac{1}{2} + j\right) d_1$$
$$z_2 = \left(\tfrac{1}{2}e^{j\pi\alpha} + j\right) d_1$$
$$d_2/d_1 = 6$$

FIGURE 6.26 Constant lines of force for two different values of α. When $\alpha = 1/2$, the field lines crowd near the corner of the gate indicating a large charge density and electric field. The length of surface near the gate edge located between z_1 and z_2 has about 2.7 equivalent squares of flux. For $\alpha = 1/12$, the field lines are much more uniform indicating a lower charge density and electric field compared to $\alpha = 1/2$. Less than 2 equivalent square of flux are located between the same distance from the edge of the gate defined by z_1 and z_2.

The total charge q_{tot} between z_1 and z_2 is proportional the number of $\Delta\Psi_n$ contained between z_1 and z_2. For Fig. 6.26 (a), $q_{tot} \propto 2.7\Delta\Psi_n$ approximatley. For Fig. 6.26 (b), $\alpha = 1/12$ and field lines are much more evenly distributed even on the transition region between gate and field plate. The region on the metal between z_1 and z_2 has approximately $q_{tot} \propto 1.5\Delta\Psi_n$ for Fig. 6.26 (b). By decreasing the transition angle, charge that was induced on the corner of the gate is now induced on the transition region between the gate and field plate instead. This lowers the charge density and decreases the electric field at the gate edge. The total charge between z_1 and z_2 as a function of transition angle is shown in Fig. 6.27 (a). The amount of charge induced on the corner of the gate is reduced by 40 % when α is decreased from $1/2$ to $1/10$. The amount of charge induced on the corner of the gate is also a function of d_2/d_1. The total charge between z_1 and z_2 as a function of d_2/d_1 for $\alpha = 1/6$ is shown in Fig. 6.27 (b). A drop in the total charge between z_1 and z_2 also occurs when the field plate to channel distance is decreased.

6.4.6 Key Findings and Summary

Conformal mapping was used to create a model for the electrostatic potential and electric field in the transition region below a transistor's gate and field plate. With the conformal mapping model, a quantitative understanding of how field plates influence charge imaging on the gate

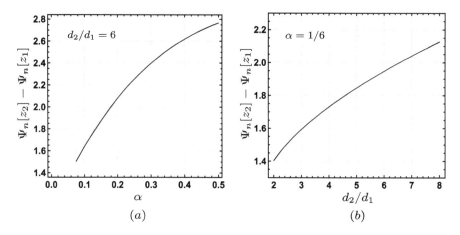

FIGURE 6.27 Total charge between z_1 and z_2 defined in Fig. 6.26 as a function of (a) α for $d_2/d_1 = 6$ and (b) as a function of d_2/d_1 for $\alpha = 1/6$. Decreasing d_2/d_1 or α leads to less charge induced on the corner of the gate and lower electric field.

corner based on design parameters (field plate to charge distance and transition angle) was obtained. Additionally, it was found that the true variable is not the field plate to charge distance d_2, but rather d_2 divided by the gate to charge distance d_1 and the natural unit of length is d_1. This type of insight is difficult to obtain with finite element simulations, but easily obtained with the closed form solution. To obtain a closed form solution, an infinite geometry was used. The infinite geometry not only simplified the mathematics for the conformal transformation, but also allowed using the method of images to obtain Green's function for the problem. Additional analysis for field plates using the conformal mapping model can be found in [3, 4].

6.5 TRIGATE FINFETS

6.5.1 Introduction

The next case study also has its origin from transistor design. To increase control of current flow for short gate lengths, 3D transistor designs have emerged where current flows through fins of conducting semiconductor and the gate metal covers three sides of the fins (see Fig. 6.28 (a)). These types of transistors are often called FinFETs. In this case study, we analyze a portion of the gate that is responsible for determining the voltage required for shutting off current flow called the pinch-off voltage ($V_{p,fin}$). The model should capture the dependence of $V_{p,fin}$ on

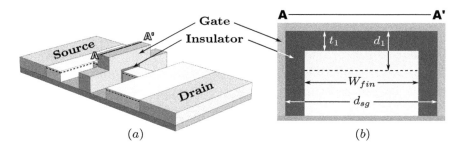

FIGURE 6.28 (a) 3D schematic of a FinFET structure. The location of charge is indicated by the dashed line. (b) Cross-sectional view of the gated channel region.

the geometry of the fin. The case study shows how to handle regions of charge where the complex potential is no longer valid.

In the off-state, there is no current flow resulting in an electrostatic environment. If the gate length is long enough to avoid what are known as short channel effects, then the region near the source side of the gate can be analyzed in 2D. The cross-section of the gate is shown in Fig. 6.28 (b). The gate metal covers three sides of the semiconductor fin and there are fixed charges located in the semiconductor fin that lead to a potential drop from gate to the fixed charges. The maximum voltage drop from gate to semiconductor is the pinch-off voltage.

6.5.2 Conformal Mapping to a Flat Plate

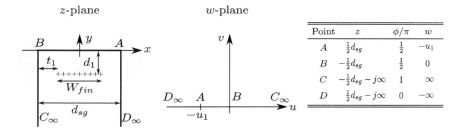

Point	z	ϕ/π	w
A	$\frac{1}{2}d_{sg}$	$\frac{1}{2}$	$-u_1$
B	$-\frac{1}{2}d_{sg}$	$\frac{1}{2}$	0
C	$-\frac{1}{2}d_{sg} - j\infty$	1	∞
D	$\frac{1}{2}d_{sg} - j\infty$	0	$-\infty$

FIGURE 6.29 S-C transformation of the gated region of a FinFET in the z-plane onto the upper half of the w-plane.

The cross-section of the gate shown in Fig. 6.28 (b) is approximated by the geometry in Fig. 6.29 that places C at $z = -\frac{1}{2}d_{sg} - j\infty$ and D at $z = \frac{1}{2}d_{sg} - j\infty$. The error associated with assuming the gate goes infinitely deep into the semiconductor is quantified later. The model assumes a uniform permittivity ϵ_u for the region confined by the gate metal. Green's function for our model is obtained by mapping the region confined by the gate into the upper half of the w-plane. If we rotate the geometry of Fig. 6.29 by $\frac{1}{2}\pi$ and translate it by $j\frac{1}{2}d_{sg}$, then we have the degenerate triangle of Fig. 5.25 (a). Therefore, the mapping is obtained with (5.137) by replacing z with $jz + j\frac{1}{2}d_{sg}$ and b with d_{sg}

$$z = -j\left(2d_{sg}/\pi\right)\sinh^{-1}[\sqrt{w_1}] - \tfrac{1}{2}d_{sg} \tag{6.112}$$

Solving for w_1 in (6.112) gives

$$w_1 = -\sin^2\left[\frac{\pi z}{2d_{sg}} + \frac{\pi}{4}\right] = -\frac{1}{2} - \cos\left[\frac{\pi z}{2d_{sg}}\right]\sin\left[\frac{\pi z}{2d_{sg}}\right] \tag{6.113}$$

6.5.3 Electrostatic Potential, ρ not specified

Green's function in the w_1-plane is

$$G_D[w_1, w_{1,0}] = \frac{1}{2\pi\epsilon_u}\ln\left[\left|\frac{w_1 - w_{1,0}^*}{w_1 - w_{1,0}}\right|\right] \tag{6.114}$$

To transform Green's function to the z-plane, we replace w_1 with $w_1[z]$ from (6.113) and $w_{1,0}$ with $w_1[z_0]$ to obtain

$$G_D[z, z_0] = \frac{1}{2\pi\epsilon_u}\ln\left[\left|\frac{\sin[\pi z/d_{sg}] - \sin[\pi z_0^*/d_{sg}]}{\sin[\pi z/d_{sg}] - \sin[\pi z_0/d_{sg}]}\right|\right] \tag{6.115}$$

Using (3.74), the electrostatic potential in the z-plane is

$$V[z] = \int_{\text{charge}} \rho[x_0, y_0]\, G_D[z, z_0]\, dx_0\, dy_0 \tag{6.116}$$

In order to evaluate (6.116), the charge distribution $\rho[x_0, y_0]$ needs to be specified. We consider two types of charge distributions. Both types of charge distributions are symmetric with respect to the center of the gate (the imaginary axis in Fig. 6.29) and only vary in the y-direction. Due to symmetry in the charge distribution, the maximum potential drop from gate to semiconductor occurs along the imaginary axis.

6.5.4 Electrostatic Potential for Sheets of Charge

First we consider a charge distribution consisting of individual sheets of charge. The complex potential for this case can be computed, but we use the Green's function approach. The charge distribution for a single sheet charge that extends from $-\frac{1}{2}W_{fin}$ to $\frac{1}{2}W_{fin}$ in the x-direction located at $y = -d_1$ with sheet charge density σ_z is

$$\rho[x_0, y_0] = \sigma_z \delta[y + d_1], \quad -\tfrac{1}{2}W_{fin} \le x_0 \le \tfrac{1}{2}W_{fin} \tag{6.117}$$

where $\delta[\ ]$ is the Dirac delta function. The electrostatic potential can now be evaluated as

$$V[z] = \frac{\sigma_z}{2\pi\epsilon_u}\mathrm{Re}\left[\int_{-\infty}^{0}\int_{-\frac{1}{2}W_{fin}}^{\frac{1}{2}W_{fin}}\delta[y+d_1]\ln\left[\frac{\sin[\alpha z]-\sin[\alpha z_0^*]}{\sin[\alpha z]-\sin[\alpha z_0]}\right]da_0\right] \tag{6.118}$$

where $\alpha = \pi/d_{sg}$ and $da_0 = dx_0\,dy_0$. With the identities

$$\frac{\sin[\alpha z]-\sin[\alpha z_0^*]}{\sin[\alpha z]-\sin[\alpha z_0]} = \frac{\left(1+e^{j\alpha(z+z_0^*)}\right)\left(1-e^{j\alpha(z-z_0^*)}\right)}{\left(1+e^{j\alpha(z+z_0)}\right)\left(1-e^{j\alpha(z-z_0)}\right)} \tag{6.119}$$

and

$$\int \ln\left[1\pm\beta e^{\alpha j x_0}\right]dx_0 = \frac{j}{\alpha}\mathrm{Li}_2\left[\mp\beta e^{\alpha j x_0}\right] \tag{6.120}$$

where $\mathrm{Li}_2[\ ]$ is the dilogarithm function, the normalized electrostatic potential $(V_n = V\epsilon_u/(d_{sg}\sigma_z))$ can be evaluated from (6.118) as

$$\begin{aligned}
\mathcal{V}_n[z] = -\frac{1}{2\pi^2}\mathrm{Im}\bigg[&\mathrm{Li}_2\left[-je^{\alpha(-jt_1-d_1+jz)}\right] - \mathrm{Li}_2\left[-je^{\alpha(-jt_1+d_1+jz)}\right]\\
&- \mathrm{Li}_2\left[-je^{\alpha(jt_1+d_1+jz)}\right] + \mathrm{Li}_2\left[-je^{\alpha(jt_1-d_1+jz)}\right]\\
&- \mathrm{Li}_2\left[je^{\alpha(jt_1-d_1+jz)}\right] + \mathrm{Li}_2\left[je^{\alpha(jt_1+d_1+jz)}\right]\\
&+ \mathrm{Li}_2\left[je^{\alpha(-jt_1+d_1+jz)}\right] - \mathrm{Li}_2\left[je^{\alpha(-jt_1-d_1+jz)}\right]\bigg]
\end{aligned} \tag{6.121}$$

Note $\mathcal{V}[z]$ has been used instead $V[z]$ to indicate modifications may be needed to prevent the potential function from crossing a branch cut. The $\mathrm{Li}_2[\ ]$ function has a branch cut along the real axis from $x = 1$ to $x = \infty$. If the potential function crosses this branch cut, a discontinuity in the electrostatic potential will occur without any physical justification such as the presence of a double layer. To check for a branch cut crossing, the arguments for the eight $\mathrm{Li}_2[\ ]$ terms in (6.121) need to be plotted along a constant y line for $y > -d_1$ and $y < -d_1$ with $-\frac{1}{2}d_{sg} < x < \frac{1}{2}d_{sg}$. Performing this check reveals that terms 3 and 7 cross the branch cut for $0 > y > -d_1$ and terms 3, 4, 7, and 8 cross the branch cut for $-d_1 > y$. Thus the region between $0 > y > -d_1$ is described by one function and the region between $-d_1 > y$ by another function. To eliminate the branch cut crossing, we use the identity

$$\text{Li}_2[-x] + \text{Li}_2[-1/x] = -\frac{\pi^2}{6} - \frac{1}{2}\ln[x]^2 \qquad (6.122)$$

on the terms that cross the branch cut and choose the appropriate phase for $\ln[-j]$. For $y > -d_1$, we have

$$
\begin{aligned}
V_n[z] = \; & \frac{2t_1 - d_{sg}}{2d_{sg}}\left(\frac{y - d_1}{d_{sg}}\right) \\
& - \frac{1}{2\pi^2}\text{Im}\Big[\text{Li}_2\Big[-je^{\alpha(-jt_1 - d_1 + jz)}\Big] - \text{Li}_2\Big[-je^{\alpha(-jt_1 + d_1 + jz)}\Big] \\
& \qquad + \text{Li}_2\Big[je^{-\alpha(jt_1 + d_1 + jz)}\Big] + \text{Li}_2\Big[-je^{\alpha(jt_1 - d_1 + jz)}\Big] \\
& \qquad - \text{Li}_2\Big[je^{\alpha(jt_1 - d_1 + jz)}\Big] + \text{Li}_2\Big[je^{\alpha(jt_1 + d_1 + jz)}\Big] \\
& \qquad - \text{Li}_2\Big[-je^{-\alpha(-jt_1 + d_1 + jz)}\Big] - \text{Li}_2\Big[je^{\alpha(-jt_1 - d_1 + jz)}\Big]\Big]
\end{aligned}
\qquad (6.123)
$$

The first term of (6.123) can be written in terms of dilogarithms that do not cross the branch cut for $y > -d_1$ as

$$
\begin{aligned}
\frac{2t_1 - d_{sg}}{2d_{sg}}&\left(\frac{y - d_1}{d_{sg}}\right) \\
&= \frac{1}{2\pi^2}\text{Im}\Big[\text{Li}_2\Big[-je^{\alpha(-jt_1 - d_1 - jz)}\Big] - \text{Li}_2\Big[je^{\alpha(jt_1 - d_1 - jz)}\Big] \\
&\qquad - \text{Li}_2\Big[-je^{\alpha(-jt_1 + d_1 + jz)}\Big] + \text{Li}_2\Big[je^{\alpha(jt_1 + d_1 + jz)}\Big]\Big]
\end{aligned}
\qquad (6.124)
$$

Using (6.124) allows writing (6.123) as

$$
\begin{aligned}
V_n[z] = \; & \frac{1}{2\pi^2}\text{Im}\Big[\text{Li}_2\Big[-je^{\alpha(-jt_1 - d_1 - jz)}\Big] - \text{Li}_2\Big[je^{\alpha(-jt_1 - d_1 - jz)}\Big] \\
& \qquad + \text{Li}_2\Big[je^{\alpha(jt_1 - d_1 + jz)}\Big] - \text{Li}_2\Big[-je^{\alpha(jt_1 - d_1 + jz)}\Big] \\
& \qquad + \text{Li}_2\Big[-je^{\alpha(jt_1 - d_1 - jz)}\Big] - \text{Li}_2\Big[je^{\alpha(jt_1 - d_1 - jz)}\Big] \\
& \qquad + \text{Li}_2\Big[je^{\alpha(-jt_1 - d_1 + jz)}\Big] - \text{Li}_2\Big[-je^{\alpha(-jt_1 - d_1 + jz)}\Big]\Big]
\end{aligned}
\qquad (6.125)
$$

Similar analysis for $-d_1 > y$ and the identity

$$\frac{(d_{sg} - 2t_1)}{d_{sg}^2 d_1^{-1}} = \frac{1}{2\pi^2}\text{Im}\left[\text{Li}_2\left[-je^{\alpha(-jt_1-d_1+jz)}\right] + \text{Li}_2\left[je^{\alpha(jt_1+d_1-jz)}\right]\right.$$
$$+ \text{Li}_2\left[-je^{\alpha(-jt_1-d_1-jz)}\right] + \text{Li}_2\left[je^{\alpha(jt_1+d_1+jz)}\right] \quad (6.126)$$
$$- \text{Li}_2\left[-je^{\alpha(-jt_1+d_1+jz)}\right] - \text{Li}_2\left[je^{\alpha(jt_1-d_1-jz)}\right]$$
$$\left.- \text{Li}_2\left[-je^{\alpha(-jt_1+d_1-jz)}\right] - \text{Li}_2\left[je^{\alpha(jt_1-d_1+jz)}\right]\right]$$

give

$$V_n[z] = \frac{1}{2\pi^2}\text{Im}\left[\text{Li}_2\left[-je^{\alpha(-jt_1-d_1-jz)}\right] - \text{Li}_2\left[je^{\alpha(-jt_1-d_1-jz)}\right]\right.$$
$$+ \text{Li}_2\left[je^{\alpha(-jt_1+d_1-jz)}\right] - \text{Li}_2\left[-je^{\alpha(-jt_1+d_1-jz)}\right] \quad (6.127)$$
$$+ \text{Li}_2\left[-je^{\alpha(jt_1-d_1-jz)}\right] - \text{Li}_2\left[je^{\alpha(jt_1-d_1-jz)}\right]$$
$$\left.+ \text{Li}_2\left[je^{\alpha(jt_1+d_1-jz)}\right] - \text{Li}_2\left[-je^{\alpha(jt_1+d_1-jz)}\right]\right]$$

We combine (6.125) and (6.127) into a single expression

$$V_n[z] = \frac{1}{2\pi^2}\text{Im}\left[\text{Li}_2\left[-je^{\alpha(-jt_1-d_1-jz)}\right] - \text{Li}_2\left[je^{\alpha(-jt_1-d_1-jz)}\right]\right.$$
$$+ \text{Li}_2\left[je^{\pm\alpha(jt_1-d_1+jz)}\right] - \text{Li}_2\left[-je^{\pm\alpha(jt_1-d_1+jz)}\right] \quad (6.128)$$
$$+ \text{Li}_2\left[-je^{\alpha(jt_1-d_1-jz)}\right] - \text{Li}_2\left[je^{\alpha(jt_1-d_1-jz)}\right]$$
$$\left.+ \text{Li}_2\left[je^{\pm\alpha(-jt_1-d_1+jz)}\right] - \text{Li}_2\left[-je^{\pm\alpha(-jt_1-d_1+jz)}\right]\right]$$

where the upper sign is used when $y > -d_1$ and the lower sign is used when $y < -d_1$.

6.5.5 Pinch-off Voltage for a Single Sheet of Charge

We can calculate the normalized pinch-off voltage for a single sheet charge as

$$V_{p,fin,n} = V_n[-jd_1]$$
$$= \frac{1}{\pi^2}\text{Im}\left[\text{Li}_2\left[je^{-\alpha jt_1}\right] + \text{Li}_2\left[je^{\alpha jt_1}\right]\right. \quad (6.129)$$
$$\left.+ \text{Li}_2\left[-je^{\alpha(jt_1-2d_1)}\right] + \text{Li}_2\left[-je^{\alpha(-jt_1-2d_1)}\right]\right]$$

If the sheet of charge extends all the way to the metal ($t_1 = 0$), then the normalized pinch-off voltage reduces to

$$V_n[-jd_1] = \frac{2}{\pi^2}\left(G - \text{Im}\left[\text{Li}_2\left[je^{-2d_1\alpha}\right]\right]\right) \tag{6.130}$$

where $G \approx 0.9160$ is Catalan's constant.

The pinch-off voltage has three main regimes depending on the proximity of the sheet charge to the different portions of the gate (top and sides). The first regime occurs when $d_{sg} \gg d_1$ and the pinch-off voltage is primarily determined by the planar top-gate instead of the side-gates. This regime is labeled the planar regime. The second regime occurs when $d_{sg} < d_1$ and the pinch-off voltage is primarily determined by the side-gates. This regime is labeled the side-gate regime. The final regime is where the entire gate structure determines the pinch-off voltage. This regime is labeled the mixed regime. From the field plate analysis in Section 6.4, the pinch-off voltage due to only a top gate was determined to be

$$V_{p,pl} = \sigma_z d_1/\epsilon_u \tag{6.131}$$

The pinch-off voltage due to only side gates can be obtained from (6.129) by letting $d_1 \to \infty$

$$V_{p,sg} = \frac{\sigma_z d_{sg}}{\epsilon_u \pi^2}\text{Im}\left[\text{Li}_2\left[je^{-\alpha jt_1}\right] + \text{Li}_2\left[je^{\alpha jt_1}\right]\right] \tag{6.132}$$

The three regimes can be observed by plotting $V_{p,fin}/V_{p,pl}$ and $V_{p,fin}/V_{p,sg}$ as a function of d_1/d_{sg} with logarithmic scales on both the horizontal and vertical axes. An example of this type of plot when $t_1 = 0$ is shown in Fig. 6.30 (a). The different regimes are easily identified. The transition from one regime to another depends on the value of t_1/d_{sg}. We define the boundary of the planar regime (B_{top}) as the value of d_1/d_{sg} corresponding to $V_{p,fin} = 0.9V_{p,pl}$ and the boundary of the side-gate regime (B_{sg}) as the value of d_1/d_{sg} corresponding to $V_{p,fin} = 0.9V_{p,sg}$. The region between these two boundaries is considered the mixed regime where the top gate and side-gates both play a role in determining the pinch-off voltage.

A plot of B_{top} and B_{sg} as a function of t_1/d_{sg} is shown in Fig. 6.30 (b). If the dielectric layer that defines t_1 is assumed to be conformal over the entire semiconductor fin, then there is a minimum d_1 even if the sheet of charge is located at the surface of the semiconductor. The minimum d_1 prevents a planar regime for $t_1/d_{sg} > 0.05$. In order for a planar regime to exist for $t_1/d_{sg} > 0.05$, the dielectric thickness on the top portion of the semiconductor fin must be thinner than the dielectric thickness on the sides of the fin. Therefore, this portion of the curve is dashed instead of solid to represent this requirement. The minimum d_1 thickness also has an impact on the side-gate regime. For $t_1/d_{sg} > 0.34$, the top gate is far enough away from the sheet of charge that the side-gate regime condition is met even if the sheet of charge is at the surface of the semiconductor. This is represented in Fig. 6.30 (b) by changing the curve for B_{sg} from solid to dashed at this point. From Fig. 6.30 (b), a good rule of thumb for the side-gates to be the primary factors determining the pinch-off voltage is to have $d_1 > 0.38d_{sg}$.

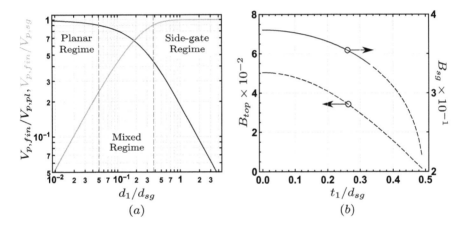

FIGURE 6.30 (a) $V_{p,fin}/V_{p,pl}$ and $V_{p,fin}/V_{p,sg}$ as a function of d_1/d_{sg} with logarithmic scales on both the horizontal and vertical axes for a FinFET with $t_1 = 0$. The planar regime, mixed regime, and side-gate regime are clearly marked. (b) The boundary values for the planar regime B_{top} and the side-gate regime B_{sg} as a function of t_1/d_{sg}.

6.5.6 Pinch-off Voltage for Multiple Sheets of Charge

The previous analysis can be extended to multiple sheets of charge using the principle of superposition. For example, let all the sheets of charge extend from $-\frac{1}{2}W_{fin}$ to $\frac{1}{2}W_{fin}$ and be symmetric with respect to the imaginary axis. Let the ith sheet have a sheet charge σ_i and be locate at $y = -d_i$. The pinch-off voltage for this configuration of sheets charges is

$$V_{p,fin} = \text{Max}[V[jy]] \qquad (6.133)$$

where

$$V[jy]$$
$$= \frac{d_{sg}}{\pi^2 \epsilon_u} \sum_i \sigma_i \text{Im} \left[\text{Li}_2 \left[-j e^{\alpha(jt_1 - d_i + y)} \right] + \text{Li}_2 \left[-j e^{\alpha(-jt_1 - d_i + y)} \right] \right. \tag{6.134}$$
$$\left. + \text{Li}_2 \left[j e^{\alpha(jt_1 - |d_i + y|)} \right] + \text{Li}_2 \left[j e^{\alpha(-jt_1 - |d_i + y|)} \right] \right]$$

6.5.7 Electrostatic Potential for a Region of Charge

Now consider a continuous distribution of charge. For example, the charge distribution for a semiconductor that is doped uniformly (Fig. 6.31) between $-\frac{1}{2}W_{fin} \leq x \leq \frac{1}{2}W_{fin}$ and $-y_2 \leq y \leq -y_1$ is

$$\rho[x_0, y_0] = q N_d, \quad -\frac{1}{2}W_{fin} \leq x_0 \leq \frac{1}{2}W_{fin}, \quad -y_2 \leq y_0 \leq -y_1 \tag{6.135}$$

Using the transform

$$V[x, y] = V_0[x, y] - \frac{1}{\epsilon_u} \int_{-y_1}^{y} dy' \int_{-y_1}^{y'} \rho[y''] \, dy'' \tag{6.136}$$

where $V_0[x, y]$ is a solution to Laplace's equation, the problem is changed from finding the solution to Poisson's equation to finding the solution to Laplace's equation. The transform of (6.136) typically results in a change in the boundary conditions. Since the region of charge does not extend to the boundaries, there is no change to the boundary conditions and $V[x, y] = V_0[x, y]$ in the charge free regions. The Green's function approach allows solving for $V_0[x, y]$ and then (6.136) can be used to obtain $V[x, y]$ in the region containing charge.

Evaluation of (6.116) with the uniform doping described by (6.135) results in the same integral over x_0 as with sheet charges and with

$$\int \text{Li}_2[\beta e^{\alpha y_0}] \, dy_0 = \frac{1}{\alpha} \text{Li}_3[\beta e^{\alpha y_0}] \tag{6.137}$$

where $\text{Li}_3[\ ]$ is the trilogarithm function, the resulting integral over y_0 gives

$$\mathcal{V}[z] = -\frac{q N_d d_{sg}^2}{2\pi^3 \epsilon_u} \text{Im}[f_3[y_1] - f_3[y_2]] \tag{6.138}$$

where

$$
\begin{aligned}
f_3[\zeta] = {} & \mathrm{Li}_3\left[-je^{\alpha(-jt_1-\zeta+jz)}\right] + \mathrm{Li}_3\left[-je^{\alpha(-jt_1+\zeta+jz)}\right] \\
& + \mathrm{Li}_3\left[-je^{\alpha(jt_1+\zeta+jz)}\right] + \mathrm{Li}_3\left[-je^{\alpha(jt_1-\zeta+jz)}\right] \\
& - \mathrm{Li}_3\left[je^{\alpha(jt_1-\zeta+jz)}\right] - \mathrm{Li}_3\left[je^{\alpha(jt_1+\zeta+jz)}\right] \\
& - \mathrm{Li}_3\left[je^{\alpha(-jt_1+\zeta+jz)}\right] - \mathrm{Li}_3\left[je^{\alpha(-jt_1-\zeta+jz)}\right]
\end{aligned}
\tag{6.139}
$$

We now have three regions to consider: $y > -y_1$, $-y_1 > y > -y_2$, and $-y_2 > y$. Since $\mathrm{Li}_3[\ ]$ and $\mathrm{Li}_2[\ ]$ have the same branch cut, we can use the previous analysis (or one can plot the arguments for the 16 $\mathrm{Li}_3[\ ]$ terms of (6.139)) to determine terms 3, 7, 11, and 15 cross the branch cut for $y > -y_1$, terms 3, 4, 7, 8, 11, and 15 cross the branch cut for $-y_1 > y > -y_2$, and terms 3, 4, 7, 8, 11, 12, 15, and 16 cross the branch cut for $-y_2 > y$. To eliminate the branch cut crossing, we use the identity

$$
\mathrm{Li}_3[-x] - \mathrm{Li}_3[-1/x] = -\frac{\pi^2}{6}\ln[x] - \frac{1}{6}\ln[x]^3
\tag{6.140}
$$

on the terms that cross the branch cut and choose the appropriate phase for $\ln[-j]$. With the identities

$$
\begin{aligned}
-\frac{(2t_1 - d_{sg})}{4d_{sg}} & (y_1 - y_2)(2y - y_1 - y_2) = \\
& \frac{d_{sg}^2}{2\pi^3}\mathrm{Im}\Big[\mathrm{Li}_3\left[-je^{\alpha(-jt_1-y_1-jz)}\right] - \mathrm{Li}_3\left[je^{\alpha(jt_1-y_1-jz)}\right] \\
& + \mathrm{Li}_3\left[-je^{\alpha(-jt_1+y_1+jz)}\right] - \mathrm{Li}_3\left[je^{\alpha(jt_1+y_1+jz)}\right] \\
& + \mathrm{Li}_3\left[je^{\alpha(jt_1-y_2-jz)}\right] - \mathrm{Li}_3\left[-je^{\alpha(-jt_1-y_2-jz)}\right] \\
& - \mathrm{Li}_3\left[-je^{\alpha(-jt_1+y_2+jz)}\right] + \mathrm{Li}_3\left[je^{\alpha(jt_1+y_2+jz)}\right]\Big]
\end{aligned}
\tag{6.141}
$$

and

$$
y_1^2 - y_2^2 = \tfrac{1}{2}(y_1 - y_2)(2y + y_1 + y_2 - (2y - y_1 - y_2))
\tag{6.142}
$$

the electrostatic potential in the charge free regions can be written as

$$
V[z] = \frac{qN_d d_{sg}^2}{2\pi^3\epsilon_u}\mathrm{Im}[f_{3a}[y_1] - f_{3a}[y_2]]
\tag{6.143}
$$

with

$$
\begin{aligned}
f_{3a}[\zeta] = & \operatorname{Li}_3\left[-je^{-\alpha(jt_1+\zeta+jz)}\right] - \operatorname{Li}_3\left[je^{-\alpha(jt_1+\zeta+jz)}\right] \\
& + \operatorname{Li}_3\left[\pm je^{\pm\alpha(jt_1-\zeta+jz)}\right] - \operatorname{Li}_3\left[\mp je^{\pm\alpha(jt_1-\zeta+jz)}\right] \\
& + \operatorname{Li}_3\left[-je^{\alpha(jt_1-\zeta-jz)}\right] - \operatorname{Li}_3\left[je^{\alpha(jt_1-y_1-jz)}\right] \\
& + \operatorname{Li}_3\left[\pm je^{\pm\alpha(-jt_1-\zeta+jz)}\right] - \operatorname{Li}_3\left[\mp je^{\pm\alpha(-jt_1-\zeta+jz)}\right]
\end{aligned}
\tag{6.144}
$$

The upper sign is for $y > -y_2$ and the lower sign is for $y < -y_2$.
To obtain $V[x, y]$ in the region containing charge, we use (6.136)

$$
V[x, y] = -\frac{qN_d\,(y+y_1)^2}{2\epsilon_u} + V_0[x, y]
\tag{6.145}
$$

where $V_0[x, y]$ is given by (6.143) with the upper sign since $V[x, y] = V_0[x, y]$ at $y = -y_1$.

6.5.8 Pinch-off Voltage for a Region of Charge

The pinch-off voltage for a symmetric region of charge is still the maximum voltage along the imaginary axis. The impact of spreading the charge over a volume vs. an area can be obtained by comparing the pinch-off voltage of a single sheet of charge to a single region of charge. To simplify the analysis, we assume $d_1 \gg d_{sg}$ allowing side-gate only comparisons. Additionally, the location for the centroid of charge and the total sheet charge

$$
\sigma_z = qN_d\,(y_2 - y_1) = 2qN_d\Delta
\tag{6.146}
$$

is kept constant where 2Δ is the thickness of the charged region in the y-direction. Under these assumptions, the pinch-off voltage for the volume distribution of charge is the potential at $z = -y_1 - \Delta$

$$
\begin{aligned}
V_{p,\rho} = & -\frac{\sigma_z\Delta}{4\epsilon_u} + \frac{\sigma_z d_{sg}^2}{2\pi^3\epsilon_u\Delta}\operatorname{Im}\left[\operatorname{Li}_3\left[je^{\alpha(-jt_1+\Delta)}\right] + \operatorname{Li}_3\left[je^{\alpha(jt_1+\Delta)}\right]\right] \\
& + \frac{\sigma_z d_{sg}^2}{2\pi^3\epsilon_u\Delta}\operatorname{Im}\left[\operatorname{Li}_3\left[-je^{\alpha(jt_1-\Delta)}\right] + \operatorname{Li}_3\left[-je^{\alpha(-jt_1-\Delta)}\right]\right]
\end{aligned}
\tag{6.147}
$$

A plot of $V_{p,\rho}/V_{p,sg}$ vs Δ/d_{sg} is shown in Fig. 6.31. Spreading the charge decreases the electric since the induced charge on the side gates has a lower peak. This results in the decrease in pinch-off voltage as Δ increases shown in Fig. 6.31.

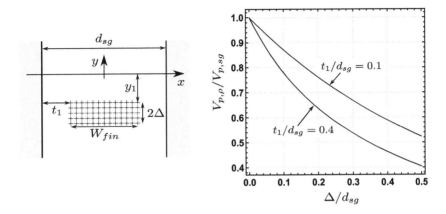

FIGURE 6.31 Comparison of the pinch-off voltage for a uniformly distributed layer of charge and when the same amount of charge is compressed into a sheet. The thickness of the uniformly distributed charge layer is 2Δ. For a fixed amount of charge, spreading the charge over a larger distance reduces the pinch-off voltage.

6.5.9 Infinite Depth Assumption

A key assumption in the FinFET model is that the gate metal extends infinitely deep into the semiconductor. This assumption allows the method of images to be used to determine Green's function in the w-plane. We have already shown as the distance between the top-gate and the channel is reduced, the side-gates play a smaller role in determining the pinch-off voltage. Therefore, the side-gate only structure requires the deepest depth (d_{max}) for the model to be valid and this depth also serves as a conservative estimate for all FinFET structures.

A cross-section of the gated region of a FinFET with side-gates that have a finite depth into the semiconductor is shown in Fig. 6.32 (a). The vertical distance from the corners of the side-gates (points A and D in Fig. 6.32 (a)) to the nearest sheet of charge (y_0 in Fig. 6.32 (a)) is defined as the depth of the side-gates. To quantify the error in the pinch-off voltage caused by the infinite depth assumption and to establish a value for d_{max}, the difference between the pinch-off voltage calculated by (6.129) and the pinch-off voltage obtained by 2D finite element electrostatic simulations for side-gates with corners is used. The error in the pinch-off voltage depends on y_0/d_{sg} and $W_{fin}/(2d_{sg})$. Values for y_0/d_{sg} of 0.4, 0.5, 0.6 and $W_{fin}/(2d_{sg})$ values ranging from 0.05 to 0.49 are used for the comparison. The relative error is plotted in Fig. 6.32 (b) for the three values of y_0/d_{sg}. As y_0/d_{sg} increases (sheet charge further away from the corners), the approximation becomes more accurate and the error decreases. Defining d_{max} as the depth of the side-gates past the last sheet charge that gives less than 1 % error results in $d_{max} = 0.424d_{sg}$. Thus, the depth depends on the distance between sides gate. Most Fin-FETs have a depth larger than d_{max} making the pinch-off voltage given by (6.129) valid for most FinFETs.

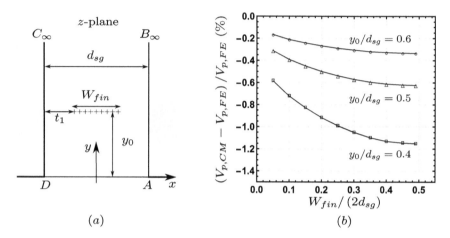

(a)   (b)

FIGURE 6.32   (a) Cross-section of the gated region of a FinFET with side-gates that have a finite depth into the semiconductor. (b) Relative error between the pinch-off voltage calculated using the conformal mapping model ($V_{p,CM}$) and finite element simulations ($V_{p,FE}$) as a function of fin width to twice the distance between side-gates. The further the sheet charge is from the corner (larger y_0/d_{sg}), the smaller the error.

6.5.10   Key Findings and Summary

Conformal mapping was used to model the electrostatic potential and electric field in the region enclosed by the gate of a FinFET. The cases of n discrete sheets of charge and for a region of uniformly distributed charge were analyzed. Closed form expressions for the pinch-off voltage of a FinFET for discrete sheets of charge and for a region of uniformly distributed charge were obtained. From the closed form solution, the natural unit of length for the problem is the side-gate to side-gate distance d_{sg}. It was also found that for a top gate to sheet charge distance (d_1) greater than $0.38d_{sg}$, the side gates were primarily responsible for determining the pinch-off voltage. To obtain a closed form solution, the gate was assumed to go infinitely deep into the semiconductor. Using finite element simulations, the model is shown to be accurate to better than

1 % if the gate metal goes into the semiconductor by at least $0.424d_{sg}$ past the last sheet of charge.

6.6 UNIFORM ELECTRIC FIELD

Inserting an object into a uniform electric field causes distortions to the electric field. In some cases, the method of images can be used to obtain the complex potential that allows calculating the distortion. To use the method of images, first we determine a charge configuration that generates a uniform electric field by a limiting process when no objects are present. Then, the object is introduced before performing the limiting process to determine the image charges needed to satisfy the boundary conditions. With the image charges, the electrostatic potential (or electric field) of the system is now determined. The limiting process is then performed and the distortion determined by the resulting electrostatic potential (or electric field) of the system in this limit. Thus, we need to obtain a charge configuration that generates a uniform electric field by a limiting process.

We start with a dipole produced by a line charge q_0 at $z = -x_0$ and a line charge $-q_0$ at $z = x_0$. The complex potential and electric field for the dipole are

$$\Phi[z] = \frac{q_0}{2\pi\epsilon} \ln\left[\frac{z - x_0}{z + x_0}\right] \qquad (6.148)$$

and

$$\mathcal{E}[z] = -\left(\frac{d\Phi}{dz}\right)^* = -\frac{q_0}{\pi\epsilon} \frac{x_0}{(z^* - x_0)(z^* + x_0)} \qquad (6.149)$$

For $x_0 \gg |z|$, a second order Taylor polynomial can be used to approximate the electric field as

$$\mathcal{E}[z] \approx \frac{q_0}{\pi\epsilon x_0}\left(1 + (z^*/x_0)^2\right) \qquad (6.150)$$

If we now let $x_0 \to \infty$ while holding q_0/x_0 constant, the electric field in finite space becomes uniform and directed in the positive x-direction

$$\mathcal{E}[z] = \mathcal{E}_0 = \frac{q_0}{\pi\epsilon x_0} \qquad (6.151)$$

When an object is introduced, the image charges due to $\pm q_0$ located at $\mp x_0$ need to be determined and the resulting electrostatic potential (or electric field) calculated with x_0 finite. Then the limit of $x_0 \to \infty$ while holding q_0/x_0 constant for the electrostatic potential (or electric field) of the system is electrostatic potential of the object inserted in a uniform electric field directed along the positive real axis.

6.7 CIRCULAR CONDUCTING OR DIELECTRIC CYLINDER IN A UNIFORM ELECTRIC FIELD

The problem of a conductive circular cylinder coated with a conformal dielectric immersed in a uniform electric field is often solved through the use of separation of variables [14]. We tackle the problem with conformal mapping. Consider a conductive cylinder with radius r_0 centered at the origin of the w-plane. Let the cylinder have a conformal dielectric coating of thickness $r_1 - r_0$ and permittivity ϵ_1 as shown in Fig. 6.34. Let the material surrounding the dielectric coated cylinder have a permittivity of ϵ_0. To obtain the complex potential for the dielectric coated cylinder in a uniform electric field, we first determine the complex potential for the coated cylinder due to a charge q_0 at $w = -x_0$ and $-q_0$ at $w = x_0$ and then let $x_0 \to \infty$ while holding q_0/x_0 constant. To use the method of images, we need to determine the locations and values of the image charge. This is done by using (4.45) to transform the concentric circular boundaries in the w-plane into planar boundaries in the z-plane. The two charges in the w-plane result in an infinite array of line charges of opposite sign separated by $\Delta y = \pi$ in the z-plane. The image charge locations and values in the z-plane can be obtained using the ray tracing method described in Section 3.10. In fact, the problem in the z-plane is almost the same as the problem depicted by Fig. 3.19 and solved in Section 3.13. To make the two problems correspond, we have

$$a = \ln[r_1/r_0], \quad b = \ln[x_0/r_1] \tag{6.152}$$

To obtain conductor boundary conditions at $z = -a$, we let $\epsilon_2 \to \infty$. The image charge locations and values for a charge of 1 at $z = b$ are given in Fig. 6.33. All the image charges can be determined in the z-plane using Fig. 6.33 and transformed into the w-plane with

$$w[z] = r_1 e^z \tag{6.153}$$

The image charge values and locations are tabulated in Table 6.1. In the z-plane, no dielectric interface exist at infinity and therefore no polarization charge at infinity. Also, for every positive line charge, there is a negative line charge of the same magnitude. Therefore, there is no charge at infinity that is transformed to finite space when transforming between the z-plane and w-plane.

Taking into account all image and true charges, applying the complex potential dielectric boundary conditions ((2.136) and (2.139)) with $\Psi =$

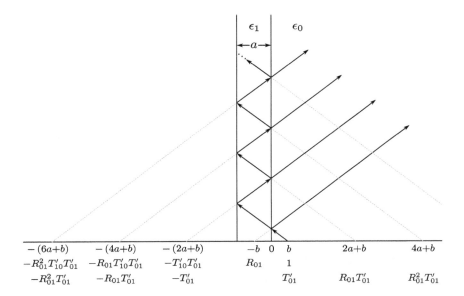

FIGURE 6.33 Location of image charges and their values for a line charge located in front of an infinite planar conductor coated with a dielectric of thickness a and permittivity ϵ_1.

0 along the real axis for $-x_0 < u < x_0$ and continuous gives

$$
\begin{aligned}
\Phi_1 = -\frac{q_0}{2\pi\epsilon_0} \bigg(& \ln\left[\frac{x_0 + w}{x_0 - w}\right] + R_{01} \ln\left[\frac{(r_1^2/x_0) + w}{(r_1^2/x_0) - w}\right] \\
& + T_{01}' T_{10}' \sum_{n=1}^{\infty} R_{01}^{n-1} \ln\left[\frac{(r_1^2/x_0)\,(r_0/r_1)^{2n} - w}{(r_1^2/x_0)\,(r_0/r_1)^{2n} + w}\right]\bigg)
\end{aligned}
\tag{6.154}
$$

for region 1 and

$$
\begin{aligned}
\Phi_2 = & -\frac{T_{01}' q_0}{2\pi\epsilon_1} \sum_{n=1}^{\infty} R_{01}^{n-1} \ln\left[\frac{(r_1^2/x_0)\,(r_0/r_1)^{2n} - w}{(r_1^2/x_0)\,(r_0/r_1)^{2n} + w}\right] \\
& - \frac{T_{01}' q_0}{2\pi\epsilon_1} \sum_{n=1}^{\infty} R_{01}^{n-1} \ln\left[\frac{x_0(r_1/r_0)^{2n-2} + w}{x_0(r_1/r_0)^{2n-2} - w}\right]
\end{aligned}
\tag{6.155}
$$

for region 2. Approximating the $\ln[\ ]$ terms with first order Taylor polynomials in $1/x_0$ and then taking the limit as $x_0 \to \infty$ with q_0/x_0 held constant gives

$$
\Phi_1 = -\mathcal{E}\left(w + \frac{R_{01} r_1^2}{w} - \frac{T_{01}' T_{10}' r_1^2}{w} \sum_{n=1}^{\infty} R_{01}^{n-1} \frac{r_0^{2n}}{r_1^{2n}}\right)
\tag{6.156}
$$

TABLE 6.1 Image Charge in w-plane

Region	Value	Location	
1	$\pm R_{01} q_0$	$\mp r_1^2/x_0$	
	$\pm T'_{01} T'_{10} R_{01}^{n-1} q_0$	$\pm \dfrac{r_1^2}{x_0}\left(\dfrac{r_0}{r_1}\right)^{2n}$	$n = 1, 2, 3, ...$
2	$\pm T'_{01} R_{01}^{n-1} q_0$	$\pm \dfrac{r_1^2}{x_0}\left(\dfrac{r_0}{r_1}\right)^{2n}$	$n = 1, 2, 3, ...$
	$\pm T'_{01} R_{01}^{n-1} q_0$	$\mp x_0\left(\dfrac{r_1}{r_0}\right)^{2n-2}$	$n = 1, 2, 3, ...$

and

$$\Phi_2 = -\frac{T'_{01}\epsilon_0 \mathcal{E}}{\epsilon_1} \sum_{n=1}^{\infty} R_{01}^{n-1} \frac{r_0^{2n}}{r_1^{2n}}\left(-\frac{r_1^2}{w} + w\frac{r_1^2}{r_0^2}\right) \tag{6.157}$$

where $\mathcal{E} = q_0/(\pi\epsilon_0 x_0)$ is the uniform electric field that would exist if the coated cylinder was not present. The infinite sum converges to

$$\sum_{n=1}^{\infty} R_{01}^{n-1} \frac{r_0^{2n}}{r_1^{2n}} = \frac{r_0^2(\epsilon_1 + \epsilon_0)}{r_0^2(\epsilon_1 - \epsilon_0) + r_1^2(\epsilon_1 + \epsilon_0)} \tag{6.158}$$

resulting in

$$\Phi_1 = -\mathcal{E}w + \mathcal{E}\frac{r_1^2}{w}\left(\frac{r_0^2(\epsilon_1 + \epsilon_0) + r_1^2(\epsilon_1 - \epsilon_0)}{r_0^2(\epsilon_1 - \epsilon_0) + r_1^2(\epsilon_1 + \epsilon_0)}\right) \tag{6.159}$$

and

$$\Phi_2 = \frac{2\epsilon_0 \mathcal{E} r_1^2 r_0^2}{r_0^2(\epsilon_1 - \epsilon_0) + r_1^2(\epsilon_1 + \epsilon_0)}\left(\frac{1}{w} - \frac{w}{r_0^2}\right) \tag{6.160}$$

Constant flux contours for a dielectric coated conductive cylinder with $\epsilon_1 = 5\epsilon_0$ are shown in Fig. 6.34.

The case for a conducting cylinder of radius r_0 is obtained by letting $\epsilon_1 = \epsilon_0$ in (6.159)

$$\Phi = -\mathcal{E}w + \mathcal{E}\frac{r_0^2}{w} \tag{6.161}$$

The case for a dielectric cylinder of radius r_1 is obtained by letting $r_0 = 0$ in (6.159) and (6.160)

$$\Phi_1 = -\mathcal{E}w + \mathcal{E}\frac{r_1^2}{w}\left(\frac{\epsilon_1 - \epsilon_0}{\epsilon_1 + \epsilon_0}\right), \quad \Phi_2 = -\frac{2\epsilon_0 \mathcal{E}w}{\epsilon_1 + \epsilon_0} \tag{6.162}$$

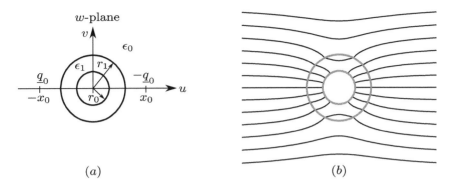

FIGURE 6.34 (a) Conductive cylinder of radius r_0 coated with a conformal dielectric coating of thickness $r_1 - r_0$. In the limit $x_0 \to \infty$, the line charges produce a constant electric field directed along the real axis. (b) Constant flux contours for a dielectric coated conductive cylinder inserted into a constant electric field.

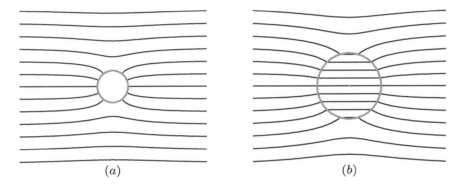

FIGURE 6.35 Constant flux contours for a (a) conductive cylinder and (b) dielectric cylinder inserted into a constant electric field.

Note that the field lines are parallel inside the dielectric cylinder. Constant flux contours for conducting and dielectric cylinders are shown in Fig. 6.35. The dielectric cylinder in Fig. 6.35 has a permittivity of $\epsilon_1 = 5\epsilon_0$.

6.8 ELLIPTIC DIELECTRIC CYLINDER IN UNIFORM ELECTRIC FIELD

Let us now determine the complex potential for an elliptic dielectric cylinder in a uniform electric field directed at an angle α with respect to the positive real axis. Far from the cylinder, the electric field approaches

$$\mathcal{E}_z = |\mathcal{E}_z| e^{j\alpha} \tag{6.163}$$

Let the major axis of the cylinder lie on the real axis centered at the origin of the z-plane. The principle of superposition is used to combine the solutions of an elliptic dielectric cylinder in a vertical field $\mathcal{E}_v = |\mathcal{E}_z|\sin[\alpha]$ and horizontal field $\mathcal{E}_h = |\mathcal{E}_z|\cos[\alpha]$ to obtain the complex potential.

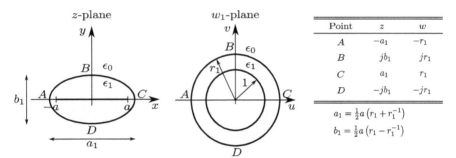

Point	z	w
A	$-a_1$	$-r_1$
B	jb_1	jr_1
C	a_1	r_1
D	$-jb_1$	$-jr_1$

$$a_1 = \tfrac{1}{2}a\left(r_1 + r_1^{-1}\right)$$
$$b_1 = \tfrac{1}{2}a\left(r_1 - r_1^{-1}\right)$$

FIGURE 6.36 Using the logarithm transformation, a dielectric coated conductive cylinder can be mapped to a dielectric elliptical cylinder. The region between $z = \pm a$ must be a constant potential contour for this mapping to be correct.

For the vertical electric field case, the portion of the real axis between the foci of the ellipse is an equipotential. Rotating the results of Section 6.7 by $\pi/2$ gives the complex potential for a circular conducting cylinder coated with a conformal dielectric in a vertical field $\mathcal{E}_v = j|\mathcal{E}_{w,v}|$ as

$$\Phi_{1,v} = j\,|\mathcal{E}_{w,v}|\left(w + \frac{r_1^2}{w}\left(\frac{r_0^2\left(\epsilon_{r,1}+1\right)+r_1^2\left(\epsilon_{r,1}-1\right)}{r_0^2\left(\epsilon_{r,1}-1\right)+r_1^2\left(\epsilon_{r,1}+1\right)}\right)\right) \quad (6.164)$$

and

$$\Phi_{2,v} = \frac{2j\,|\mathcal{E}_{w,v}|\,r_1^2 r_0^2}{r_0^2\left(\epsilon_{r,1}-1\right)+r_1^2\left(\epsilon_{r,1}+1\right)}\left(\frac{1}{w}+\frac{w}{r_0^2}\right) \quad (6.165)$$

where $\epsilon_{r,1} = \epsilon_1/\epsilon_0$ is the relative permittivity of the dielectric coating. To transform (6.164) and (6.165) into complex potentials for a dielectric elliptic cylinder centered at the origin of the z-plane of major axis $2a_1$ (aligned with the real axis) and minor axis $2b_1$ in a vertical electric field, we first must make the w-plane of Fig. 6.34 match the w_1-plane of Fig. 5.33 by setting $r_0 = 1$ as shown in Fig. 6.36. Then we replace w in (6.164) and (6.165) with $w_1[z]$ from (5.155) and r_1 with (5.183) to obtain

$$\Phi_{1,v} = j\,|\mathcal{E}_{w,v}|\left(w_1[z] + \frac{1}{w_1[z]}\left(\frac{a_1+b_1}{a_1-b_1}\right)\left(\frac{a_1\epsilon_{r,1}-b_1}{a_1\epsilon_{r,1}+b_1}\right)\right) \quad (6.166)$$

and

$$\Phi_{2,v} = \frac{j\,|\mathcal{E}_{w,v}|\,(a_1 + b_1)}{a_1 \epsilon_{r,1} + b_1}\left(\frac{1}{w_1[z]} + w_1[z]\right) \qquad (6.167)$$

We can use the value of $\mathcal{E}_z$ far away from the cylinder to establish the value of $|\mathcal{E}_{w,v}|$. Using (4.9) to relate the $\mathcal{E}_z$ and $\mathcal{E}_w$ gives

$$\lim_{|z|\to\infty} [\mathcal{E}_z] = \lim_{|z|\to\infty}\left[\mathcal{E}_w\left(\frac{dw}{dz}\right)^*\right] = 2\mathcal{E}_w/a \qquad (6.168)$$

Far away from the cylinder in the z-plane, $\mathcal{E}_z \to \mathcal{E}_v = |\mathcal{E}_z| \sin[\alpha]$ which requires

$$|\mathcal{E}_{w,v}| = \tfrac{1}{2}a\,|\mathcal{E}_z| \sin[\alpha] \qquad (6.169)$$

Constant flux contours for a dielectric elliptic cylinder in the z-plane and a dielectric coated conductive cylinder in the w-plane are shown in Fig. 6.37. The dielectric elliptic cylinder and dielectric coating have $\epsilon_1 = 7\epsilon_0$.

z-plane

w-plane

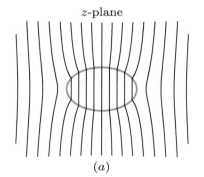
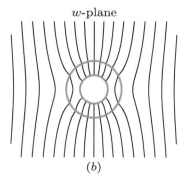

(a)

(b)

FIGURE 6.37 (a) Constant flux lines for an elliptic cylinder with major axis perpendicular to a constant electric field. (b) Constant flux lines for a dielectric coated circular conductive cylinder that were transformed using the logarithm transformation to obtain the constant flux lines of (a).

To obtain the solution for the horizontal field, the same procedure described in Section 6.7 can be used except now the circular boundary of radius r_0 is a line of force instead of a conductor. The image locations are the same as shown in Fig. 6.33, but the image charges located at $x < -b$ are positive instead of negative. All the image charge locations and values are given in Table 6.2. Taking into account all image and true charges, applying the complex potential dielectric boundary conditions ((2.136) and (2.139)) with $\Psi = 0$ along the real axis for $-x_0 < u < x_0$

and continuous gives

$$\Phi_{1,h} = -\frac{q_0}{2\pi\epsilon_0} \left(\ln\left[\frac{x_0 + w}{x_0 - w}\right] + R_{01}\ln\left[\frac{(r_1^2/x_0) + w}{(r_1^2/x_0) - w}\right] \right.$$
$$\left. + T_{01}'T_{10}' \sum_{n=1}^{\infty} R_{10}^{n-1}\ln\left[\frac{(r_1^2/x_0)(r_0/r_1)^{2n} + w}{(r_1^2/x_0)(r_0/r_1)^{2n} - w}\right] \right) \tag{6.170}$$

for region 1 and

$$\Phi_{2,h} = -\frac{T_{01}'q_0}{2\pi\epsilon_1} \sum_{n=1}^{\infty} R_{10}^{n-1}\ln\left[\frac{(r_1^2/x_0)(r_0/r_1)^{2n} + w}{(r_1^2/x_0)(r_0/r_1)^{2n} - w}\right]$$
$$-\frac{T_{01}'q_0}{2\pi\epsilon_1} \sum_{n=1}^{\infty} R_{10}^{n-1}\ln\left[\frac{x_0(r_1/r_0)^{2n-2} + w}{x_0(r_1/r_0)^{2n-2} - w}\right] \tag{6.171}$$

for region 2. Approximating the ln[ ] terms with first order Taylor polynomials in $1/x_0$, then taking the limit as $x_0 \to \infty$ with q_0/x_0 held constant, and using

$$\sum_{n=1}^{\infty} R_{10}^{n-1}\frac{r_0^{2n}}{r_1^{2n}} = \frac{r_0^2(\epsilon_{r,1} + 1)}{r_1^2(\epsilon_{r,1} + 1) - r_0^2(\epsilon_{r,1} - 1)} \tag{6.172}$$

gives

$$\Phi_{1,h} = |\mathcal{E}_{w,h}|\left(-w + \frac{r_1^2}{w}\left(\frac{r_1^2(\epsilon_{r,1} - 1) - r_0^2(\epsilon_{r,1} + 1)}{r_1^2(\epsilon_{r,1} + 1) - r_0^2(\epsilon_{r,1} - 1)}\right)\right) \tag{6.173}$$

and

$$\Phi_{2,h} = -\frac{2|\mathcal{E}_{w,h}|r_1^2r_0^2}{r_1^2(\epsilon_{r,1} + 1) - r_0^2(\epsilon_{r,1} - 1)}\left(\frac{1}{w} + \frac{w}{r_0^2}\right) \tag{6.174}$$

where $|\mathcal{E}_{w,h}| = q_0/(\pi\epsilon_0 x_0)$. To transform (6.173) and (6.174) into complex potentials for a dielectric elliptic cylinder centered at the origin of the z-plane of major axis $2a_1$ (aligned with the real axis) and minor axis $2b_1$ in a horizontal electric field, we first must make the w-plane of Fig. 6.34 match the w_1-plane of Fig. 5.33 by setting $r_0 = 1$. Then we replace w in (6.173) and (6.174) with $w_1[z]$ from (5.155) and r_1 with (5.183) to obtain

$$\Phi_{1,h} = -|\mathcal{E}_{w,h}|\left(w_1[z] + \frac{1}{w_1[z]}\left(\frac{a_1 + b_1}{a_1 - b_1}\right)\left(\frac{a_1 - b_1\epsilon_{r,1}}{a_1 + b_1\epsilon_{r,1}}\right)\right) \tag{6.175}$$

and

$$\Phi_{2,h} = -\frac{|\mathcal{E}_{w,h}|(a_1 + b_1)}{(a_1 + b_1\epsilon_{r,1})}\left(\frac{1}{w_1[z]} + w_1[z]\right) \tag{6.176}$$

Requiring $\mathcal{E}_z \to \mathcal{E}_h = |\mathcal{E}_z|\cos[\alpha]$ far away from the cylinder gives

$$|\mathcal{E}_{w,h}| = \tfrac{1}{2}a\,|\mathcal{E}_z|\cos[\alpha] \tag{6.177}$$

Constant flux contours for a dielectric elliptic cylinder in the z-plane and a dielectric coated circular cylinder with constant flux line boundaries in the w-plane are shown in Fig. 6.38. The dielectric elliptic cylinder and dielectric coating have $\epsilon_1 = 7\epsilon_0$.

TABLE 6.2   Image Charge in w-plane

Region	Value	Location	
1	$\pm R_{01}q_0$	$\mp r_1^2/x_0$	
	$\pm T'_{01}T'_{10}R_{10}^{n-1}q_0$	$\mp \dfrac{r_1^2}{x_0}\left(\dfrac{r_0}{r_1}\right)^{2n}$	$n = 1,2,3,\ldots$
2	$\pm T'_{01}R_{10}^{n-1}q_0$	$\mp \dfrac{r_1^2}{x_0}\left(\dfrac{r_0}{r_1}\right)^{2n}$	$n = 1,2,3,\ldots$
	$\pm T'_{01}R_{10}^{n-1}q_0$	$\mp x_0\left(\dfrac{r_1}{r_0}\right)^{2n-2}$	$n = 1,2,3,\ldots$

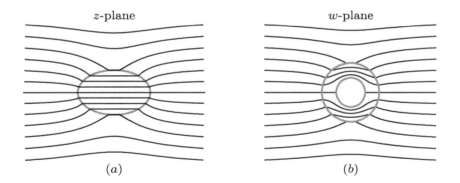

z-plane w-plane

(a)   (b)

FIGURE 6.38   (a) Constant flux lines for an elliptic cylinder with major axis parallel to a constant electric field. (b) Constant flux lines for a dielectric coated circular cylinder that were transformed using the logarithm transformation to obtain the constant flux lines of (a). The boundary conditions on the circular cylinder are now constant flux instead of constant potential.

The principle of superposition allows combining the horizontal and vertical solutions to obtain the complex potential for an elliptic dielectric cylinder in a uniform field at an angle α with respect to the horizontal

$$\Phi_1 = \Phi_{1,h} + \Phi_{1,v}$$
$$= -\tfrac{1}{2}a|\mathcal{E}_z|\left(e^{-j\alpha}w_1[z] + \frac{1}{w_1[z]}\left(A'\cos[\alpha] - jA''\sin[\alpha]\right)\right) \quad (6.178)$$

and

$$\Phi_2 = \Phi_{2,h} + \Phi_{2,v} = -\left(\frac{\cos[\alpha]}{a_1 + b_1\epsilon_{r,1}} - \frac{j\sin[\alpha]}{a_1\epsilon_{r,1} + b_1}\right)|\mathcal{E}_z|(a_1 + b_1)z \quad (6.179)$$

where $(aw_1[z]) = z + \sqrt{z^2 - a^2}$, $a = \sqrt{a_1^2 - b_1^2}$, and

$$A' = \left(\frac{a_1 + b_1}{a_1 - b_1}\right)\left(\frac{a_1 - b_1\epsilon_{r,1}}{a_1 + b_1\epsilon_{r,1}}\right), \quad A'' = \left(\frac{a_1 + b_1}{a_1 - b_1}\right)\left(\frac{a_1\epsilon_{r,1} - b_1}{a_1\epsilon_{r,1} + b_1}\right) \quad (6.180)$$

Constant flux contours for an elliptic dielectric cylinder with major axis parallel to the real axis in a constant electric field directed at an angle $\alpha = \pi/6$ with respect to the positive real axis is shown in Fig. 6.39.

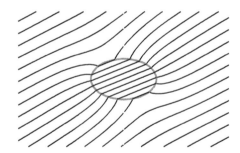

FIGURE 6.39 Constant flux contours for an elliptic dielectric cylinder with major axis parallel to the real axis in a constant electric field directed at an angle $\alpha = \pi/6$ with respect to the positive real axis.

6.9 LIMITATIONS FOR CONFORMAL MAPPING

There are two key limitations for conformal mapping. First is the mathematical complexity for polygons with greater than 4 finite vertices or smooth shapes that are not circular or oval. Numerical techniques must be used for these cases and the physical insight into the problem becomes similar to that obtained with finite element simulations.

Despite this limitation, the S-C transformation can be modified to obtain a smooth transition between the sides of a polygon instead of the sharp corner and closed form solutions can be obtained for simply polygons. This is physically important due to the non-physical infinite electric field on conductors containing sharp corners with negative external angles. Rounding such a corner results in a finite electric field.

To round a sharp corner on a polygon, the S-C transformation in integral form is determined with the sharp corner. Next the factor $(w - b_i)^{-\phi_i/\pi}$ associated with the corner to be rounded is replaced with

$$(w - b_i)^{-\phi_i/\pi} \rightarrow (w - b_i')^{-\phi_i/\pi} + \lambda \, (w - b_i'')^{-\phi_i/\pi} \qquad (6.181)$$

where b_i', b_i'' and λ are all real numbers and $b_{i-1} < b_i' < b_i'' < b_{i+1}$. The S-C integral is then computed and certain constraints are used to determine b_i', b_i'', and λ. Although a smooth curve has replaced the sharp corner, the curve can only be made approximately circular. To obtain a good approximation to a circular arc, b_i', b_i'' and λ are chosen so that

$$|z[b_i'] - z[b_i]| = |z[b_i''] - z[b_i]| = r_c \qquad (6.182)$$

where r_c is the radius for true circular rounding of the sharp corner and $z[b_i]$ is the location of the sharp corner before any rounding.

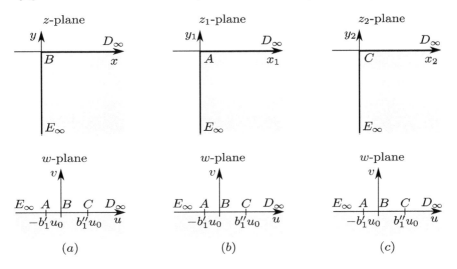

FIGURE 6.40 (a) Polygon with a single right angle corner. (b) Translated mapping of (a) with the corner mapped to $w = -b_1' u_0$. (c) Translated mapping of (a) with the corner mapped to $w = b_1'' u_0$.

To visualize the rounding process first consider a circle centered at the origin with a radius r. The circle is the sum of two orthogonal vectors

(x and y) in a certain way. When the magnitude of one of the vectors is increasing the magnitude of the other vector is decreasing. When the magnitude of one vector is maximum, the other vector is zero. In order to be a true circle, the vectors must follow a certain relationship between them. The relationship between x and y for the portion of the circle in the second quadrant in parametric form is

$$x = -\sqrt{r^2 - t^2}\hat{x}, \quad y = t\hat{y}, \quad 0 \leq t \leq r \tag{6.183}$$

Now consider the single corner polygon shown in Fig. 6.40 (a). The S-C transformation is

$$z = A_1 \int \sqrt{w}dw + B_1 = \tfrac{2}{3}A_1 w^{3/2} + B_1 \tag{6.184}$$

The boundary condition $z = 0$ when $w = 0$ requires $B_1 = 0$. If we let $z = x_0$ when $w = u_0$, then A_1 can be determined and the transformation becomes

$$z = x_0 w_1^{3/2} \tag{6.185}$$

where $w_1 = w/u_0$. Let z_1 correspond to a translated (6.185) so that $z[-b_1'] = 0$

$$z_1 = x_0 \left(w_1 + b_1'\right)^{3/2} \tag{6.186}$$

Let z_2 correspond to a translated (6.185) so that $z[b_1''] = 0$

$$z_2 = x_0 \left(w_1 - b_1''\right)^{3/2} \tag{6.187}$$

As w_1 is increased along the real axis from $-\infty$ to $-b_1'$, both z_1 and z_2 correspond to increasing values along the negative imaginary axis as seen in Fig. 6.40 (b) and (c). At $w_1 = -b_1'$, w_1 begins mapping to positive real values in the z_1-plane, but continues mapping to increasing values along the negative imaginary in the z_2-plane. For $w_1 > b_1''$, z_1 and z_2 correspond to positive real values. Adding z_1 and z_2 results in a curve that conforms to the general characteristics of a circle in the region $-b_1' < w < b_1''$, but does not have the exact circle relationship between the orthogonal vectors. To help the resulting curve resemble more closely a circle, z_2 is multiplied by an adjustment factor λ. The resulting transformation is

$$z_3/x_0 = A_2 \left((w_1 + b_1')^{3/2} + \lambda\,(w_1 - b_1'')^{3/2}\right) \tag{6.188}$$

where A_2 is needed to rescale the transformation back to its original scale. Applying requirement (6.182) gives

$$z_3[-b_1'] = -jr_c, \quad z_3[b_1''] = r_c \tag{6.189}$$

which leads to

$$r_c = A_2 x_0 (b_1' + b_1'')^{3/2}, \quad \lambda = 1 \qquad (6.190)$$

The transformation for the rounded corner written in terms of the true rounding radius r_c is

$$z_3 = r_c \left(\left(\frac{w_1 + 1}{1 + \eta} \right)^{3/2} + \left(\frac{w_1 - \eta}{1 + \eta} \right)^{3/2} \right) \qquad (6.191)$$

where $\eta = b_1''/b_1'$. No further boundary conditions need to be met and we are free to choose $b_1' = b_1''$ resulting in

$$z_3 = r_c \left(\left(\tfrac{1}{2} (w_1 + 1) \right)^{3/2} + \left(\tfrac{1}{2} (w_1 - 1) \right)^{3/2} \right) \qquad (6.192)$$

The resulting curve (C_1) and true circular rounding (C_c) are shown in Fig. 6.41 (a) and (b). Although C_1 and C_c do not overlap over most of the region that was rounded, many similar features are present. For example, both curves are smooth with continuously decreasing slopes and are symmetric with respect to the line $\alpha = 3\pi/4$. For this example, we can obtain the equation of the resulting curve as

$$x^{2/3} + (-y)^{2/3} = r^{2/3} \qquad (6.193)$$

The distance r_1 from C_1 to the center of C_c differs from the radius r_c of C_c by less than 9 % over the entire range (see Fig. 6.41 (c)). Other examples using this technique for rounding can be found in [12, 2, 5].

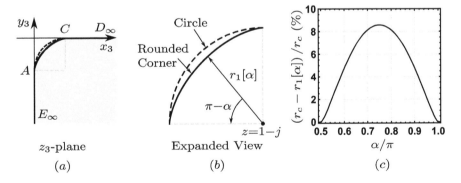

z_3-plane

(a)

Expanded View

(b)

α/π

(c)

FIGURE 6.41 (a) Polygon of Fig. 6.40 (a) after the corner has been rounded with $r_c = 1$. (b) Magnified view of the rounded corner. Circular rounding of radius r_c included for comparison. (c) Percent difference of rounded corner "radius" $r_1[\alpha]$ and true circular rounding radius r_c as a function of α.

The second key limitation for conformal mapping is the difficulty in handling regions where the permittivity is a function of position. When the permittivity is a function of position, Poisson's equation becomes

$$\nabla^2 V = -\frac{\rho}{\epsilon} - \nabla V \cdot \nabla \epsilon \qquad (6.194)$$

In order to transform (6.194) back into Laplace's equation, a function that eliminates both terms on the right is needed. Such a function is difficult to obtain. In general, when the permittivity is a function of position or the dielectric boundary has a shape that is not easily conformally mapped, then it is often better to tackle the problem by other means. Even when this is the case, useful physical insight is often gained by assuming a uniform permittivity and then solving the problem by conformal mapping. The insight gained can then be used to guide finite element simulations.

6.10 CONCLUSIONS

In this chapter, problems associated with plates at different potentials, fixed charges imaging on plates, and dielectric boundaries have been solved. We used infinite geometries to reduce mathematical complexities which in turn allowed for closed form solutions. The natural unit of length and true variables are typically easily identified with the closed form solutions. Although models that use infinite geometries to approximate finite geometries have a limited geometric range where they are valid/accurate, the insight into the physics of the problem gained by the closed form solution is often well worth the trade-off in the geometric range of accuracy in the model.

EXERCISES

6.1 A microstrip tranmission line has a rectangular signal conductor a distance d above the ground plane. Determine the width of the signal conductor that produces a 50 Ω characteristic impedance when the thickness of the signal conductor is 4 μm and d is 100 μm. Assume a uniform permittivity $\epsilon = 10\epsilon_0$ for all space not occupied by a conductor.

6.2 a. Calculate the characteristic impedance for a coaxial transmission line where the centered inner circular conductor has a radius r_1 and the outer circular conductor has a radius r_2.

b. Any misalignment of the inner conductor from the center of the outer conductor modifies the characteristic impedance. Calculate the characteristic impedance for the misaligned case. If $r_2 = 10r_1$, how much misalignment (in unit of r_2) is needed to change Z_0 by 5 %.

6.3 A transmission line consisting of a round signal wire centered inside a grounded square cylinder has been studied many times over the years. In [11], the characteristic impedance of the transmission line is given as

$$Z_0 = \frac{1}{2\pi c\epsilon_0} \ln\left[R_{eff}\frac{s}{d}\right]$$

where c is the speed of light in vacuum, ϵ_0 is the permittivity of vacuum, s is the length of a side for the square outer conductor, d is the diameter of the inner conductor, and R_{eff} is a parameter that depends on s/d. Several values for R_{eff} are given in Table 6.3. Use

TABLE 6.3   Values of R_{eff} for Given d/s

d/s	0.5	0.6	0.7	0.8	0.9
R_{eff}	1.07861	1.07831	1.07731	1.07439	1.06568

the small diameter wire approximation to determine Z_0 for the d/s values listed in Table 6.3. Assuming Z_0 given above is exact, when does the error for the small diameter wire approximation exceed 1 %?

6.4 Round the gate corner (vertex D) for a field plated gate with $\alpha = 1/2$. If the rounding radius is r_c show

a. the new transform between the w-plane and the z-plane is

$$z = jb - \frac{2b}{\pi(1+\lambda)}\left(\sinh^{-1}[f_1[\beta]] - \sqrt{\frac{\beta}{\eta}}\tanh^{-1}\left[f_1[w_1]\sqrt{\frac{\eta}{\beta}}\right]\right)$$
$$- \frac{2b}{\pi(1+\lambda)}\left(\lambda\sinh^{-1}[f_2[1]] - \frac{\lambda}{\sqrt{\eta}}\tanh^{-1}[f_2[w_1]\sqrt{\eta}]\right)$$

where

$$f_1[\zeta] = \sqrt{\frac{w_1-\beta}{\zeta-\eta}}, \quad f_2[\zeta] = \sqrt{\frac{w_1-1}{\zeta-\eta}},$$

$w_1 = w/u_1$. The location where the field plate and gate meet (vertex C) is mapped to ηu_1, where the rounding starts and ends are mapped to βu_1 and u_1, respectively, and

$$\lambda = -\frac{\sinh^{-1}\left[\sqrt{\frac{1-\beta}{\beta-\eta}}\right] - \sqrt{\frac{\beta}{\eta}}\tanh^{-1}\left[\sqrt{\frac{\eta}{\beta}}\sqrt{\frac{1-\beta}{1-\eta}}\right]}{\sin^{-1}\left[\sqrt{\frac{1-\beta}{1-\eta}}\right] - \sqrt{\frac{1}{\eta}}\tan^{-1}\left[\sqrt{\eta}\sqrt{\frac{1-\beta}{\beta-\eta}}\right]}$$

The values of β and η are found by simultaneously solving

$$\frac{a}{b} = \frac{\sqrt{\beta}+\lambda}{\sqrt{\eta}\,(1+\lambda)}$$

and

$$r_c = \frac{2b}{\pi\,(1+\lambda)}\left(\sinh^{-1}\left[\sqrt{\frac{1-\beta}{\beta-\eta}}\right] - \sqrt{\frac{\beta}{\eta}}\tanh^{-1}\left[\sqrt{\frac{\eta}{\beta}}\sqrt{\frac{1-\beta}{1-\eta}}\right]\right)$$

b. Let $a/b = 6$ and $r_c = 2b$. Compare the electric field components along the sheet of charge to a field plate configuration with $\alpha = 1/2$ but no rounding. How much has the peak $\mathcal{E}_x$ and $\mathcal{E}_y$ been reduced by introducing the rounding?

6.5 Let a grounded conductor bound the region defined by $|x| \leq x_1$ and $y \geq 0$ along with portion of the y-axis between $0 \leq y \leq a$ also consisting of a grounded conductor. Let the bounded region have a uniform charge density ρ. Show that the electrostatic potential is given by

$$V[x,y] = \frac{\rho x_1^2}{\pi\epsilon}\mathrm{Im}\left[\tanh^{-1}\left[\frac{1}{w_1[z]^*}\right] - \frac{4w_1[z]^*}{\pi^2}I[w_1[z]]\right] + \frac{\rho\left(x_1^2 - x^2\right)}{2\epsilon}$$

where

$$I[w_1] = \int_\eta^1 \left(\sin^{-1}\left[\frac{\sqrt{u_0^2 - \eta^2}}{\sqrt{1-\eta^2}}\right]\right)^2 \left(\frac{1}{(w_1^*)^2 - u_0^2}\right)du_0,$$

$$w_1 = \sqrt{(1-\eta^2)\sin^2\left[\frac{\pi z}{2x_1}\right] + \eta^2}, \quad \text{and} \quad \eta = \tanh\left[\frac{\pi a}{2x_1}\right]$$

6.6 Show the complex potential for a side-gate only FinFET with a single sheet charge of charge density σ_z is

$$\Phi[z] = \frac{-j\sigma_z d_{sg}}{2\pi^2\epsilon}\left(\mathrm{Li}_2\left[je^{\pm\alpha(jt_1+jz)}\right] - \mathrm{Li}_2\left[-je^{\pm\alpha(jt_1+jz)}\right]\right.$$
$$\left. + \mathrm{Li}_2\left[je^{\pm\alpha(-jt_1+jz)}\right] - \mathrm{Li}_2\left[-je^{\pm\alpha(-jt_1+jz)}\right]\right)$$

The coordinate axes orientation is such that the sheet charge is symmetric with respect to the imaginary axis and coincides with the real axis.

6.7 Show the induced charge density along the right side-gate for the FinFET in problem 6.6 is

$$\sigma = \frac{\sigma_z}{\pi} \tanh^{-1} \left[\frac{\cos[\pi t_1/d_{sg}]}{\cosh[\pi y/d_{sg}]} \right]$$

6.8 A circular dielectric cylinder of radius r_1 and permittivity ϵ has its center at the origin of the z-plane with the portion of the surface defined by $-\alpha \leq \theta \leq \alpha$ covered by a grounded conductor. Let the region around the cylinder be vacuum. If a line charge q_0 is located at z_0 where $|z_0| > r_1$,

a. Show the complex potential is

$$\Phi_1[z] = -\frac{q_0}{2\pi\epsilon_0} \ln\left[\frac{f[z/r_1] - f[z_0/r_1]}{f[z/r_1] - (f[z_0/r_1])^*} \right]$$
$$-\frac{Rq_0}{2\pi\epsilon_0} \ln\left[\frac{f[z/r_1] + (f[z_0/r_1])^*}{f[z/r_1] + f[z_0/r_1]} \right]$$

outside the cylinder and

$$\Phi_2[z] = -\frac{T'q_0}{2\pi\epsilon} \ln\left[\frac{f[z/r_1] - f[z_0/r_1]}{f[z/r_1] - (f[z_0/r_1])^*} \right]$$

inside the cylinder where

$$R = \frac{\epsilon_0 - \epsilon}{\epsilon_0 + \epsilon}, \quad T' = \frac{2\epsilon}{\epsilon_0 + \epsilon}, \quad f[z] = \sqrt{\frac{1 - ze^{j\alpha}}{z - e^{j\alpha}}}$$

b. Determine the total charge induced on the side of the conductor that touches dielectric as a function of line charge position z_0.

REFERENCES

[1] W.H. Chang. Analytical IC metal-line capacitance formulas (short papers). *IEEE Transactions on Microwave Theory and Techniques*, 24(9):608–611, Sep 1976.

[2] J.D. Cockcroft. The effect of curved boundaries on the distribution of electrical stress round conductors. *Journal of the Institution of Electrical Engineers*, 66(376):385–409, Apr 1928.

[3] Robert Coffie. Analytical field plate model for field effect transistors. *IEEE Transactions on Electron Devices*, 61(3):878–883, Mar 2014.

[4] Robert Coffie. Slant field plate model for field-effect transistors. *IEEE Transactions on Electron Devices*, 61(8):2867–2872, Aug 2014.

[5] Laszlo Fogaras and Wolfgang Lampe. Calculation of electrical field strength around transformer winding corners. *IEEE Transactions on Power Apparatus and Systems*, PAS-87(2):399–405, Feb 1968.

[6] S. Frankel. Characteristic impedance of parallel wires in rectangular troughs. *Proceedings of the IRE*, 30(4):182–190, Apr 1942.

[7] Sidney Frankel. *Mutliconductor Transmission Line Analysis*. Artech House, Inc, 1977.

[8] A. E. H. Love. Some electrostatic distributions in two dimensions. *Proceedings of the London Mathematical Society*, s2-22(1):337–369, 1924.

[9] James Clerk Maxwell. *A Treatise on Electricity and Magnetism*, volume Vol. I. Oxford, 3rd edition, 1904.

[10] Harlan B. Palmer. The capacitance of a parallel-plate capacitor by the schwartz-christoffel transformation. *Electrical Engineering*, 56(3):363–368, Mar 1937.

[11] H.J. Riblet. An accurate approximation of the impedance of a circular cylinder concentric with an external square tube. *IEEE Transactions on Microwave Theory and Techniques*, 31(10):841–844, Oct 1983.

[12] H. W. Richmond. On the electrostatic field of a plane or circular grating formed of thick rounded bars. *Proceedings of the London Mathematical Society*, s2-22(1):389–403, 1924.

[13] A. Ringhandt and H.G. Wagemann. An exact calculation of the two-dimensional capacitance of a wire and a new approximation

formula. *IEEE Transactions on Electron Devices*, 40(5):1028–1032, may 1993.

[14] William R. Smythe. *Static and Dynamic Electricity*. McGraw-Hill Book Company, 3rd edition, 1968.

[15] H. Wheeler. The transmission-line properties of a round wire between parallel planes. *IRE Transactions on Antennas and Propagation*, 3(4):203–207, Oct 1955.

[16] H.A. Wheeler. Transmission-line conductors of various cross sections. *IEEE Transactions on Microwave Theory and Techniques*, 28(2):73–83, Feb 1980.

CHAPTER 7

Other Fields of Physics

Although the focus up to now has been on 2D electrostatic analysis, the problem solving approaches developed so far apply to any 2D analysis that obeys Laplace's equation. This is important for two reasons. First, it allows obtaining solutions to problems in other areas of physics by simply translating the solution of the equivalent electrostatic problem. Second, solutions to electrostatic problems may be obtained by translating the solutions of equivalent problems in other areas of physics. This chapter describes how to translate electrostatic analysis into other areas of physics. Although there are many areas of physics that obey Laplace's equation, we limit our discussion to the areas of steady electric currents, magnetostatics, steady heat power flow, and steady fluid flow.

7.1 TRANSLATING TO OTHER AREAS OF PHYSICS

Table 7.1 is a summary of the key relationships between the different areas of physics to be discussed. Note that all the relationships given in Table 7.1 assume 2D analysis. Although some additional information is needed, most of the information required for translating a solution from electrostatics to one of the areas listed in Table 7.1 (or vice versa) is provided. The additional information needed for translation is provided in the sections of this chapter.

7.2 STEADY ELECTRIC CURRENT

Conductive materials that obey Ohm's law (ohmic materials) have current densities J (unit: A/m^2) that are proportional to electric field

$$J = \sigma_c \boldsymbol{E} \qquad (7.1)$$

TABLE 7.1 Translation between Different Areas of Physics

Quantity	Electrostatics	Steady Electric Current	Magnetostatics	Steady Heat Power Flow	Steady Fluid Flow
Potential	V	V	Ω	T	ϕ
Medium Constant	ϵ	σ_c	μ	k_{th}	ρ_l
Field Vector	$\boldsymbol{\mathcal{D}} = -\epsilon\nabla V$	$\boldsymbol{J} = -\sigma_c\nabla V$	$\boldsymbol{\mathcal{B}} = -\mu\nabla\Omega$	$\boldsymbol{J}_{th} = -k_{th}\nabla T$	$\boldsymbol{\mathcal{F}} = -\rho_l\nabla\phi$
d(Field Vector Flux)	$d\underline{\psi} = \boldsymbol{\mathcal{D}}\cdot\hat{n}\,ds$	$d\underline{I} = \boldsymbol{J}\cdot\hat{n}\,ds$	$d\underline{\psi}_m = \boldsymbol{\mathcal{B}}\cdot\hat{n}\,ds$	$d\underline{Q} = \boldsymbol{J}_{th}\cdot\hat{n}\,ds$	$d\underline{\psi}_l = \boldsymbol{\mathcal{F}}\cdot\hat{n}\,ds$
Gauss's Law	$\oint d\underline{\psi} = \underline{q}_{enc}$	$\oint d\underline{I} = \underline{I}_{enc}$	$\oint d\underline{\psi}_m = 0$	$\oint d\underline{Q} = \underline{g}_{enc}$	$\oint d\underline{\psi}_l = \underline{e}$
Field Vector Divergence	$\nabla\cdot\boldsymbol{\mathcal{D}} = \rho_f$	$\nabla\cdot\boldsymbol{J} = \rho_I$	$\nabla\cdot\boldsymbol{\mathcal{B}} = 0$	$\nabla\cdot\boldsymbol{J}_{th} = g_v$	$\nabla\cdot\boldsymbol{\mathcal{F}} = e_v$
Possion's Equation†	$\nabla^2 V = -\rho/\epsilon$	$\nabla^2 V = -\rho_l/\sigma_c$	$\nabla^2\Omega = 0$	$\nabla^2 T = -g_v/k_{th}$	$\nabla^2\phi = -e_v/\rho_l$
Field transmittance	$\underline{C} = \underline{\psi}/\Delta V$	$\underline{G} = \underline{I}/\Delta V$	$\underline{P} = \underline{\psi}_m/\Delta\Omega$	$\underline{G} = \underline{Q}/\Delta T$	$\underline{G} = \underline{\psi}_l/\Delta\phi$

† The medium constant is assumed isotropic.

where σ_c (unit: S/m) is the electrical conductivity of the material and in general is a tensor. For regions of steady current flow and free of current sources, the divergence of $\boldsymbol{J}$ is zero

$$\nabla \cdot \boldsymbol{J} = 0 \qquad (7.2)$$

This type of current flow is known as Direct Current (DC). If the region contains current sources, then a term ρ_I (unit: A/m^3) that describes the density of current sources needs to be included in (7.2). For example, the divergence of $\boldsymbol{J}$ for a region that has a point current source located at $z = z_0$ is

$$\nabla \cdot \boldsymbol{J} = \underline{I}\delta[z - z_0] \qquad (7.3)$$

Combining (7.1) and (7.2) gives

$$\nabla \cdot (\sigma_c \boldsymbol{E}) = 0 \qquad (7.4)$$

Therefore, ohmic materials with direct current and uniform σ_c have zero charge density in regions free of current sources. A common question at this point is how can an electric field exist inside a conductor? Up to now we have only considered ideal conductors that have infinite electrical conductivity. From (7.1), the electric field in the ohmic material is inversely proportional to the electrical conductivity. For an ideal conductor, the electric field is zero due to its infinite electrical conductivity, but for conductors of finite electrical conductivity, the electric field is finite.

Now consider two conductors with an electrostatic potential difference $\Delta V = V_2 - V_1$ between them (see Fig. 7.1). Assume the region between the conductors is filled with a material of uniform permittivity ϵ_0 and zero electrical conductivity. Let Φ be the complex potential calculated for this case. As discussed in Section 2.21, the capacitance per

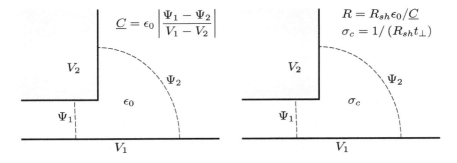

FIGURE 7.1 When the capacitance per unit length $\underline{C}$ of a region filled with permittivity ϵ_0 is known, the resistance of the same region filled with ohmic material of electrical conductivity σ_c is $R = R_{sh}\epsilon_0/\underline{C}$ where R_{sh} is the sheet resistance of the ohmic material.

unit length for the region enclosed by two lines of force Ψ_1 and Ψ_2 and the boundary of the two conductors is

$$\underline{C} = \epsilon_0 \frac{(\Psi_1 - \Psi_2)}{\Delta V} \tag{7.5}$$

Now replace the insulating material in the region that defines the capacitor of (7.5) with ohmic material of electrical conductivity σ_c (see Fig. 7.1). Due to the potential difference between the conductors, a current flows in the conductive material from one conductor to the other. In both cases, the charge density between the conductors is zero and the boundary conditions are the same. Therefore, the electric field is still given by

$$\boldsymbol{\mathcal{E}} = - \left(\frac{d\Phi}{dz} \right)^* \tag{7.6}$$

where Φ was determined when the region of interest was filled with the insulating material. From (7.1), the current density is

$$\boldsymbol{J} = \sigma_c \boldsymbol{\mathcal{E}} = -\sigma_c \left(\frac{d\Phi}{dz} \right)^* \tag{7.7}$$

In 2D, the current per unit length is

$$\underline{I} = \int \boldsymbol{J} \cdot \hat{n}\, ds = \int \mathrm{Re} \left[-\sigma_c \left(\frac{d\Phi}{dz} \right) (-j\,dz) \right] = \sigma_c \left(\Psi_1 - \Psi_2 \right) \tag{7.8}$$

Let the conductive material have a thickness (perpendicular to the z-plane) of $t_\perp$. The resistance R (unit: Ω) times thickness is given by

$$Rt_\perp = \Delta V / \underline{I} = \frac{\Delta V}{\sigma_c \left(\Psi_1 - \Psi_2 \right)} = \frac{\epsilon_0}{\sigma_c \underline{C}} \tag{7.9}$$

Defining the sheet resistance R_{sh} (unit: $\Omega/\square$) of the conducting material as

$$R_{sh} = 1/\left(\sigma_c t_\perp \right) \tag{7.10}$$

gives the resistance between the conductors as

$$R = \Delta V / I = \frac{\Delta V}{\sigma_c t_\perp \left(\Psi_1 - \Psi_2 \right)} = R_{sh} \left(\epsilon_0 / \underline{C} \right) \tag{7.11}$$

The last form of (7.11) allows calculating the resistance between two conductors if the capacitance for the region between the two conductors is already known. As a very simple example, the infinite horizontal parallel plate capacitor has lines of force that are vertical. Let the distance between plates be δ. The capacitance per unit length defined by two lines of force a distance l apart is

$$\underline{C} = \epsilon_0 l/\delta \qquad (7.12)$$

If the region between the two lines of force is replaced with ohmic material with sheet resistance R_{sh}, the resistance between the top and bottom plates is

$$R = R_{sh}\delta/l \qquad (7.13)$$

Note that resistance and capacitance have the opposite dependence on the ratio of the number of curvilinear squares along the perimeter of the conductor to the number of curvilinear squares between conductors.

Based on Table 7.1, a point current source $\underline{I}$ (unit: A/m) is analogous to a line charge and the electrical conductivity of the material is analogous to the permittivity. With these analogies, we can convert the complex potential due to n lines charges given by (2.89) into the complex potential for an infinite sheet of ohmic material with electrical conductivity σ_c and n point current sources by replacing $\underline{q}_i$ with $\underline{I}_i$ and ϵ with σ_c

$$\Phi[z] = -\frac{1}{2\pi\sigma_c}\sum_{i=1}^{n}\underline{I}_i \ln[z - z_i] + C_0 \qquad (7.14)$$

where z_i is the location of the ith current source of value $\underline{I}_i$.

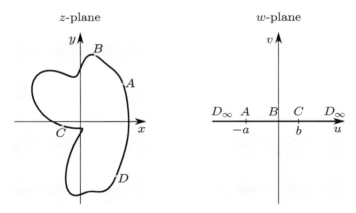

FIGURE 7.2 Diagram for the van der Pauw derivation. Four contacts (A, B, C, D) exist on the edge of the sample. Current is forced and removed between two adjacent contacts and the voltage drop is measured between the two remaining contacts.

With the translation used to generate (7.14), we can derive van der Pauw's theorem in terms of cross-resistances as it was originally done [8].

Consider the general geometry shown in Fig. 7.2. Riemann's mapping theorem states the general geometry can be conformally mapped onto the upper half-plane of the w-plane. Let points A, B, C, and D in the z-plane map to $w = -a$, $w = 0$, $w = b$, and $w = \pm\infty$ on the real axis with $a > 0$ and $b > 0$. Let point currents sources $\underline{I}_0$ and $-\underline{I}_0$ be located at $z = A$ and $z = D$, respectively. Using (7.14), the complex potential in the w-plane is

$$\Phi_1[w] = -\frac{\underline{I}_0}{\pi\sigma_c}\ln[w + a] \tag{7.15}$$

Note (7.15) is the complex potential for the whole w-plane. In order to have $\underline{I}_0$ in the upper half plane, $2\underline{I}_0$ is used for $|\underline{I}_i|$ in (7.14). We define the cross-resistance R_1 as the voltage difference between $w = 0$ and $w = b$ divided by value of the current in the upper half plane

$$R_1 = \frac{V_0 - V_b}{\underline{I}_0 t_\perp} = -\frac{R_{sh}}{\pi}\left(\ln[a] - \ln[b + a]\right) \tag{7.16}$$

Now let the point currents sources $\underline{I}_0$ and $-\underline{I}_0$ be located at $z = B$ and $z = A$, respectively. The complex potential in the w-plane is

$$\Phi_2[w] = -\frac{1}{\pi\sigma_c}\underline{I}_0\left(\ln[w] - \ln[w + a]\right) \tag{7.17}$$

The cross-resistance R_2 is defined as the voltage difference between $w = b$ and $w = \infty$

$$R_2 = \frac{V_b - V_\infty}{\underline{I}_0 t_\perp} = -\frac{R_{sh}}{\pi}\left(\ln[b] - \ln[a + b]\right) \tag{7.18}$$

Rearranging and combining (7.16) and (7.18) gives

$$e^{-\pi R_1/R_{sh}} + e^{-\pi R_2/R_{sh}} = 1 \tag{7.19}$$

which is van der Pauw's theorem in terms of cross-resistances and the sheet resistance of the ohmic material.

Based on the above analysis, all results previously derived for electrostatics can be directly applied to ohmic material with direct current for either calculating resistances or determining the complex potential when current sources are present. This includes the method of images and Green's function in terms of current sources based on the translation used to obtain (7.14).

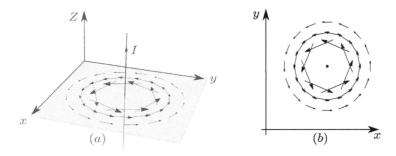

FIGURE 7.3 Vector field plot of $\mathcal{H}$ in (a) 3D and (b) 2D for a wire with current I flowing in the positive Z-direction.

7.3 MAGNETOSTATICS

When current flows, a magnetic field is generated. In order for the magnetic field to be effectively two-dimensional, current flow must extend infinitely in the direction perpendicular to the complex plane (similar to a line charge). For example, let an infinitely long wire that is perpendicular to the z-plane go through $z = z_0$ and have a current $I > 0$ flowing through it. Let the positive direction of current flow be in the Z-direction in Cartesian coordinates as shown in Fig. 7.3. The magnetic field intensity vector $\mathcal{H}$ (unit: A/m) in the complex plane due to the current flow in the wire is

$$\mathcal{H} = \frac{I}{2\pi r}\hat{\boldsymbol{\theta}} = \frac{jI}{2\pi (z^* - z_0^*)} \tag{7.20}$$

where $\hat{\boldsymbol{\theta}}$ is a unit vector perpendicular to the radial direction with the positive direction as counterclockwise. In current free regions, the curl of (7.20) is zero. Therefore, $\mathcal{H}$ can be obtained from the gradient of a scalar magnetic potential Ω (unit: A) as

$$\mathcal{H} = -\nabla\Omega \tag{7.21}$$

To ensure the integral of $\nabla \times \mathcal{H}$ over the area of any closed path is zero and to keep Ω single valued, a suitable barrier must be inserted [7]. From Table 7.1, the divergence of the magnetic flux density vector $\mathcal{B}$ (unit: T) is zero

$$\nabla \cdot \mathcal{B} = 0 \tag{7.22}$$

$\mathcal{B}$ and $\mathcal{H}$ are related through the permeability μ (unit: H/m) of the material as

$$\mathcal{B} = \mu\mathcal{H} \tag{7.23}$$

where μ in general is a tensor. Combining (7.21) through (7.23) gives

$$\nabla \cdot (-\mu \nabla \Omega) = 0 \tag{7.24}$$

If μ is a constant, then Ω obeys Laplace's equation. It should then be possible to obtain $\mathcal{H}$ and the flux contours of $\mathcal{B}$ from a complex potential $\Phi_{\mathcal{H}}$ (unit: A).

Comparing (2.4) and (7.20), $\mathcal{H}$ has the same form as $\mathcal{E}$ due to an infinite line charge and can be made identical by multiplying (2.4) by $j\epsilon I_i/\underline{q}_i$. We can then convert the complex potential due to n lines charges given by (2.89) into the magnetic complex potential for n wires of current by multiplying (2.89) by $\left(j\epsilon I_i/\underline{q}_i \right)^*$

$$\Phi_{\mathcal{H}}[z] = \frac{j}{2\pi} \sum_{i=1}^{n} I_i \ln[z - z_i] + C_0 \tag{7.25}$$

where z_i is the location of the ith wire with current I_i.

From the above analysis, a constant flux boundary in electrostatics is analogous to a constant magnetic potential boundary and a constant electrostatic potential boundary is analogous to a constant magnetic flux boundary. Based on these analogies, the image current source required to produce a constant magnetic potential planar boundary has the same sign as the true current source and the image current source required to produce a constant magnetic flux planar boundary has the opposite sign as the true current source. Additionally, the boundary conditions at the interface between materials with different permeability values (assuming no current flow along the interface) are

$$\Omega_1 = \Omega_2, \quad \mu_1 \frac{\partial \Omega_1}{\partial n} = \mu_2 \frac{\partial \Omega_2}{\partial n} \tag{7.26}$$

To demonstrate the method of magnetic images, consider the problem of a wire with current I in the space between two concentric circular cylinders of infinite permeability. Let the cylinders have their centers at the origin of the w-plane and the location of the current source be on the real axis at $w = c$. Let cylinder 1 have a radius b and cylinder 2 have a radius $a > b$ as shown in Fig. 7.5 (a). According to [3] (and after adjusting for the different unit and coordinate axes conventions), the boundary conditions for the magnetic potential are

$$\Omega[be^{j\theta}] = -I/2, \quad \Omega[ae^{j\theta}] = -I\theta/(2\pi) \tag{7.27}$$

If we change the zero of the magnetic potential to $w = b$, the boundary conditions become

$$\Omega\left[be^{j\theta}\right] = 0, \quad \Omega\left[ae^{j\theta}\right] = -I\left(\theta - \pi\right)/\left(2\pi\right) \tag{7.28}$$

Instead of solving this problem directly, we use the method of images to solve the equivalent electrostatic problem of a line charge between two conductive concentric circular cylinders with a constant flux boundary for cylinder 1 and a flux boundary that is proportional to θ on cylinder 2. First we determine the images required to achieve a circular constant flux boundary of radius r_1 centered at the origin for a true line charge q_0 located at $w = w_0$. From Section 3.7, a circular constant electrostatic potential boundary is achieved by placing an image charge $-q_0$ at $w = r_1^2/w_0^*$. If this image charge is replaced by an image line charge q_0, then the imaginary part of the complex potential is

$$\text{Im}[\Phi] = \frac{q_0}{2\pi\epsilon_0}\text{Im}\left[\ln\left[\left(w - w_0\right)\left(w - \frac{r_1^2}{w_0^*}\right)\right]\right] = \frac{q_0\theta}{2\pi\epsilon_0} \tag{7.29}$$

when $w = re^{j\theta}$. This is not the desired boundary condition for cylinder 1, but it is the desired boundary condition for cylinder 2. In order to eliminate the θ dependence, an image charge of $-q_0$ needs to be placed at the origin. Then $\text{Im}\left[\Phi\left[r_1e^{j\theta}\right]\right] = 0$. These results are summarized in Fig. 7.4.

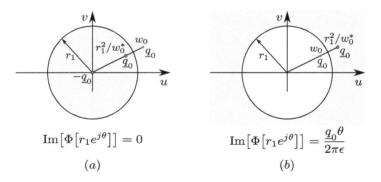

FIGURE 7.4 (a) A line charge q_0 located at w_0, an image charge q_0 at $w = r_1^2/w_0^*$ and an image charge $-q_0$ at $w = 0$ are required to make $\text{Im}\left[\Phi\left[r_1e^{j\theta}\right]\right] = 0$. (b) A line charge q_0 located at w_0 and an image charge q_0 at $w = r_1^2/w_0^*$ are required to make $\text{Im}\left[\Phi\left[r_1e^{j\theta}\right]\right] = q_0\theta/\left(2\pi\epsilon\right)$.

From Fig. 7.4, we need to place an image charge of $-q_0$ at $w = 0$ and q_0 at $w = b^2/c$ to make cylinder 1 a constant flux boundary and an image charge q_0 at $w = a^2/c$ to make cylinder 2 have a flux that is proportional to θ. These image charges then require additional image charges to satisfy the boundary conditions leading to an infinite array of image charges. If we let one iteration consist of three images (two to make cylinder 1 a constant flux boundary and one to make cylinder 2 have a flux that varies proportionally to θ), then after $2p + 1$ iterations, the complex potential is

$$\Phi = -\frac{q_0}{2\pi\epsilon} \ln\left[\frac{1}{w^{2p+1}}\left(w - \frac{a^{2p+2}}{b^{2p}c}\right)\prod_{n=-p}^{n=p}\left(w - \frac{b^{2n+2}}{a^{2n}c}\right)\left(w - \frac{b^{2n}c}{a^{2n}}\right)\right] \quad (7.30)$$

Setting the zero of the complex potential at $z = b$ and letting $p \to \infty$ in (7.30) gives the complete complex potential as

$$\Phi = -\frac{q_0}{2\pi\epsilon} \ln\left[\prod_{n=-\infty}^{n=\infty}\left(\frac{ca^{2n} - (b^{2n+2}/w)}{ca^{2n} - b^{2n+1}}\right)\left(\frac{a^{2n}w - cb^{2n}}{a^{2n}b - cb^{2n}}\right)\right] \quad (7.31)$$

To translate this solution into the magnetic complex potential, we multiply (7.31) by $\left(j\epsilon I/q_0\right)^*$ to obtain

$$\Phi_{\mathcal{H}} = \frac{jI}{2\pi} \ln\left[\prod_{n=-\infty}^{n=\infty}\left(\frac{ca^{2n} - (b^{2n+2}/w)}{ca^{2n} - b^{2n+1}}\right)\left(\frac{a^{2n}w - cb^{2n}}{a^{2n}b - cb^{2n}}\right)\right] \quad (7.32)$$

A field map of (7.32) is shown in Fig. 7.5. Although Fig. 7.5 has been drawn with $\Delta\Omega = \Delta\text{Im}[\Phi_{\mathcal{H}}] = 0.05I$, $w = ae^{j\theta}$ is not a constant magnetic flux boundary or a constant magnetic potential boundary. Therefore, the regions with a portion of their boundaries composed of $w = ae^{j\theta}$ do not form curvilinear squares.

As a second example, we calculate the complex magnetic potential for a wire located at $z = -x_0 < 0$ with current I and permeability μ_1 for $x > 0$ and μ_0 for $x < 0$. In terms of electrostatics, this problem is equivalent to a line charge q_0 at $z = -x_0$ with permittivity ϵ_1 for $x > 0$ and ϵ_0 for $x < 0$. The electrostatic complex potential for this problem was given in Section 3.10 as

$$\Phi[z] = \begin{cases} -\dfrac{q_0}{2\pi\epsilon_0}\left(\ln[z + x_0] + \dfrac{\epsilon_0 - \epsilon_1}{\epsilon_0 + \epsilon_1}\ln[x_0 - z]\right), & x < 0 \\[4mm] -\dfrac{q_0}{\pi\left(\epsilon_0 + \epsilon_1\right)}\ln[z + x_0] & , \quad x > 0 \end{cases} \quad (7.33)$$

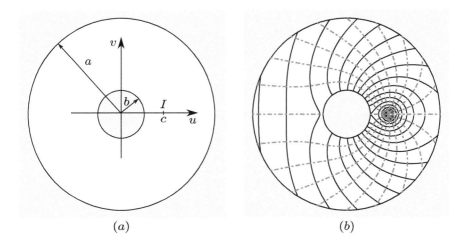

(a) (b)

FIGURE 7.5 Magnetic field map for a wire of current I located in the region between two concentric circular cylinders of infinite permeability. Dashed lines are constant magnetic potential contours and solid lines are constant magnetic flux contours. $\Delta\Omega = \Delta\text{Im}[\Phi_{\mathcal{H}}] = 0.05I$, $a = 4b$, and $c = 1.8b$.

To obtain the magnetic complex potential, we replace ϵ_i with μ_i and $\underline{q_0}$ with $(j\mu_0 I)^*$ in (7.33)

$$\Phi_{\mathcal{H}}[z] = \begin{cases} \dfrac{jI}{2\pi}\left(\ln[z+x_0] + \dfrac{\mu_0 - \mu_1}{\mu_0 + \mu_1}\ln[x_0 - z]\right), & x < 0 \\[2ex] \dfrac{jI\mu_0}{\pi(\mu_0 + \mu_1)}\ln[z+x_0] & , & x > 0 \end{cases} \tag{7.34}$$

7.4 STEADY HEAT POWER FLOW

Similar to charges that are infinite in one direction that result in a 2D electrostatic potential, a heat source that is uniform and infinite in one direction creates a temperature distribution that is two-dimensional. The value of a heat source depends on its dimensions and rate of heat energy it generates. For example, the value of a 3D heat source (g_v (unit: W/m^3)) depends on the rate per unit time of heat energy that is generated per unit volume. 2D and 1D heat sources are labeled g_s (unit: W/m^2) and g (unit: W/m), respectively. Heat sources are analogous to charge density in electrostatics. The heat power flow density vector $\boldsymbol{J}_{th}$ (unit: W/m^2) describes the flow of heat energy per unit time per unit area. The flow of heat energy results in a change in temperature in the region surrounding the heat source. The ability of a material to transfer heat energy is called the thermal conductivity (k_{th}) (unit: W/m·K) of the material. In general k_{th} is a tensor. Experimentally, $\boldsymbol{J}_{th}$ is found to be proportional to the gradient of a scalar potential T (unit: K) better known as temperature

$$\boldsymbol{J}_{th} = -k_{th}\boldsymbol{\nabla}T \tag{7.35}$$

In regions free of heat sources, the divergence of $\boldsymbol{J}_{th}$ is zero

$$\boldsymbol{\nabla}\cdot\boldsymbol{J}_{th} = 0 \tag{7.36}$$

Plugging (7.35) into (7.36) gives

$$\boldsymbol{\nabla}\cdot(-k_{th}\boldsymbol{\nabla}T) = 0 \tag{7.37}$$

If k_{th} is a constant, then the temperature function obeys Laplace's equation and a complex potential can be defined for the region. The mathematical relationships between T and $\boldsymbol{J}_{th}$ are exactly the same as the mathematical relationships between V and $\boldsymbol{D}$ in electrostatics. This relationship allows converting an electrostatic complex potential into the equivalent thermal complex potential by replacing ϵ with k_{th}, letting constant voltages represent constant temperatures, and charges represent heat sources. For example, the conversion of the complex potential due to n lines charges given by (2.89) into the complex potential for an infinite material of thermal conductivity k_{th} with n heat line sources is

$$\Phi[z] = -\frac{1}{2\pi k_{th}}\sum_{i=1}^{n}\underline{g}_{i}\ln[z - z_{i}] + C_{0} \tag{7.38}$$

where z_i is the location of the ith heat line source of value $\underline{g}_i$.

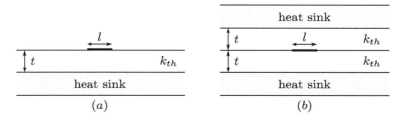

FIGURE 7.6 (a) Cross-section of a material of thickness t and thermal conductivity k_{th} with backside mounted to a heat sink. The surface has a heat source of length l. (b) Placing the mirror image of the structure at the surface results in the desired boundary condition of a constant heat flux contour at the surface outside of the heat source.

As an example, let us determine the temperature distribution in an infinite slab of material with thermal conductivity k_{th} and thickness t with a heat source of width l at the surface and the bottom attached to an ideal infinite heat sink (see Fig. 7.6). Let the heat sink have a temperature T_A which results in the boundary condition

$$T|_{y=0} = T_A \tag{7.39}$$

Assuming heat can only be removed from the backside, the surface boundary conditions are

$$k_{th} \frac{\partial T}{\partial y}\bigg|_{y=t} = \begin{cases} 0 & |x| > \frac{1}{2}l \\ g_s, & |x| < \frac{1}{2}l \end{cases} \tag{7.40}$$

where g_s is the power per unit area the surface heat source produces. Outside the heat source, the surface corresponds to a constant heat flux contour. To meet this boundary condition, an infinite heat sink can be placed at distance t above the original surface with the material between the surface and the top heat sink having the same thermal conductivity. Due to symmetry, half of the heat energy flows to the top heat sink and half to the bottom. This requires the heat source value to be doubled in order to achieve the same heat energy flow to the bottom heat as in the original problem. The corresponding electrostatic problem is a region of sheet charge of width l midway between two infinite conductors at the same potential a distance $2t$ apart. The complex potential for a line charge q_0 between grounded conductors was given in Section 5.11 which we can now translate into the complex thermal potential as

$$\Phi[z] = \frac{g}{\pi k_{th}} \ln\left[\frac{e^{\pi z/(2t)} - e^{\pi z_0^*/(2t)}}{e^{\pi z/(2t)} - e^{\pi z_0/(2t)}}\right] \tag{7.41}$$

where g is a line heat source located at z_0. For this problem, $z_0 = x_0 + jt$. To obtain the complex potential due to the entire heat source, we use the principle of superposition and sum up all the line charge contributions that make up the heat source

$$\Phi[z] = \frac{g_s}{\pi k_{th}} \int_{-\frac{1}{2}l}^{\frac{1}{2}l} \ln\left[\frac{1 + je^{\pi(x_0-z)/(2t)}}{1 - je^{\pi(x_0-z)/(2t)}}\right] dx_0 \tag{7.42}$$

Using (6.120), we can integrate (7.42) as

$$\Phi[z] = \frac{2tg_s}{\pi^2 k_{th}} \left(\text{Li}_2\left[je^{\pi(l-2z)/(4t)}\right] + \text{Li}_2\left[-je^{-\pi(l+2z)/(4t)}\right] \right.$$
$$\left. - \text{Li}_2\left[-je^{\pi(l-2z)/(4t)}\right] - \text{Li}_2\left[je^{-\pi(l+2z)/(4t)}\right] \right) \tag{7.43}$$

In order for the heat sink to be at temperature T_A, we must add T_A to (7.43). The temperature field map for (7.43) is shown in Fig. 7.7. Due to symmetry, the peak temperature occurs at $z = jt$. Plugging $z = jt$ into (7.43) and taking the real part gives the peak temperature as

$$T_{peak} = \frac{2tg_s}{\pi^2 k_{th}} \operatorname{Re}\left[\operatorname{Li}_2[e^\alpha] + \operatorname{Li}_2\left[-e^{-\alpha}\right] - \operatorname{Li}_2[-e^\alpha] - \operatorname{Li}_2\left[e^{-\alpha}\right]\right] \quad (7.44)$$

where $\alpha = \pi l / (4t)$.

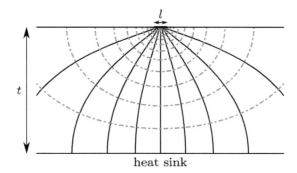

FIGURE 7.7 Temperature field map for a heat source l at the surface of material of thickness t and backside attached to a heat sink. Dashed lines are constant temperature contours and solid lines are constant heat flux contours.

7.5 FLUID DYNAMICS

Fluid dynamics has many examples of 2D problems that conform to Laplace's equation [6]. We limit our discussion to irrotational, incompressible 2D fluid flow. If we assume the 2D fluid flow is irrotational, then the velocity of the fluid has a curl of zero

$$\nabla \times \boldsymbol{v} = 0 \quad (7.45)$$

From Section 2.9, any vector with a curl of zero can be represented by the gradient of a scalar potential. This allows writing

$$\boldsymbol{v} = -\nabla \phi \quad (7.46)$$

where ϕ (unit: m^2/s) is the scalar velocity potential. As discussed in the preface of [4], the minus sign in the definition of the velocity potential reflects it is the potential of an impulsive force that would start the flow and has the added advantage of being consistent with of other fields of

Mathematical Physics. If the fluid is also incompressible then the density of the fluid particle does not change [6] requiring

$$\nabla \cdot \boldsymbol{v} = 0 \tag{7.47}$$

Combining (7.46) and (7.47) gives

$$\nabla^2 \phi = 0 \tag{7.48}$$

Since ϕ obeys Laplace's equation, we can define a complex velocity potential

$$\Phi_v = \phi + j\psi \tag{7.49}$$

where ψ is known as the streamline function. The tangent at each point of a constant streamline contour is in the direction of the fluid velocity. The fluid velocity can be obtained directly from the complex velocity potential as

$$\boldsymbol{v} = -\left(\frac{d\Phi_v}{dz}\right)^* \tag{7.50}$$

We are now ready to derive the Milne-Thomson circle theorem for fluid dynamics [5]. Let there be no rigid boundaries and let the complex velocity potential of fluid flow be $f[z]$, where the singularities of $f[z]$ are all at a distance greater than a from the origin. If a circular cylinder of radius a centered at the origin in introduced into the field of flow, the complex velocity potential becomes

$$\Phi_v[z] = f[z] + \left(f\left[a^2/z^*\right]\right)^* \tag{MTCT}$$

To prove this theorem, we need to show that the boundary of the cylinder is a streamline and only the singularities of $f[z]$ exist in the region $|z| > a$. Along the cylinder's boundary, $z = ae^{j\theta}$ and the complex potential is

$$\Phi_v[z] = f\left[ae^{j\theta}\right] + \left(f\left[ae^{j\theta}\right]\right)^* = 2\mathrm{Re}\left[f\left[ae^{j\theta}\right]\right] \tag{7.51}$$

Therefore, the boundary of the circle is a streamline with value $\psi = 0$. Next, let z_0 be a singularity for $f[z]$. The location of this singularity for the function $(f[a^2/z^*])^*$ is

$$z_s = a^2/z_0^* \tag{7.52}$$

which has a magnitude less than a since $|z_0| > a$ by assumption. Therefore, all the singularities of $(f[a^2/z^*])^*$ are located within the circle $|z| = a$ and the theorem is proven.

From Table 7.1, the electric field is analogous to fluid velocity and a streamline is analogous to a constant line of force. Bases on these analogies, we can translate the circle theorem into an electrostatic circle theorem. Let there be no boundaries and let the complex electrostatic potential be $f[z]$, where the singularities of $f[z]$ are all at a distance greater than a from the origin. If a circular line of force boundary condition of radius a centered at the origin is introduced into the electric field, the complex electrostatic potential is given by (MTCT).

When there are no boundaries, the complex electrostatic potential has the form of (2.89) and a singularity in the complex potential corresponds to either a positive or negative line charge. Now consider the complex potential of a line charge q_0 located at $z = z_0$ given as

$$f[z] = -\frac{q_0}{2\pi\epsilon} \ln[z - z_0] \tag{7.53}$$

When a circular line of force boundary of radius $a < |z_0|$ centered at the origin is introduced, the complex electrostatic potential is obtained from (MTCT) as

$$
\begin{aligned}
\Phi[z] &= -\frac{q_0}{2\pi\epsilon} \ln\left[(z - z_0) \left(\frac{a^2}{z} - z_0^* \right) \right] \\
&= -\frac{q_0}{2\pi\epsilon} \ln\left[\frac{z_0^* (z_0 - z)}{z} \left(z - \frac{a^2}{z_0^*} \right) \right]
\end{aligned}
\tag{7.54}
$$

As shown in Fig. 7.8, (7.54) is equivalent to the potential due to a negative line charge located at the origin and positive line charges located at $z = z_0$ and $z = a^2/z_0^*$. All the line charges have the same magnitude. Thus, the circle theorem is the method of images in disguise. It obtains the complex potential by including the image charges due to true charges in finite space and the charge at infinity.

As a second demonstration of the theorem, consider the complex electrostatic potential for a uniform electric field of magnitude $|\mathcal{E}_0|$ in the direction of the positive real axis given as

$$f[z] = -|\mathcal{E}_0|z \tag{7.55}$$

When a circular line of force boundary of radius a centered at the origin is introduced, the complex electrostatic potential is obtained from (MTCT) as

$$\Phi[z] = -|\mathcal{E}_0|z - |\mathcal{E}_0|\frac{a^2}{z} \tag{7.56}$$

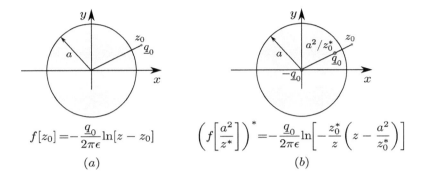

$$f[z_0] = -\frac{q_0}{2\pi\epsilon}\ln[z - z_0]$$

(a)

$$\left(f\left[\frac{a^2}{z^*}\right]\right)^* = -\frac{q_0}{2\pi\epsilon}\ln\left[-\frac{z_0^*}{z}\left(z - \frac{a^2}{z_0^*}\right)\right]$$

(b)

FIGURE 7.8 (a) Complex potential due to a single line charge q_0 located at $z = z_0$. (b) Location of line charges introduced by the circle theorem for the complex potential in (a). The circle theorem introduces image charges due to the true charges in finite space and any charges at infinity.

From Section 6.6, a uniform field can be viewed as originating from a positive line charge and a negative line charge that go to infinity in opposite directions. The ratio of charge to distance between charges is kept constant as the distance between charges goes to infinity. The circle theorem places a point dipole at the origin that has a dipole moment $\underline{M} = -2\pi\epsilon|E_0|a^2$ as the image charge for the charges at infinity.

Fluid sources are defined by the volume of fluid flow through a boundary that encloses them. If more volume of fluid is flowing out of the bounded region than entering, the net volume of fluid flow through the boundary is considered positive and it is labeled a positive source. If more volume of fluid is entering the bounded region than flowing out, the net volume of flow fluid through the boundary is considered negative and it is labeled a negative source. Negative sources are also called sinks. In 2D, a point fluid source is really a 3D fluid line source that is defined by the volume of fluid flow per unit length through the boundary. Thus, 2D point positive and negative sources are analogous to positive and negative line charges in electrostatics. The complex velocity potential due to n point sources is

$$\Phi_v[z] = -\frac{1}{2\pi}\sum_{i=1}^{n} m_i \ln[z - z_i] + C_0 \qquad (7.57)$$

where z_i is the location of the ith point source and $\underline{m_i}$ (unit: m^2/s) is the strength of the point source. Translating a complex velocity potential due to point sources into an complex electrostatic potential is performed by letting $\underline{m}_i$ represent the line charge value at z_i and then dividing the complex velocity potential by the permittivity of the region.

To complete the translation between electrostatics and steady fluid flow, we introduce the flow rate density vector $\boldsymbol{\mathcal{F}}$ (unit: kg/m²·s) given by

$$\boldsymbol{\mathcal{F}} = \rho_l \boldsymbol{v} \tag{7.58}$$

where ρ_l (unit: kg/m³) is the density of the fluid. Flow rate density is analogous to the electric flux density and ρ_l is analogous to permittivity. The equivalent of Gauss's law for fluid flow is

$$\oint \boldsymbol{\mathcal{F}} \cdot \hat{n}\, ds = \underline{e} \tag{7.59}$$

which states that the efflux (flowing out) of fluid per unit length ($\underline{e}$ (unit: kg/m·s)) from an enclosed region is the integral of the normal component of the flow rate density vector along the boundary. Taking the divergence of (7.58) and assuming an irrotational fluid gives the differential form of (7.59) as

$$\nabla \cdot \boldsymbol{\mathcal{F}} = \nabla \cdot (-\rho_l \nabla \phi) = e_v \tag{7.60}$$

where e_v (unit: kg/m³·s) is the space density of efflux.

EXERCISES

7.1 In the derivation of van der Pauw's theorem, point contacts were assumed. This theorem has been extended to finite contacts [2, 1].

   a. Show four resistance measurements are needed to determine the sheet resistance when two of the contacts have finite width and two contacts are points.

   b. Use the results from Section 5.18.8 to show the sheet resistance is

$$R_{sh} = R_{qs,qs} K[1 - m] / K[m]$$

   for two point contacts and two finite contacts where a point contact is always followed by a finite contact in either direction along the boundary. The value of m is determined by

$$R_{pq,rq} = \frac{(R_{qs,qs} - R_{qs,ps})(R_{qs,qs} - R_{qs,rs})}{R_{qs,qs}}$$

$$+ \frac{2R_{qs,qs}K[1-m]}{\pi K[m]}\sqrt{\frac{(n-1)(n-m)}{n}}$$

$$\times \mathrm{Re}\left[\Pi[n,\theta_1[x_r],m] + \Pi[n,m]\left(1 - \frac{R_{qs,rs}}{R_{qs,qs}}\right)\right]$$

with

$$\theta_1[x_r] = \sin^{-1}\left[\mathrm{sn}\left[\left(\frac{R_{qs,qs} - R_{qs,rs}}{R_{qs,qs}} + j\frac{K[1-m]}{K[m]}\right)K[m],m\right]\right],$$

$$n = 1/\mathrm{sn}^2\left[K[m]\left(\frac{R_{qs,qs} - R_{qs,ps}}{R_{qs,qs}}\right),m\right],$$

and

$$R_{ij,mn} = \frac{V_m - V_n}{I_{ij}}$$

is defined as the voltage drop between contacts m and n divided by the current being forced from contact i to contact j. For the case of four finite contacts see [1].

7.2 Plate 1 has zero thickness, width w_1, and lies along the real axis. Plate 2 has zero thickness, width w_2, and lies along the real. If the material surrounding the plates is ohmic, has a sheet resistance R_{sh}, and the distance between plates is g, what is the resistance between the plates?

7.3 An ohmic resistor has the shape shown in Fig. 7.9 with the contacts along the boundaries between BC and EF. If the resistive material has a sheet resistance R_{sh} and $b = 3a$, what is the resistance between the contacts?

7.4 A uniform magnetic field exist along the horizontal direction ($\mathcal{H} = |\mathcal{H}_0|\hat{x}$). A circular cylinder with infinite permeability of radius r_0 is conformally coated with a magnetic material of permeability μ_1 and thickness $r_1 - r_0$. The coated cylinder is placed in the uniform magnetic field. What is complex potential inside and outside the coated magnetic material? Draw the constant magnetic flux contours near the cylinder.

z-plane

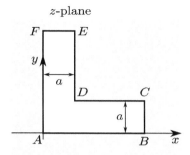

Point	z	ϕ/π
A	0	1/2
B	b	1/2
C	$b + ja$	1/2
D	$a + ja$	$-1/2$
E	$a + jb$	1/2
F	jb	1/2

FIGURE 7.9 Geometry of resistor for problem 7.3.

7.5 A wire with current I is located at $z = b > 0$. A circular cylinder with permeability μ_1 and radius $r_1 < b$ has its center at the origin of the z-plane. Assume the material for $|z| > r_1$ has a permeability μ_0 and determine the magnetic complex potential.

7.6 A rectangular enclosure of width $2l$ and height h has the two vertical sides and the top at temperature $0\,°C$. The bottom of the enclosure is kept at T_0. What is the temperature distribution throughout the enclosure? Draw the temperature field map for $h = l$.

7.7 A plate of length $2a$ is centered inside a circular cylinder or radius $r_1 > a$. The material between the plate and cylinder has a uniform thermal conductivity k_{th}. If the plate is heated to a temperature T_0 and cylinder is kept at $0\,°C$, what is the temperature distribution between the plate and cylinder?

7.8 Uniform fluid flow with speed $|v_0|$ in the horizontal direction has the complex velocity potential

$$\Phi_v = -|v_0|z$$

a. If an obstruction of elliptical cross-section with major axis $2a$ and minor axis $2b$ is inserted into the flow with the major axis aligned with the direction of fluid flow, determine the complex velocity potential and draw a streamline map.

b. If the obstruction has the shape of flat plate with length $2l$ at an angle θ_1 to the direction of fluid flow, determine the complex velocity potential and draw a streamline map.

7.9 The flow of fluid begins in a single main channel and then splits into three channels (see Fig. 7.10). Let the main channel have a width $2h_1$, the two angled channels each have a width h_2 and make an angle $\pm\alpha$ with respect to the horizontal, and the horizontal channel between the two angled channels have a width $2h_3$.

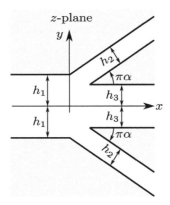

FIGURE 7.10 Single channel splitting into three channels.

a. Determine the S-C transformation that maps the region of fluid flow above the line of symmetry onto the upper half of the w-plane.

b. If the velocity at infinity upstream in the main channel is $2|v_1|$ directed along the horizontal, determine the velocity in each of the channels and draw and a streamline map. Remember the boundaries (and line of symmetry) are equivalent to lines of force in electrostatics. The fluid can be viewed as originating from vertex GH and being removed at vertices AB and DE (which are equivalent to line charges). Finally, note vertex C is a stagnation point which means velocity is zero there.

REFERENCES

[1] M. Cornils and O. Paul. Beyond van der pauw: Sheet resistance determination from arbitrarily shaped planar four-terminal devices with extended contacts. In *2008 IEEE International Conference on Microelectronic Test Structures*. IEEE, mar 2008.

[2] Martin Cornils and Oliver Paul. Sensor calibration of planar four-contact devices with up to two extended contacts. In *2007 IEEE Sensors*. IEEE, 2007.

[3] B. Hague. *The Principles of Electromagnetism*. Dover Publications, Inc., 1962.

[4] Horace Lamb. *Hydrodynamics*. Cambridge University Press, 5th edition, 1924.

[5] L. M. Milne-Thomson. Hydrodynamical images. *Mathematical Proceedings of the Cambridge Philosophical Society*, 36(2):246–247, April 1940.

[6] L. M. Milne-Thomson. *Theoretical Hydrodynamics*. The Macmillan Company, 4th edition, 1960.

[7] William R. Smythe. *Static and Dynamic Electricity*. McGraw-Hill Book Company, 3rd edition, 1968.

[8] L. J. van der Pauw. A method of measuring the resistivity and hall coefficient of lamellae of arbitrary shape. *Philips Technical Review*, 20:220–224, 1958.

Differentiating an Integral

Many advanced functions are effectively integrals with names. Understanding how to differentiate these types of functions is important. In this section we derive Leibniz's formula for differentiating an integral with variable limits of integration. Let

$$g[y] = \int_a^b f[x, y]\, dx = F[b, y] - F[a, y] \qquad (A.1)$$

where

$$F[x, y] = \int f[x, y]\, dx + C, \quad \frac{\partial F[x, y]}{\partial x} = f[x, y] \qquad (A.2)$$

Assuming a and b are independent of y, the limits do not change when y is increased by Δy and we have

$$\frac{g[y + \Delta y] - g[y]}{\Delta y} = \int_a^b \frac{f[x, y + \Delta y] - f[x, y]}{\Delta y}\, dx \qquad (A.3)$$

Taking the limit of both sides for $\Delta y \to 0$ gives

$$\frac{dg}{dy} = \frac{d}{dy}\left(\int_a^b f[x, y]\, dx\right) = \int_a^b \frac{\partial f}{\partial y}\, dx, \quad a, b \text{ are constants} \qquad (A.4)$$

If the integration limits a and b are not constants, then (A.1) and (A.2) give

$$\frac{\partial g}{\partial b} = \frac{\partial F[b, y]}{\partial b} = f[b, y], \quad \frac{\partial g}{\partial a} = -\frac{\partial F[a, y]}{\partial a} = -f[a, y] \qquad (A.5)$$

The complete derivative of the integral with respect to y when the limits a and b depend on y is

$$\frac{dg}{dy} = \int_a^b \frac{\partial f[x,y]}{\partial y} dx + \frac{\partial g}{\partial b} \frac{db}{dy} + \frac{\partial g}{\partial a} \frac{da}{dy}$$

$$= \int_a^b \frac{\partial f[x,y]}{\partial y} dx + f[b,y] \frac{db}{dy} - f[a,y] \frac{da}{dy}$$

(A.6)

which is known as Leibniz's formula for differentiating an integral. A special case of (A.6) that is used often with functions defined in terms of integrals is

$$h[y] = \int_a^y f[x] \, dx, \quad \frac{dh}{dy} = f[y]$$

(A.7)

Dirac δ-Function

In electrostatics, discrete charge distributions are often specified using the Dirac δ-function. The δ-function is not an ordinary function with a function value assigned to each argument value, but instead is categorized as a generalized function.

In 1D, the δ-function may be defined by the requirement

$$\int_a^b f[x]\,\delta[x - x_0]\,dx = \begin{cases} f[x_0], & a < x_0 < b \\ 0, & \text{otherwise} \end{cases} \tag{B.1}$$

This definition implies

$$\delta[x] = 0 \quad \text{for} \quad x \neq 0 \tag{B.2}$$

but no function value can be assigned to $\delta[0]$. Instead, $\delta[0]$ is defined by the integral of (B.1).

Although no ordinary function satisfies (B.1), it is possible to specify ordinary functions that approach the δ-function through the limit process

$$\lim_{n \to \infty} \left[\int_a^b f[x]\,\delta_n[x - x_0]dx \right] = f[x_0] \tag{B.3}$$

for any interval $a < x_0 < b$ and for any function $f[x]$ that is continuous and finite in that interval.

The ordinary function $\delta_n[x]$ is not a unique function. Two examples of $\delta_n[x]$ are

$$\delta_n[x] = \frac{n}{\pi\left(1 + (nx)^2\right)} \quad \text{and} \quad \delta_n[x] = \frac{n}{\sqrt{\pi}}\exp\left[-(nx)^2\right] \tag{B.4}$$

Using the first representation of δ_n gives

$$\lim_{n \to \infty} \left[\frac{1}{\pi} \int_a^b \frac{n}{1 + (nx)^2} dx \right] = \frac{1}{\pi} \int_{-\infty}^{\infty} \frac{1}{1 + s^2} ds = 1 \tag{B.5}$$

assuming $a < 0 < b$.

Therefore, the δ-function can be viewed as representing the limit

$$\delta[x] = \underset{n \to \infty}{\text{ilim}} [\delta_n[x]] \tag{B.6}$$

where the notation "ilim" rather than "lim" denotes that for $x = 0$, $\delta[x]$ has meaning only under an integral sign, and the limit is to be performed after integration over the argument x in $\delta[x]$ [4].

These results are easily generalized to 2D. The 2D δ-function $\delta[z]$ can be defined analogously to (B.6) as

$$\delta[z] = \underset{n \to \infty}{\text{ilim}} [\delta_n[z]] \tag{B.7}$$

and δ_n satisfies

$$\lim_{n \to \infty} \left[\int_{area} f[z] \, \delta_n[z - z_0] \, da \right] = f[z_0] \tag{B.8}$$

where the integration is over any area that contains z_0 and $f[z]$ is any function that is continuous and finite in that area. In terms of Cartesian coordinates, the 2D δ-function is just the product of the δ-function in each direction

$$\delta[z - z_0] = \delta[x - x_0]\delta[y - y_0] \tag{B.9}$$

In general, the δ-function is not the product of δ-functions for each coordinate direction. For example, if we let

$$\delta[z] = \delta[r]\delta[\theta], \quad \text{incorrect} \tag{B.10}$$

in polar coordinates, we obtain

$$\int_{area} \delta[r]\,\delta[\theta]\, r dr d\theta = 0 \tag{B.11}$$

even when the area encloses the origin. Clearly (B.11) does not fulfill the requirement of (B.1). To satisfy (B.1), we must divide $\delta[r]\delta[\theta]$ by r to obtain the δ-function in polar coordinates

$$\delta[z] = \delta[r]\delta[\theta]/r \tag{B.12}$$

Another consequence of (B.1) is that $\delta[x]$ must have the inverse units of the integration variable. Thus in 1D with the integration variable representing length, the units of $\delta[x]$ are inverse length. This is important for dimensional analysis with functions described in δ-function notation. For example, volume charge density ρ_f has unit of C/m^3. A line charge density q with units C/m located at (x_0, y_0) in the z-plane can be represented in terms of a volume charge density as

$$\rho_f = q\delta[x - x_0]\delta[y - y_0] \tag{B.13}$$

Since each δ-function in (B.13) has units of $1/m$, (B.13) has units of C/m^3 as desired.

Further details on generalized functions and the δ-function can be found in [6].

Elliptic Integrals

First exposure to elliptic integrals can often result in confusion. Common sources of confusion are the various notations in the literature for the same elliptic integrals, complete versus incomplete elliptic integrals, and the complementary modulus forms. Thus, elliptic integrals should be clearly defined when used. In this section, we confine our discussion to elliptic integrals of the first, second, and third kind.

The incomplete elliptic integral of the first kind is defined as

$$F[\phi, m] = \int_0^\phi \frac{dt}{\sqrt{1 - m\sin^2[t]}} = \int_0^{\sin[\phi]} \frac{dw}{\sqrt{1 - w^2}\sqrt{1 - mw^2}} \qquad (C.1)$$

Note that two parameters are required to obtain a value for F. The parameter ϕ is called the amplitude of u (see Appendix D). The parameter $m = k^2$ is the elliptic module squared and k is the elliptic module. Many books write $F[\phi, k]$ instead of $F[\phi, m]$ with m replaced on the right hand side of (C.1) by k^2. The reason for using m instead of k and k^2 is that the two main mathematical software programs (Mathematica and Matlab) used by scientist and engineers use m. Although most software programs can handle almost any value of m, it is standard to have $0 < m < 1$. In addition to the elliptic module k there is a complementary module k' defined as

$$k' = \sqrt{1 - k^2} = \sqrt{1 - m} \qquad (C.2)$$

This leads to a complementary module squared defined as

$$m' = (k')^2 = 1 - m \qquad (C.3)$$

When $\phi = \frac{1}{2}\pi$, the incomplete elliptic integral of the first kind becomes the complete elliptic integral of the first kind defined as

$$K[m] = F[\tfrac{1}{2}\pi, m] = \int_0^1 \frac{dw}{\sqrt{1 - w^2}\sqrt{1 - mw^2}} \tag{C.4}$$

Note that the symbol K has been used to distinguish between the complete and incomplete elliptic integrals of the first kind. This is not the case for the elliptic integrals of the second and third kind. Often the elliptic module is dropped and the complete elliptic integrals of the first kind becomes K with the module squared m being understood. To distinguish between $K[m]$ and $K[m']$ when dropping the elliptic module, the notation

$$K[m'] = K' \tag{C.5}$$

is used.

The incomplete elliptic integral of the second kind is defined as

$$E[\phi, m] = \int_0^\phi \sqrt{1 - m \sin^2[t]}\, dt = \int_0^{\sin[\phi]} \frac{\sqrt{1 - mw^2}}{\sqrt{1 - w^2}}\, dw \tag{C.6}$$

When $\phi = \tfrac{1}{2}\pi$, the incomplete elliptic integral of the second kind becomes the complete elliptic integral of the second kind defined as

$$E[m] = E[\tfrac{1}{2}\pi, m] = \int_0^1 \frac{\sqrt{1 - mw^2}}{\sqrt{1 - w^2}}\, dw \tag{C.7}$$

Unlike the complete elliptic integral of the first kind, the same symbol is used for both the incomplete and complete integrals of the second kind. The only difference is the number of arguments for E. Similar to the complete elliptic integrals of the first kind, the complete elliptic integral of the second kind is often written as E with the module squared m being understood. To distinguish between $E[m]$ and $E[m']$ when dropping the elliptic module, the notation

$$E[m'] = E' \tag{C.8}$$

is used.

The incomplete elliptic integral of the third kind is defined as

$$
\begin{aligned}
\Pi[n, \phi, m] &= \int_0^\phi \frac{dt}{(1 - n \sin^2[t])\sqrt{1 - m \sin^2[t]}} \\
&= \int_0^{\sin[\phi]} \frac{dw}{(1 - nw^2)\sqrt{1 - w^2}\sqrt{1 - mw^2}}
\end{aligned} \tag{C.9}
$$

where n is the characteristic parameter. When $\phi = \frac{1}{2}\pi$, the incomplete elliptic integral of the third kind becomes the complete elliptic integral of the third kind defined as

$$\Pi[n, m] = \Pi\left[n, \tfrac{1}{2}\pi, m\right] = \int_0^1 \frac{dw}{(1 - nw^2)\sqrt{1 - w^2}\sqrt{1 - mw^2}} \quad \text{(C.10)}$$

Similar to the complete elliptic integral of the second kind, the same symbol is used for both the incomplete and complete integrals of the third kind. The only difference is the number of arguments for Π.

Two relationships that come up often when using elliptic integrals are now derived. Since

$$K[m'] = \int_0^1 \frac{dw}{\sqrt{1 - w^2}\sqrt{1 - m'w^2}} = \int_1^{1/\sqrt{m}} \frac{du}{\sqrt{u^2 - 1}\sqrt{1 - mu^2}} \quad \text{(C.11)}$$

where $u^2 = (1 - m'w^2)^{-1}$ and

$$F\left[\sin^{-1}\left[1/\sqrt{m}\right], m\right] = \int_0^1 \frac{dw}{\sqrt{1 - w^2}\sqrt{1 - mw^2}} \\ + \int_1^{1/\sqrt{m}} \frac{dw}{\sqrt{1 - w^2}\sqrt{1 - mw^2}} \quad \text{(C.12)}$$

we have

$$F\left[\sin^{-1}\left[1/\sqrt{m}\right], m\right] = K[m] + jK[m'] \quad \text{(C.13)}$$

Similar analysis with $u^2 = (1 - m'w^2)/m$ leads to

$$E[m'] = \int_0^1 \frac{\sqrt{1 - m'w^2}}{\sqrt{1 - w^2}} dw = \int_1^{1/\sqrt{m}} \frac{mu^2}{\sqrt{u^2 - 1}} \frac{du}{\sqrt{1 - mu^2}} \quad \text{(C.14)}$$

and

$$E\left[\sin^{-1}\left[1/\sqrt{m}, m\right]\right] = E[m] + j\left(K[m'] - E[m']\right) \quad \text{(C.15)}$$

The last relationship we discuss for elliptic integrals is known as Legendre's relation for complete elliptic integrals given as

$$E[m]\,K[m'] + E[m']\,K[m] - K[m]\,K[m'] = \tfrac{1}{2}\pi \quad \text{(C.16)}$$

To prove (C.16), we first show derivative of the left hand side of (C.16) with respect to k is zero, which means the left hand side of (C.16) is a constant. Then we show the constant is $\frac{1}{2}\pi$. The identities

$$\frac{\partial E[m]}{\partial k} = \int_0^1 \frac{-kw^2}{\sqrt{1 - mw^2}\sqrt{1 - w^2}} dw = \frac{E[m] - K[m]}{k} \quad \text{(C.17)}$$

and

$$\frac{\partial E[m']}{\partial k} = \frac{\partial E[m']}{\partial k'}\frac{dk'}{dk} = \frac{(K[m'] - E[m'])\,k}{m'} \tag{C.18}$$

are easily obtained by differentiating (C.7). The identity

$$\frac{m'w^2}{(1-mw^2)^{3/2}\sqrt{1-w^2}} = \frac{\sqrt{1-w^2}}{\sqrt{1-mw^2}} - \frac{\partial}{\partial w}\left(\frac{w\sqrt{1-w^2}}{\sqrt{1-mw^2}}\right) \tag{C.19}$$

allows obtaining

$$km'\frac{\partial K[m]}{\partial k} = m\int_0^1 \frac{\sqrt{1-w^2}}{\sqrt{1-mw^2}}dw = E[m] - m'K[m] \tag{C.20}$$

and

$$km'\frac{\partial K[m']}{\partial k} = km'\frac{\partial K[m']}{\partial k'}\frac{dk'}{dk} = mK[m'] - E[m'] \tag{C.21}$$

by differentiating (C.4). Taking the derivative of the left hand side of (C.16) and then using identities (C.17)–(C.21) results in zero. Taking the limit as $k \to 0$ of the left hand side of (C.16) gives $\frac{1}{2}\pi$, proving Legendre's relationship.

Further identities and details regarding elliptic integrals can be found in [1].

Jacobi's Elliptic Functions

The trigonometric functions sine and cosine are sometime referred to as circular functions. This is because they can be viewed as special cases of the Jacobi's elliptic functions sn[ ] and cn[ ]. The definition of sn[ ] is obtained by first defining the inverse of sn[ ] equal to the incomplete elliptic integral of the first kind

$$\text{sn}^{-1}[\sin[\phi],m] = u = \int_0^{\sin[\phi]} \frac{dw}{\sqrt{1-w^2}\sqrt{1-mw^2}} \tag{D.1}$$

Based on the definition of an inverse, we have

$$\sin[\phi] = \text{sn}[u,m] \tag{D.2}$$

Two additional Jacobi elliptic functions are defined in terms of sn[ ] as

$$\text{cn}[u,m] = \sqrt{1-\text{sn}^2[u,m]} \tag{D.3}$$

and

$$\text{dn}[u,m] = \sqrt{1-m\,\text{sn}^2[u,m]} \tag{D.4}$$

Based on (D.1) and (D.3), we have

$$\text{sn}[u,0] = \sin[u], \quad \text{cn}[u,0] = \cos[u] \tag{D.5}$$

Since $u = 0$ when $\phi = 0$ and $u = K[m]$ when $\phi = \pi/2$, we have

$$\text{sn}[0,m] = 0, \quad \text{cn}[0,m] = 1, \quad \text{dn}[0,m] = 1$$
$$\text{sn}[K[m],m] = 1, \quad \text{cn}[K[m],m] = 0, \quad \text{dn}[K[m],m] = k' \tag{D.6}$$

Typically, the elliptic module squared m is understood and left out of the notation. Quotients and reciprocals of $sn[u]$, $cn[u]$, and $dn[u]$ lead to an addition nine Jacobi elliptic functions

$$ns[u] = \frac{1}{sn[u]}, \quad tn[u] = sc[u] = \frac{sn[u]}{cn[u]}, \quad sd[u] = \frac{sn[u]}{dn[u]}$$

$$nc[u] = \frac{1}{cn[u]}, \quad \frac{1}{tn[u]} = cs[u] = \frac{cn[u]}{sn[u]}, \quad cd[u] = \frac{cn[u]}{dn[u]} \qquad (D.7)$$

$$nd[u] = \frac{1}{dn[u]}, \quad ds[u] = \frac{dn[u]}{sn[u]}, \quad dc[u] = \frac{dn[u]}{cn[u]}$$

Further identities and details regarding Jacobi's elliptic functions can be found in [1].

Gamma and Beta Functions

The gamma function is defined for $\mathrm{Re}[z] > 0$ and $s > 0$ as

$$\Gamma[z] = \int_0^\infty t^{z-1}e^{-t}dt = s^z \int_0^\infty t^{z-1}e^{-st}dt \qquad \text{(E.1)}$$

Integrating (E.1) by parts with $u = e^{-t}$ and $v = t^z/z$ gives

$$\Gamma[z] = z^{-1} \int_0^\infty e^{-t}t^z du = z^{-1}\Gamma[z+1] \qquad \text{(E.2)}$$

Since

$$\Gamma[1] = \int_0^\infty e^{-t}dt = 1 \qquad \text{(E.3)}$$

we have

$$\Gamma[n+1] = n\,(n-1)\cdots(2)\,(1) = n! \qquad \text{(E.4)}$$

for n a positive integer.

The Beta function $B[x, y]$, can be expressed in terms of the gamma function by first integrating the left side of the equality

$$\int_0^\infty dt \int_0^\infty t^{x+y-1}e^{-(v+1)t}v^{x-1}dv$$
$$= \int_0^\infty dv \int_0^\infty t^{x+y-1}e^{-(v+1)t}v^{x-1}dt \qquad \text{(E.5)}$$

to obtain

$$\int_0^\infty dt \int_0^\infty t^{x+y-1}e^{-(v+1)t}v^{x-1}dv = \Gamma[x]\,\Gamma[y] \qquad \text{(E.6)}$$

Integrating the right hand side of (E.5) with respect to t gives

$$\int_0^\infty dv \int_0^\infty t^{x+y-1} e^{-(v+1)t} v^{x-1} dt = \Gamma[x+y] \int_0^\infty \frac{v^{x-1}}{(v+1)^{x+y}} dv \quad (\text{E.7})$$

Equating (E.6) to (E.7) leads to

$$B[x,y] = \int_0^\infty \frac{v^{x-1}}{(v+1)^{x+y}} dv = \int_0^1 t^{x-1}(1-t)^{y-1} dt = \frac{\Gamma[x]\,\Gamma[y]}{\Gamma[x+y]} \quad (\text{E.8})$$

where the middle equation of (E.8) was obtained by letting $v = t/(1-t)$. The gamma function reflection formula

$$\Gamma[z]\,\Gamma[1-z] = \frac{\pi}{\sin[\pi z]} \quad (\text{E.9})$$

can be derived using the Beta function integral

$$\Gamma[z]\,\Gamma[1-z] = B[z, 1-z] = \int_0^\infty \frac{v^{z-1}}{1+v} dv \quad (\text{E.10})$$

In order to determine the value of (E.10), let $v = y^{2n}$ and $z = \frac{2m+1}{2n}$ where m, n are integers and $n > m$

$$I = \int_0^\infty \frac{v^{z-1}}{1+v} dv = 2n \int_0^\infty \frac{y^{2m}}{1+y^{2n}} dy = n \int_{-\infty}^\infty \frac{y^{2m}}{1+y^{2n}} dy \quad (\text{E.11})$$

The roots of the denominator of (E.11) are

$$a_k = e^{j\pi(2k+1)/(2n)} \quad (\text{E.12})$$

where $-n \le k < n-1$ and k is an integer. Expanding the integrand of (E.11) into partial fractions gives

$$I = n \int_{-\infty}^\infty \sum_{k=-n}^{n-1} \frac{A_k + jB_k}{y - a_k} dy$$

$$= 2n \int_{-\infty}^\infty \sum_{k=0}^{n-1} \left(\frac{A_k\,(y - \mathrm{Re}[a_k])}{|y - a_k|^2} - \frac{B_k\,\mathrm{Im}[a_k]}{|y - a_k|^2} \right) dy \quad (\text{E.13})$$

where

$$A_k = -\frac{\cos[(2k+1)\,\theta]}{2n}, \quad B_k = -\frac{\sin[(2k+1)\,\theta]}{2n} \quad (\text{E.14})$$

and

$$\theta = \frac{(2m+1)\pi}{2n} \tag{E.15}$$

By letting $x = y - \text{Re}[a_k]$, (E.14) becomes

$$
\begin{aligned}
I &= 2n \int_{-\infty}^{\infty} \sum_{k=0}^{n-1} \left(\frac{A_k x - B_k \text{Im}[a_k]}{x^2 + \text{Im}[a_k]^2} \right) dx \\
&= 2n \int_{-\infty}^{\infty} \sum_{k=0}^{n-1} \left(\frac{-B_k \text{Im}[a_k]}{x^2 + \text{Im}[a_k]^2} \right) dx
\end{aligned}
\tag{E.16}
$$

Integrating term by term gives

$$I = 2n\pi \left(\sum_{k=0}^{n-1} -B_k \right) = \pi \sum_{k=0}^{n-1} \sin[(2k+1)\,\theta] = \frac{\pi}{\sin[\theta]} = \frac{\pi}{\sin[\pi z]} \tag{E.17}$$

Although n, m were assumed integers, we can approximate any number for z between 0 and 1 with the appropriate choice of n and m. By continuity, (E.17) holds for any value of z between 0 and 1. By analytic continuation, the result is valid for any $z \neq 0, \pm 1, \pm 2, \ldots$.

Further details regarding the gamma and beta functions can be found in [2].

Gauss's Hypergeometric Function

For $|z| < 1$ and $c \neq 0, -1, -2, \ldots$, Gauss's hypergeometric function can be defined by the infinite series

$$_2F_1[a, b, c, z] = \sum_{n=0}^{\infty} \frac{(a)_n (b)_n}{(c)_n} \frac{z^n}{n!} \tag{F.1}$$

where

$$(a)_n = \Gamma[a + n] / \Gamma[a] \tag{F.2}$$

and $\Gamma[z]$ is the gamma function given by (E.1). For $\mathrm{Re}[c] > \mathrm{Re}[b] > 0$, Gauss's hypergeometric function can be defined in integral form as

$$_2F_1[a, b, c, z] = \frac{\Gamma[c]}{\Gamma[b]\Gamma[c-b]} \int_0^1 t^{b-1}(1-t)^{c-b-1}(1-zt)^{-a}dt \tag{F.3}$$

It may be advantageous to use one form of $_2F_1$ over the other depending on the analysis being performed. For example, the equality

$$_2F_1[a, b, c, z] = {}_2F_1[b, a, c, z] \tag{F.4}$$

is obtained by inspection with the infinite series representation but not with the integral form. Setting $z = 0$ in (F.3) and using (E.8) gives

$$_2F_1[a, b, c, 0] = \frac{\Gamma[c]}{\Gamma[c-b]\,\Gamma[b]} \int_0^1 t^{b-1}(1-t)^{c-b-1}dt = 1 \tag{F.5}$$

Setting $z = 1$ in (F.3) and using (E.8) gives

$$
\begin{aligned}
{}_2F_1[a,b,c,1] &= \frac{\Gamma[c]}{\Gamma[c-b]\,\Gamma[b]} \int_0^1 t^{b-1}(1-t)^{c-b-a-1}\,dt \\
&= \frac{\Gamma[c]}{\Gamma[c-b]}\,\frac{\Gamma[c-a-b]}{\Gamma[c-a]}
\end{aligned}
\tag{F.6}
$$

For certain values of a, b, and c, Gauss's hypergeometric function can be expressed in terms of elementary functions. For example,

$$
{}_2F_1[a,b,b,z] = \sum_{n=0}^{\infty} (a)_n \frac{z^n}{n!}
\tag{F.7}
$$

The right hand side of (F.7) is the series expansion for $(1-z)^{-a}$. Therefore,

$$
{}_2F_1[a,b,b,z] = (1-z)^{-a}
\tag{F.8}
$$

Other values of a, b, and c that allow ${}_2F_1$ to be expressed in terms of elementary functions can be found in [2].

The derivative of ${}_2F_1$ in the infinite series representation is

$$
\frac{\partial\,({}_2F_1[a,b,c,z])}{\partial z} = \sum_{n=0}^{\infty} \frac{(a)_n (b)_n}{(c)_n}\,\frac{z^{n-1}}{n-1!}
\tag{F.9}
$$

By combining (F.9) with

$$
\begin{aligned}
\left(\frac{c-1}{z}\right)({}_2F_1[a,b,c-1,z]) &= \sum_{n=0}^{\infty} \frac{(c-1)(a)_n(b)_n}{z(c-1)_n}\,\frac{z^n}{n!} \\
&= \sum_{n=0}^{\infty} \frac{(a)_n(b)_n}{(c)_n}\,\frac{z^n}{n!}\left(\frac{c+n-1}{z}\right)
\end{aligned}
\tag{F.10}
$$

and

$$
\left(\frac{c-1}{z}\right)({}_2F_1[a,b,c,z]) = \sum_{n=0}^{\infty} \frac{(a)_n(b)_n}{(c)_n}\,\frac{z^n}{n!}\left(\frac{c-1}{z}\right)
\tag{F.11}
$$

we obtain the identity

$$
\frac{\partial\,({}_2F_1[a,b,c,z])}{\partial z} = \left(\frac{c-1}{z}\right)({}_2F_1[a,b,c-1,z] - {}_2F_1[a,b,c,z])
\tag{F.12}
$$

With the help of (F.12), several useful indefinite integrals can be obtained in terms of $_2F_1$. For example, using (F.12) and (F.8) allows evaluating

$$\frac{\partial}{\partial z}\left(z^{\varphi+1}{}_2F_1\left[a,b,b+1,\frac{z}{\alpha}\right]\right)$$

$$= z^{\varphi}{}_2F_1\left[a,b,b+1,\frac{z}{\alpha}\right](\varphi+1-b)+\frac{a^{a}z^{\varphi}b}{(\alpha-z)^{a}} \qquad (\text{F.13})$$

If we let $b = \varphi + 1$, the first term on the right hand side is zero and we are left with

$$\frac{\partial}{\partial z}\left(z^{\varphi+1}{}_2F_1\left[a,b,b+1,\frac{z}{\alpha}\right]\right) = z^{\varphi}\left(\varphi+1\right)a^{a}(\alpha-z)^{-a} \qquad (\text{F.14})$$

or

$$\int z^{\varphi}(z-\alpha)^{-a}dz = \frac{z^{\varphi+1}}{(\varphi+1)(-\alpha)^{a}}{}_2F_1[a,1+\varphi,2+\varphi,z/\alpha]+C \quad (\text{F.15})$$

where C is a constant. Letting $\varphi = -\beta$, $a = -\gamma$, $\alpha = b_1 - a_1$ and $z = w - a_1$ in (F.15) gives

$$\int \frac{(w-b_1)^{\gamma}}{(w-a_1)^{\beta}}dw$$

$$= \frac{(w-a_1)^{1-\beta}}{(1-\beta)(a_1-b_1)^{-\gamma}}{}_2F_1\left[-\gamma,1-\beta,2-\beta,\frac{w-a_1}{b_1-a_1}\right]+C \qquad (\text{F.16})$$

Partial fractions expansion and substitution allows several other integrals to be put into the form of (F.16). We give three examples that are used in the book. First consider the integral

$$I_1 = \int \frac{(w-b)^{a}}{w(w-a)^{a}}dw \qquad (\text{F.17})$$

Using partial fractions, we can rewrite (F.17) as

$$\int \frac{(w-b)^{a}}{w(w-a)^{a}}dw = \int\left(\frac{(w-b)^{a-1}}{(w-a)^{a}}-\frac{b(w-b)^{a-1}}{w(w-a)^{a}}\right)dw \qquad (\text{F.18})$$

The first partial fraction is already in the form of (F.16). The second partial fraction can be transformed into the form of (F.16) by letting $u = 1/w$

$$-b\int \frac{(w-b)^{a-1}}{w(w-a)^{a}}dw = -\left(\frac{b}{a}\right)^{a}\int\frac{(u-b^{-1})^{a-1}}{(u-a^{-1})^{a}}du \qquad (\text{F.19})$$

Second, consider the integral

$$I_2 = \int \frac{w(w-b)^\beta}{(w-a)^\alpha} dw \qquad \text{(F.20)}$$

Decomposition of (F.20) into partial fractions gives

$$
\begin{aligned}
I_2 = {} & \frac{1}{(2-\alpha+\beta)} \int \left(\frac{a\,(\beta+1)+(1-\alpha)\,b}{(w-a)^\alpha (w-b)^{-\beta}} \right) dw \\
& + \frac{\beta+1}{(2-\alpha+\beta)} \int \left(\frac{(w-b)^\beta}{(w-a)^{\alpha-1}} + \frac{(1-\alpha)\,(w-b)^{\beta+1}}{(\beta+1)\,(w-a)^\alpha} \right) dw
\end{aligned}
\qquad \text{(F.21)}
$$

All three terms have the appropriate form of (F.16). Note the integration of the second integral can be expressed without $_2F_1[\ ]$ as

$$\int \left(\frac{(\beta+1)\,(w-b)^\beta}{(w-a)^{\alpha-1}} + \frac{(1-\alpha)\,(w-a)^{-\alpha}}{(w-b)^{-\beta-1}} \right) dw = \frac{(w-b)^{1+\beta}}{(w-a)^{\alpha-1}} \qquad \text{(F.22)}$$

Now consider the integral

$$I_3 = \int \frac{(w-b)^\alpha}{w^2\,(w-a)} dw \qquad \text{(F.23)}$$

Decomposition of (F.23) into partial fractions gives

$$
\begin{aligned}
I_3 = {} & -\frac{1}{ab} \int \left(\frac{(w-b)^\alpha\,(w\alpha+b)}{w^2} \right) dw \\
& + \frac{1}{ab} \int \left(\frac{b(w-a)^{-1}}{a(w-b)^{-\alpha}} + \frac{(a\alpha-b)\,w^{-1}}{a(w-b)^{-\alpha}} \right) dw
\end{aligned}
\qquad \text{(F.24)}
$$

Each term in the second integral has the form of (F.16) and the first integral can be integrated without $_2F_1[\ ]$ as

$$\int \frac{(w-b)^\alpha\,(w\alpha+b)}{w^2} dw = \frac{(w-b)^{\alpha+1}}{w} \qquad \text{(F.25)}$$

Before ending this section, the incomplete beta function $B[z,a,b]$ should be introduced. It is a special case of Gauss's hypergeometric function given as

$$B[z,a,b] = (z^a/a)\,_2F_1[a,1-b,1+a,z] \qquad \text{(F.26)}$$

When $b=0$ and a is a rational number, $B[z,a,0]$ can be expressed in terms of elementary functions [3]. This property may be useful when reducing equations to simplified form.

Further details regarding the Gauss's hypergeometric function can be found in [2].

Dilogarithm and Trilogarithm Functions

The dilogarithm function is defined in integral form as

$$\text{Li}_2[z] = -\int_0^z \frac{\ln[1-t]}{t}dt \tag{G.1}$$

Since the logarithm of a complex variable is a multifunction, a branch cut from $z = 1$ to $z = \infty$ is used to keep the dilogarithm a single valued function. When $|z| \leq 1$, the $\ln[1-t]$ in the integrand can be expanded in series form and term by term integration gives

$$\text{Li}_2[z] = \sum_{n=1}^{\infty} \frac{z^n}{n^2}, \quad |z| \leq 1 \tag{G.2}$$

Note that $z = 1$ in (G.2) is the famous Basel problem [7] which gives

$$\text{Li}_2[1] = \sum_{n=1}^{\infty} \frac{1}{n^2} = \frac{\pi^2}{6} \tag{G.3}$$

Since

$$\sum_{n=1}^{\infty} \frac{1}{n^2} = \sum_{n=1}^{\infty} \left(\frac{1}{(2n)^2} + \frac{1}{(2n-1)^2} \right) \tag{G.4}$$

and

$$\sum_{n=1}^{\infty} \frac{1}{(2n)^2} = \frac{1}{4} \sum_{n=1}^{\infty} \frac{1}{n^2} = \frac{\pi^2}{24}, \quad \sum_{n=1}^{\infty} \frac{1}{(2n-1)^2} = \frac{\pi^2}{8} \tag{G.5}$$

we can also calculate

$$\text{Li}_2[-1] = \sum_{n=1}^{\infty} \frac{(-1)^n}{n^2} = \sum_{n=1}^{\infty} \left(\frac{1}{(2n)^2} - \frac{1}{(2n-1)^2} \right) = -\frac{\pi^2}{12} \qquad (\text{G.6})$$

To obtain a useful identity, we first replace z with $-1/z$ in (G.1) and then differentiate with respect to z

$$\frac{d\left(\text{Li}_2[-1/z]\right)}{dz} = \frac{1}{z}\ln\left[1 + \frac{1}{z}\right] \qquad (\text{G.7})$$

Integrating (G.7) from $z = 1$ to $z = x$ gives

$$\text{Li}_2[-1/x] - \text{Li}_2[-1] = -\text{Li}_2[-x] + \text{Li}_2[-1] - \tfrac{1}{2}\ln[x]^2 \qquad (\text{G.8})$$

Rearranging and inserting the value of $\text{Li}_2[-1]$ from (G.6) gives the identity

$$\text{Li}_2[-1/x] + \text{Li}_2[-x] = -\tfrac{1}{6}\pi^2 - \tfrac{1}{2}\ln[x]^2 \qquad (\text{G.9})$$

An integral identity used in the book is obtained by first letting $z = \mp\beta e^{\alpha j x_0}$ in (G.1) and then differentiating with respect to x_0

$$\frac{d\left(\text{Li}_2\left[\mp\beta e^{\alpha j x_0}\right]\right)}{dx_0} = -\alpha j \ln\left[1 \pm \beta e^{\alpha j x_0}\right] \qquad (\text{G.10})$$

Next, we integrate both sides to obtain

$$\int \ln\left[1 \pm \beta e^{\alpha j x_0}\right] dx_0 = \frac{j}{\alpha}\text{Li}_2\left[\mp\beta e^{\alpha j x_0}\right] + C \qquad (\text{G.11})$$

where C is a constant.

The trilogarithm is defined as

$$\text{Li}_3[z] = \int_0^z \frac{1}{t}\text{Li}_2[t]\, dt \qquad (\text{G.12})$$

When $|z| \le 1$, the $\ln[1 - t]$ in the integrand can be expanded in series form and term by term integration gives

$$\text{Li}_3[z] = \sum_{n=1}^{\infty} \frac{z^n}{n^3}, \quad |z| \le 1 \qquad (\text{G.13})$$

Dividing (G.9) by x and integrating gives

$$\int_1^z \left(\frac{\text{Li}_2[-1/x]}{x} + \frac{\text{Li}_2[-x]}{x} \right) dx = -\int_1^z \left(\frac{\pi^2}{6x} + \frac{\ln[x]^2}{2x} \right) dx \qquad (\text{G.14})$$

Performing the integration leads to the identity

$$\text{Li}_3[-z] - \text{Li}_3[-1/z] = -\tfrac{1}{6}\pi^2 \ln[z] - \tfrac{1}{6}\ln[x]^3 \qquad (G.15)$$

An integral identity used in the book is obtained by first letting $z = \beta e^{\alpha y_0}$ in (G.12) and then differentiating with respect to y_0

$$\frac{d\left(\text{Li}_3[\beta e^{\alpha y_0}]\right)}{dy_0} = \alpha \text{Li}_2[\beta e^{\alpha y_0}] \qquad (G.16)$$

Next, we integrate both sides with respect to y_0 to obtain

$$\int \text{Li}_2[\beta e^{\alpha y_0}]\, dy_0 = \frac{1}{\alpha}\text{Li}_3[\beta e^{\alpha y_0}] + C \qquad (G.17)$$

where C is a constant.

Further details regarding the Dilogarithm and Trilogarithm can be found in [5].

REFERENCES

[1] Paul F. Byrd and Morris D. Friedman. *Handbook of Elliptic Integrals for Engineers and Physicists*. Springer-Verlag, 1954.

[2] A. Erdelyi, editor. *Higher Transcendental Functions Vol 1*. McGraw-Hill Book Company, Inc., 1953.

[3] J. L. González-Santander. A note on some reduction formulas for the incomplete beta function and the lerch transcedent, 2020.

[4] Herbert Kroemer. *Quantum Mechanics*. Prentice-Hall, Inc., 1994.

[5] L. Lewin. *Dilogarithms and Associated Functions*. Macdonald & Co., 1958.

[6] M. J. Lighthill. *Introduction to Fourier Analysis and Generalised Functions*. Cambridge University Press, 1958.

[7] Paul J. Nahin. *An Imaginary Tale The Story of $\sqrt{-1}$*. Princeton University Press, 1998.

Index

For Product Safety Concerns and Information please contact our EU
representative GPSR@taylorandfrancis.com
Taylor & Francis Verlag GmbH, Kaufingerstraße 24, 80331 München, Germany

www.ingramcontent.com/pod-product-compliance
Ingram Content Group UK Ltd.
Pitfield, Milton Keynes, MK11 3LW, UK
UKHW021125180425
457613UK00006B/230